Gerhard Piekarski

Medical Parasitology

With 33 Plates, Mostly in Colour
and 32 Illustrations

Springer-Verlag Berlin Heidelberg New York
London Paris Tokyo Hong Kong

Professor Dr. GERHARD PIEKARSKI, Emeritus
Institute of Medical Parasitology
University of Bonn
Sigmund-Freud-Straße 25
5300 Bonn 1
Federal Republic of Germany

Translated from the third German edition by:

DORA WIRTH
Languages Ltd.
85 Campden Street
Kensington, London W8 7EN
Great Britain

Translation of "Medizinische Parasitologie in Tafeln"
3. Auflage, 1987

ISBN-13: 978-3-642-72950-8 e-ISBN-13: 978-3-642-72948-5
DOI: 10.1007/978-3-642-72948-5

Library of Congress Cataloging-in-Publication Data. Piekarski, Gerhard. [Medizinische Parasitologie in Tafeln. English] Medical Parasitology/G. Piekarski; [translated by Dora Wirth]. p. cm. Translation of: Medizinische Parasitologie in Tafeln. 1987. Bibliography: p. Includes index.

1. Medical Parasitology. I. Title. [DNLM: 1. Parasites. 2. Parasitic Diseases. QX 4 P613m] QR251.P5413 1989 616.9'6—dc19 DNLM/DLC

Softcover reprint of the hardcover 1st edition 1989

Printing of the colour plates: O. Brandstetter Druckerei GmbH & Co. KG, Wiesbaden.

Typesetting, printing and binding: Konrad Triltsch, Graphischer Betrieb, Würzburg.
2127/3130/543210 – Printed on acid-free paper

Preface to the Third German Edition

In the period since the publication of the second edition of this book in 1975, the importance of human parasitism for clinical medicine has continued its undiminished growth. The number of patients with parasitic infections now seeking medical advice has risen considerably. A major factor in such increases has been the ease of rapid and frequent travel for reasons of business or leisure to tropical countries with a high prevalence of parasites putting visitors at great risk of acquiring infection. In addition, western countries are visited by guests and refugees from tropical regions; such individuals are often infected with parasites and consequently are a potential source of new infections (GSELL, 1978; WEISE, 1982).
The increased interest in parasitic disorders is best demonstrated by the special programme promoted for some years by the World Health Organization (WHO) to combat the most important parasitic infections: malaria, trypanosomiasis, leishmaniasis, schistosomiasis, filariasis, as well as leprosy. Third-world inhabitants in particular suffer severely from these parasitoses. Their productivity is considerably reduced, and the economic development of their countries is markedly affected. Children and young adults are especially burdened in this respect, resulting in a delay in their mental and physical development.
Many parasites injure their hosts not only at the site of infection (e.g. intestine, lungs, liver), but also produce systemic effects through the deprivation of nourishment and through the deleterious effects of toxic metabolites affecting the entire organism, including the central nervous system.
Since the appearance of the previous edition in 1975, much new information has been acquired on various parasites in humans, for instance the leishmanias (p. 27) and the coccidia (p. 67), and there have been developments in malaria research and babesiosis (pp. 95 and 108). Free-living amoebae that invade and cause serious disease in humans have also assumed a new importance (p. 61).
One discovery, however, has proved to be of special significance: latent parasitic infections can be reactivated in cases of immunodeficiency and lead to serious illness. In connection with this is the recent discovery that, in addition to certain drugs, the human immunodeficiency virus (HIV) is able to suppress the immune system, thus playing a special role as an initiating factor in the acquired immunodeficiency syndrome (AIDS). The opportunistic parasites particularly involved in the syndrome are *Pneumocystis carinii, Toxoplasma gondii, Cryptosporidium* sp., *Giardia lamblia* and *Strongyloides stercoralis.*

Major progress has been made in the chemotherapy of parasitic infections with the discovery of new highly effective and well tolerated compounds. There have also been advances in the serodiagnosis of parasitic diseases. The complex problems associated with serological results arise from the lack of any standardization of existing methods, but intensive work is now being done in this area (JANITSCHKE and WERNER, 1983; JANITSCHKE, 1985). Only the future will show whether the present fashion in the use of monoclonal antibodies for detecting antigens or genetechnological methods will lead to more specific results.

Today we know a great deal about the life cycles of almost all the parasites of man. Consequently, many infections can be avoided by alterations in traditional diet and ways of life in those countries where parasites are endemic. The WHO has once again stressed these facts. It has emphasized the urgent need for a systematic form of health education, and has placed the onus on people themselves to decisively limit the spread of parasites. These are the ways, together with new drugs suitable for mass treatment, by which the parasitic diseases can be largely controlled and to some extent eradicated. So far, however, attainment of this goal has come up against problems of social hygiene and economic difficulties which nevertheless could be resolved with international aid.

The wealth and frequency of research publications in the specialist literature make it extremely difficult to select objectively. If readers' expectations are not entirely fulfilled, I must ask for understanding. At any rate, I have found it necessary to provide a somewhat longer text than in the last edition. In the case of clinical and chemotherapeutic information, I have received permission from the WHO to refer extensively to the latest reports of the expert commissions, which are available in, for example, the *Technical Report Series.*

I am grateful for the help provided in many ways by my colleagues Dr. W. L. CURRENT, USA, Prof. R. GEIGY, Basel, Prof. H.M. SEITZ, Bonn, Dr. W. H. WERNSDORFER, Geneva, Prof. Y. YOSHIDA, Kyoto, and the WHO, which allowed or arranged for me to use illustrations. I am also grateful to my colleagues Dr. E. GÖBEL, Munich, Prof. K. JANITSCHKE, Berlin, Prof. H.E. KRAMPITZ, Munich, Prof. W. MAIER, Bonn, Priv.-Doz. Dr. C. MEIER-BROOK, Tübingen, Dr. R. MICHEL, Coblenz, Prof. H. MÜHLPFORDT, Hamburg, Dr. H. SCHULZ-KEY, Tübingen, and Dr. G. WERNSDORFER, Erlangen, for fruitful discussions and valuable references to the literature. I am most indebted to Prof. H. MEHLHORN, Bochum, and Dr. G. SCHILLING, Bonn, who took the trouble to read the text critically, and Prof. CHOBOTAR, Berrien Springs, Mich. USA, who used his experience as a parasitologist in checking the English translation. I also thank my co-worker Frau U. MÜLLER, Bonn, whose untiring support and critical assistance in typing and correction of the manuscript were vital. Finally, I must thank Springer-Verlag for fulfilling all my wishes in regard to printing. I hope that this English edition is granted the warm reception given to previous editions.

Bonn, Spring 1989 G. PIEKARSKI

Contents

List of Plates

Nematodes

Introduction

Medical Parasitology is primarily intended to be an illustrated textbook which provides a review of the most important species of parasite which occur in man; their areas of distribution, morphology and development, the typical disease symptoms resulting from infection, epidemiology and also methods of detection and indications for therapy. The main emphasis is on the protozoan and helminthic diseases; medical entomology has only been covered in connection with the epidemiology of the diseases described here.
Parasites sometimes occur exclusively **in man** (anthropoparasites) and sometimes also in animals (anthropozoonotic parasites). The **monoxenous species** complete their development in man or in one animal alone (Scheme I). **Heteroxenous species,** which include most of the medically important parasites, develop partly in man and partly in animals in the course of their life cycle. They may even be forced to infect different species so that they can continue their development. This may sometimes be associated with a digenesis, the larval development taking place in one intermediate (Scheme II Ⓐ) or in two different intermediate hosts (Scheme III Ⓑ, Ⓒ), and the sexually mature stage developing in another host, the so-called definitive host (Scheme III Ⓐ). The importance of the intermediate hosts can vary considerably (see below).

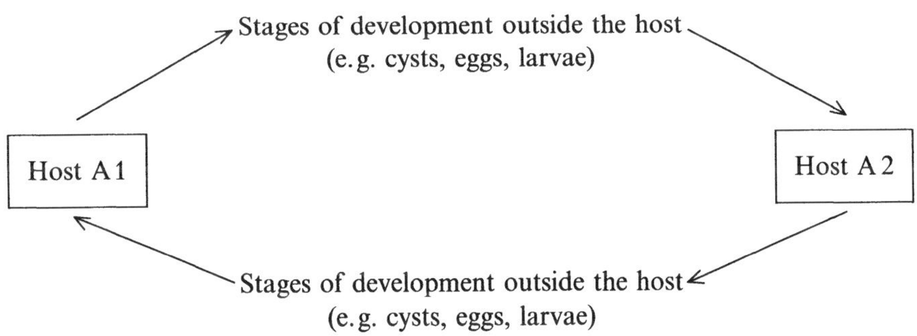

Scheme I: Monoxenous parasites, e.g. *Entamoeba histolytica, Ancylostoma, Strongyloides, Ascaris* (see pp. 49, 225, 235 and 261).

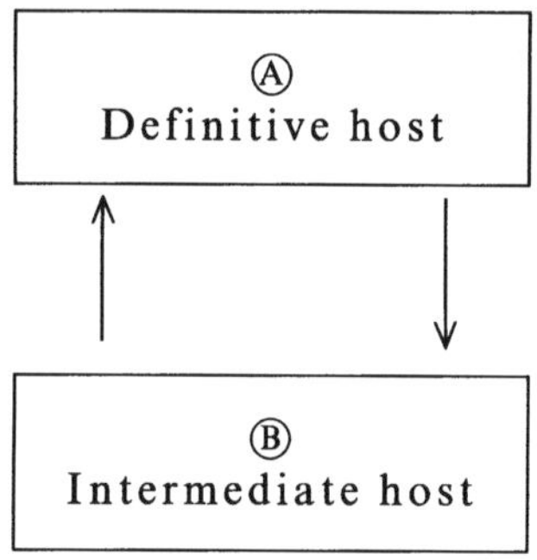

Scheme II: Dixenous parasites, e.g. (a) *Schistosoma* (man – snail – man); (b) Filariae (man – mosquito – man) (see text pp. 161 and 269)

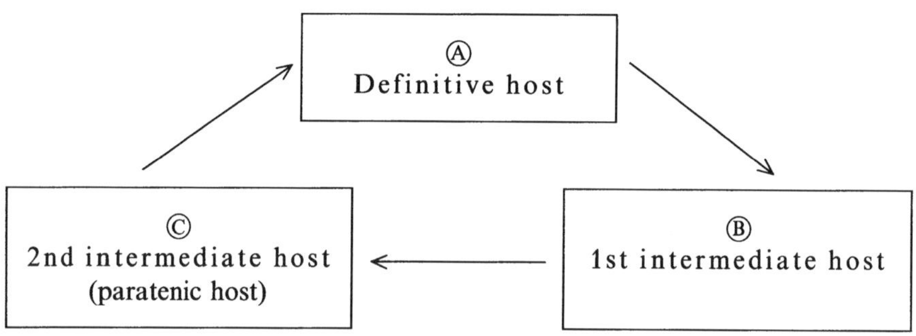

Scheme III: Trixenous parasites, e.g.
a) *Paragonimus westermani* (man Ⓐ – snail Ⓑ – fresh-water crustaceans (crabs) Ⓒ – man Ⓐ),
b) *Diphyllobothrium latum* (man Ⓐ – crustacean (copepod) Ⓑ – fresh-water fish Ⓒ – man Ⓐ) (see text pp. 153 and 175)

Parasites for which the definitive host is always man, for example *Taenia saginata* or *Trichuris trichiura,* are said to be species specific (Table 4, p. 326). If animals can also be the definitive host, the parasite may be termed non-species specific and may sometimes be capable of infecting a wide range of hosts (e.g. *Trichinella*). The parasitic stages which occur in man always differ from those of the intermediate host for a given species. Man is often the definitive host but sometimes, for example in the blood stages of malaria and the larvae of *Echinococcus,* occupies the position of an intermediate host.
For parasites that develop in only one host, in order to ensure transmission, there is the problem of disseminating cysts, eggs and in some cases larvae (see Scheme I).

Some parasites first develop outside the host to a stage where they are capable of invasion (e.g. the larvae of *Ascaris, Ancylostoma*), whilst others can, under certain conditions, develop into free-living adults which multiply in the soil (e.g. *Strongyloides*). The larvae which are produced may then go on to infect new hosts. If a parasite uses several hosts to complete development (e.g. *Schistosoma*, Plate XVIII; *Diphyllobothrium*, Plate XIX), then in the illustrations the definitive host Ⓐ has been placed opposite the one intermediate host Ⓑ (as in Scheme II) or opposite the two intermediate hosts Ⓑ and Ⓒ (1st and 2nd intermediate hosts; Scheme III). In the intermediate hosts there is either (a) an asexual multiplication of the parasites (e.g. the development of trematodes in snails) or (b) further development to an infective stage (e.g. filariae in insects, *Diphyllobothrium* in crustaceans and fish), with sexual maturity then being reached in the definitive host Ⓐ.

For some species of parasite man is the improper host resulting in incomplete or arrested development. In such cases the parasite does not become sexually mature nor can it leave the host. This happens with the larval stages of certain nematodes which normally become sexually mature in domestic animals and animals closely associated with man, e.g. dogs, cats, rats. Other parasites involved include the larvae of *Angiostrongylus cantonensis, Anisakis* species, the tapeworm larva *Cysticercus cellulosae, Echinococcus* species cysts and also the schistosomulae of certain species of schistosome, which are primarily parasites of waterbirds. All of these species have become of increasing interest because of their medical importance.

In this context questions about the mechanism responsible for host specificity are often raised. Physiological observations provide certain indications about this. It has been found that parasitic helminths – with a few exceptions – have lost the capacity for the de novo synthesis of lipid complexes. The dependency of these parasites on the lipids which they receive from the host apparently determines the host specificity. Presumably these species have become closely adapted to the available lipids, without which the parasites would be unable to survive in their host. This is perhaps the explanation for the absence of certain groups of helminths in many hosts. This problem applies to carbohydrates and proteins less, since these can be synthesized from relatively simple molecules of a non-specific nature (FRAYHA and SMYTH, 1983).

Whenever possible the development of the parasite on the plates is shown as a cycle (e.g. as for malaria and the Filariae). With monoxenous parasites, or those which are monoxenous to a certain extent, the epidemiological relationships are shown pictorially (e.g. *Entamoeba*, p. 49; *Toxoplasma*, p. 77; *Trichinella*, p. 209; *Ascaris*, p. 261). For the stages of development of trypanosomes (*Trypanosoma, Leishmania*) the epidemiological relationships among man, vectors and animal reservoirs are emphasized. For diagnosis by microscopy two pictures give a comparison of the size of the most important protozoan cysts and helminth eggs found in faeces (magnified about 500 times; Plate XXXII), and also a comparison of the size of the different species of sexually mature helminths drawn approximately life size (Plate XXXIII).

The scales of magnification have also been given in the legends, where they can assist in the diagnosis. This could not be done throughout because the clarity of the illustrations would have suffered as a result. The text illustration on p. 305 is intended to draw attention to organic structures which are frequently found in preparations of faeces and which on microscopic examination are mistaken for parasites, e.g. protozoan cysts and helminth eggs.

The Appendix consists of tabulated synopses. Tables 1 and 2 describe the pathogenic intestinal parasites, giving clinical and diagnostic data, and detection methods of practical importance including a summary of the preparation times. Table 3 gives a synopsis of the extraintestinal blood and tissue parasites of man and Table 4 a summary of the routes of infection and the development of helminth eggs in man. Tables 5 and 6 give keys for the identification by microscopy of the most important protozoan cysts and helminth eggs (see p. 327).

In the short texts accompanying the plates the names of authors have mostly not been given, because this would have lengthened this work excessively. References to original studies mostly refer to papers published in the last ten years. Older publications can be found in relevant comprehensive textbooks and handbooks. Some of the most important works from the international literature have been marked with an * in the list of references; these should be used to obtain answers to individual questions.

Detailed presentations of the clinical pictures of parasitic diseases have appeared in the handbook *Infektionskrankheiten*, edited by GSELL O. and W. MOHR, Springer-Verlag, Heidelberg, 1972; in the study by KATZ, M., D. D. DESPOMMIER and R. GWADZ on *Parasitic Diseases*, Springer-Verlag, Berlin, Heidelberg, New York, 1982, in *Clinical Parasitology*, P. C. BEAVER, R. C. JUNG and E. W. CUPP, Lea and Febiger, Philadelphia, 9th edition 1984 and MANSON-BAHR, P. E. C. and D. R. BELL: Manson's Tropical Diseases, 19th edition, Baillière Tindall, London, Philadelphia, 1987. Detailed parasitological presentations are contained in *Grundriss der Parasitologie*, 3rd edition by MEHLHORN H. and G. PIEKARSKI, Gustav Fischer, Stuttgart, 1989 and in *Parasitology in Focus*, MEHLHORN H., Springer-Verlag, Berlin, Heidelberg, New York, 1988. MEHLHORN H. and W. PETERS published a summary of the laboratory methods for the *Diagnose der Parasiten des Menschen*, Gustav Fischer Verlag, Stuttgart, 1983.

Protozoa

The pathogenic protozoa of man are unicellular parasites, which belong to four different groups: Flagellata, Amoebae, Sporozoa and Ciliata [1]. This group classification mostly relates to the locomotor organelles of these parasites. Flagellates mostly possess one or several flagella, amoebae mostly move by means of pseudopodia, ciliates by means of cilia. Most of the Sporozoa which live intracellularly do not have any external locomotor organelles, but instead have subpellicular microtubules which make rotatory movements possible. Specific anterior end organelles such as conoids and rhoptries are thought to be associated with penetration into the host cell. There are also differences with regard to sexual multiplication which occurs in the sporozoa and ciliates. Although a sexual stage is absent in amoebae, it can no longer be completely excluded from the parasitic flagellates, in the trypanosomes at least in the tsetse flies.

The **transmission** of pathogenic species can take place:

(**1**) Directly from man to man by a trophic stage (e.g. *Trichomonas vaginalis*) or by means of a resting stage (cysts; e.g. *Entamoeba histolytica*).

Most pathogenic protozoa, however, are either:

(**2**) Transmitted to people by insects (e.g. malaria parasites, trypanosomes, leishmanias) or

(**3**) Through the consumption of raw infected fish or meat (e.g. *Sarcocystis*). Toxoplasmas occupy a special position and can be acquired by humans through the consumption of raw infected meat or through the ingestion of resting stages (oocysts) from cats, and may be transmitted congenitally.

From a **clinical point of view** several species of protozoa are of world-wide importance. Large areas of the tropics and subtropics are still dominated by malaria, sleeping sickness and CHAGAS' disease, leishmaniasis and amoebic dysentery. The World Health Organisation in Geneva has therefore included the first four of these protozoan diseases in a special programme of research and eradication. Also of

[1] This "old" classification of the protozoa, which was still held by E. REICHENOW (1953), was subjected to a complete revision by LEVINE et al. (1980). New names were created for larger and smaller classification units. Thus flagellates and amoebae, because they have certain characteristics in common, have been combined into the phylum *Sarcomastigophora*, but have been retained as the subphylum *Mastigophora* (Flagellata) and the subphylum *Sarcodina* (amoebae in the wider sense). The *Sporozoa* are to be found as a class in the phylum *Apicomplexa*, and the only ciliate species that affects man, *Balantidium coli*, is in the phylum *Ciliophora*.

world-wide distribution are giardias, sarcosporidians, and the ciliate genus *Balantidium*. These are also potentially pathogenic but they do not occur in epidemic form like the other protozoan diseases of hot countries and their importance is more individual in nature, since the course of an acute disease mostly depends on external circumstances, e.g. nutrition and occupational activity.

Protozoan infections may also become of particular importance in the future, because a virus-induced infectious immunosuppression (AIDS; acquired immunodeficiency syndrome) is rapidly spreading. This causes an exacerbation of latent infections and leads to severe diseases (amoebiasis, giardiasis, pneumocystosis, toxoplasmosis etc.). In AIDS one is dealing with a viral infection, primarily acquired venereally, which apparently even on its own has life-threatening effects (see pp. V, 42, 52, 79, 87, 89, 91, 237). Haemophilic patients and intravenous drug users are considered to be at high risk, particularly the latter, because of the habit of sharing contaminated needles.

Today all protozoan infections can be treated with effective drugs. Some preparations can be used for chemoprophylaxis and offer protection against disease.

Plate I ⇨

Flagellates

Trypanosoma brucei gambiense
Trypanosoma brucei rhodesiense

Plate I

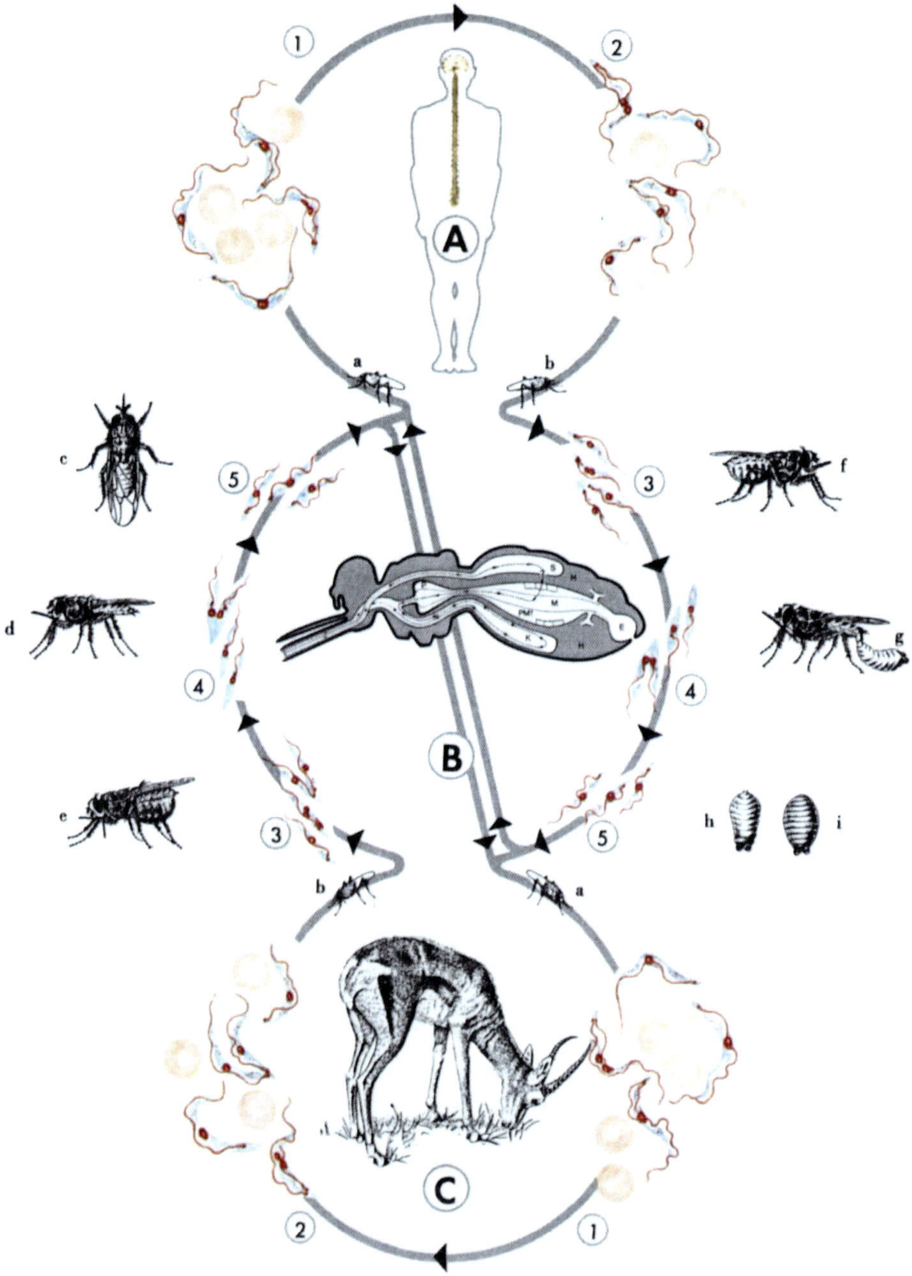

Trypanosoma brucei gambiense DUTTON, 1902
Trypanosoma brucei rhodesiense STEPHENS and FANTHAM, 1910

Pathogens of African trypanosomiasis (Sleeping sickness)

Ⓐ Development in man: trypanosomes in the peripheral blood; central nervous system affected

1, 2 Trypanosomes, partly in process of division (GIEMSA stain). In *2* there are "stumpy" forms, which alone are capable of further development in the tsetse fly (see text p. 11)

Ⓑ Development in the tsetse fly (*Glossina* species) vector

3 Trypanosomes from the fly's stomach
4 Epimastigote form from the fly's intestine
5 Metacyclic (trypomastigote) form from the salivary gland (*S*)
a, b Biting tsetse fly
c *Glossina* adult at rest
d Fasting fly
e Fly after feeding
f Gravid female *Glossina*
g *Glossina* depositing a larva
h Larva
i Pupa

Migration route of the trypanosomes in *Glossina* species:

The trypanosomes first enter the crop (*C*), then gain access to the midgut (*M*), pass through the intestinal epithelium and peritrophic membrane (*PM*) directly into the haemocoel (*H*). The predominant form in the haemolymph is the epimastigote form which migrates to the dorsal salivary glands (*S*) and transforms into an infective metacyclic trypomastigote stage. (For details see text and illustration on p. 12.)

Ⓒ Same kind of development as Ⓐ in the peripheral blood of the animal hosts (e.g. antelope; parasite reservoir).

Transmission can take place by *Glossina* species:

1. From man to man — Ⓐ → Ⓑ → Ⓐ
2. From animal to animal as parasite reservoirs — Ⓒ → Ⓑ → Ⓒ
3. From animal to man and — Ⓒ → Ⓑ → Ⓐ
   vice versa — Ⓐ → Ⓑ → Ⓒ

African trypanosomiasis, caused by flagellate protozoas *Trypanosoma brucei gambiense* and *T. b. rhodesiense*, is one of the great plagues of tropical Africa. It is essentially confined to the region of latitude between 20° north and 20° south. Its distribution is determined by the occurrence of the tsetse fly (*Glossina* species), which transmits the pathogen in its saliva when biting man during a blood meal. Following a period when the density of the flies decreased, the incidence of trypanosomiasis has increased again in the African population with the spread of *Glossina* species to the north, and the disease occurs in both Ethiopia and southern Sudan. At the present time about 45 million people living in 23 countries are at risk, with about 250,000 persons infected at present and about 10,000 new cases of the disease occurring annually (WHO, 1982).

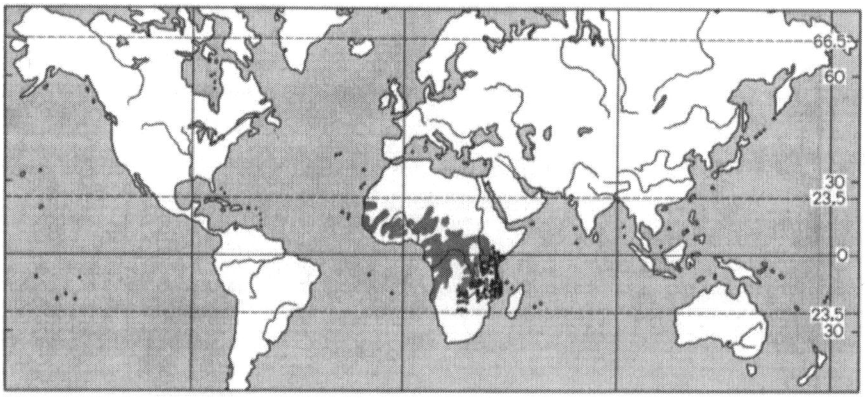

Distribution of African trypanosomes: ■ *T. b. gambiense* ▩ *T. b. rhodesiense*

The two members of the *Trypanosoma brucei* group of species that infect man, *T. b. gambiense* and *T. b. rhodesiense*, occur in geographically widely separate regions and basically have their own species of *Glossina* as vectors. *T. b. gambiense*, the more frequent, is mainly to be found in West and Central Africa, in Senegal, in Gambia, Guinea, on the Ivory Coast and in Liberia, Ghana, Nigeria, Equatorial Africa, Congo and Zaire. *T. b. rhodesiense*, the more dangerous of the two, is, on the other hand, mostly to be found in East Africa, e.g. Zimbabwe, Mozambique, Tanzania, Kenya, but also in Uganda, Malawi, Zambia and Angola. As far as man is concerned, *Glossina palpalis* and *G. tachinoides* are considered to be the main vectors for *T. b. gambiense*, and the species *G. morsitans*, *G. pallidipes* and *G. swynnertoni* for *T. b. rhodesiense*.

The two trypanosomes that cause African trypanosomiasis have been grouped together with the pathogen of the Nagana epidemic of large mammals, in the *T. brucei* group; they can only be distinguished morphologically, biologically and serologically with difficulty but undergo the same cyclical development in *Glossina* leading to infection of the salivary glands (therefore they are sometimes referred too as salivaria; see below).

In order to be able to differentiate between the sub-species, RICKMAN and ROBSON (1972) have developed the blood incubation test. It is based on the fact that *T. b. brucei* is destroyed in human blood, but *T. b. rhodesiense* is not. Mice are inoculated with infected blood and from the second passage blood containing trypanosomes is taken from these experimental animals. In each case 0.25 ml is thoroughly mixed with (a) 2 ml of human blood mixed with an anticoagulant, and (b) as a control 2 ml of buffered saline solution (pH 7.4). After incubation for 5 h at 37° C in a water bath the suspensions are each injected into a rat intraperitoneally. If a *T. b. rhodesiense* infection is present, then both rats become positive, i.e. the infectivity to rats is still retained. If one is dealing with *T. b. brucei*, then only the saline control becomes positive.
A limitation of this test has been found following the observation of RICKMAN et al. (1984) that the sensitivity or resistance of *T. b. brucei* in the tsetse fly is influenced by the species of host from which the blood originated. Thus in *Glossina* species fed with human blood, metacyclic trypanosomes resistant to man develop (BRUN and JENNI, 1984). This test has been supplemented by the in vitro test of JENNI and BRUN (1982) based on the culture forms developed by BRUN et al. (1981) for blood stages of the different species of *Trypanosoma* and subspecies of *T. b. brucei* instead of obtaining the parasites from mice.
Using isoenzyme analysis as a genetic marker is another possible way of differentiating members of the *T. brucei* group. The method involves electrophoretic separation of the components of a dissolved extract of the strain. The enzymes migrate at a varying rate depending on their molecular structure. The position of the various enzymes is made visible through a suitable dye. The resulting enzyme pattern is called a zymodeme (GODFREY, 1979). Similar zymodemes suggest the same genetic origin. This method allows the development of a biological taxonomy for protozoan strains which cannot be differentiated morphologically. Enzyme electrophoresis has contributed a great deal to an understanding of the taxonomy and epidemiology of leishmanias and trypanosomes (see also p. 21, 32). This procedure is of practical use, amongst other things for the identification of strains of trypanosomes from man and wild or domestic mammals when searching for reservoirs of parasites.

Morphology and Development. The characteristic slender forms (15–30 μm; Plate I, Ⓐ *1–2*) present in blood possess a long flagellum which originates posteriorly at the basal granule close to the kinetoplast (hence Kinetoplastida). With light microscopy a fold in the cell membrane, often called the undulating membrane, runs the length of the cell with the flagellum forming the leading edge. In actual fact the freely moving flagellum adheres to the cell membrane at only a few points.
The cell nucleus generally lies in the middle of the cell. On GIEMSA-stained blood smears the cytoplasm appears blue; the nucleus, kinetoplast and basal granule are distinctly red, the flagellum often dark red. Individual red-blue granules may lie in the cytoplasm. Multiplication takes place by longitudinal binary division, where the kinetoplast and basal granule divide before the cell nucleus. Viable trypanosomes can be preserved for years in liquid nitrogen with the addition of glycerin or dimethylsulphoxide (DMSO).
In *Glossina* species, during their migration via the intestines to the salivary glands ("salivaria", 20–40 days) the trypanosomes undergo a morphological and physiological transformation into epimastigotes and finally into the infectious, metacyclic trypomastigote forms (Plate I, Ⓑ *3–5*). This transformation is only undergone, however, by the trypanosomes which have already assumed a "stumpy" form in the blood of the person. In the tsetse fly the trypanosomes are assured of survival

only if they get into the crop (*K*) of the fly. During the metabolic transformation in the fly the surface coat disappears and is redeveloped only after the migration into the salivary gland. This host-specific surface coat, consisting of glycoprotein and mucopolysaccharides, can alter and thus produce variations in the antigenic character of the trypanosomes, which protects them against the immune system of the host (VICKERMAN, 1978). This is one of the reasons why vaccination against trypanosomiasis has not yet succeeded. Ideas to date concerning the migration route of the trypanosomes in the tsetse fly seem to be in need of correction. The trypanosomes which get into the endoperitrophic space with the blood meal do not migrate downwards through the endoperitrophic space and back up through the ectoperitrophic space, but appear to penetrate the peritrophic membrane directly, and now changed in the epimastigote stage, reach the ectoperitrophic space and via the middle intestinal epithelium into the haemocoel. After that they reach the salivary glands via the haemolymph, where they transform into the metacyclic trypomastigotes (EVANS and ELLIS, 1983). This interpretation has been largely confirmed by electron microscopy (ref. in MEHLHORN and PIEKARSKI, 1985). Both sexes of the tsetse fly can transmit trypanosomes about 20–40 days after ingesting infected blood; they then remain infectious for weeks. The position of the kinetoplast in relation to the cell nucleus alters during development within the insect. In the epimastigote the kinetoplast is situated anterior to the nucleus but is in the posterior position in the metacyclic trypomastigote. When the fly feeds, then the infectious form of the parasite is transmitted with the saliva to man or to some other receptive host.
The metacyclic stages in the salivary gland of the tsetse fly, in contrast to the preceding stages, once again have a protective surface layer (surface coat) which enables the trypanosomes to survive in the bloodstream of the host.

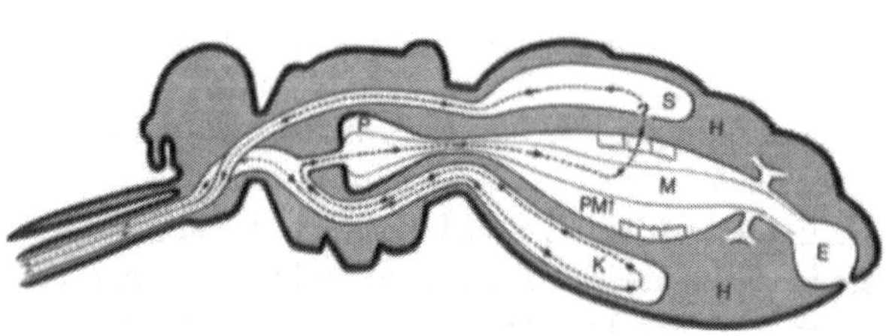

Glossina species. Migration route of the trypanosomes. After passing through the crop (*K*) the trypanosomes migrate into the midgut (*M*), penetrate the peritrophic membrane (*PM*), move into the ectoperitrophic space and via the middle intestine epithelium into the haemocoel (*H*). They reach the salivary glands (*S*) via the haemolymph

Clinical Symptoms. The site where the infected tsetse fly has fed initially becomes oedematous and the pathogens can be demonstrated within the lesion by microscopy during the first 2–3 weeks, before any general symptoms occur. It is very important to recognize this lesion so that treatment may be initiated early. After the prepatent period (about 2 weeks) the trypanosomes appear in the peripheral blood and can be demonstrated in a blood smear, or better still in a centrifuged sample of blood (p. 14). The trypanosomes are predominantly

found during attacks of fever, but then not in large numbers, in the peripheral blood and also in the lymph vessels and lymph nodes. Later on, but in many cases relatively early in the course of the disease, trypanosomes may be found in the cerebrospinal fluid.
The main feature of trypanosomiasis is that the patient develops an irregular fever. Suspicious symptomatology includes an accelerated pulse, headaches, transient local swelling or general swelling of the lymph nodes, in particular of the cervical lymph nodes, and also circumscribed reddening of individual parts of the body surface and transient localized oedema (eyelid oedema). However, these symptoms may be absent. Periods of disease with distinct clinical symptoms may be followed by general remissions even with complete freedom from fever. If untreated however, the disease always progresses. Symptoms indicating involvement of the nervous system include – severe headaches, dizziness, transient or lasting pareses, apoplectiform attacks, visual disorders, hypersensitivity of individual areas of skin, muscle weakness, convulsions, and finally gait disorders and ataxia. Along with progressive exhaustion and anaemia, somnolence becomes an increasing feature ("sleeping sickness"). The patient dies in a state of extreme cachexia. In infection due to *T. b. gambiense* these symptoms occur very slowly one after the other, with the disease taking a chronic course, whereas in infection with *T. b. rhodesiense* the disease takes an acute course ("The patient does not get to the stage of sleeping"; RODENWALDT, 1944).

The **transmission** of *T. b. gambiense* and *T. b. rhodesiense* is virtually exclusively by both sexes of *Glossina* species (Plate I, Ⓑ). Other blood-sucking insects (e.g. *Stomoxys*) can at the most effect mechanical transmission, but this route is probably of no epidemiological importance. Glossinas do not lay eggs but instead give birth to individual larvae (10–15 during their lifetime), which are ready to pupate and which complete their development in the soil (Plate I, Ⓑ *f–i*). The collecting of pupae and flies and also the use of fly traps in combination with insecticides are therefore effective control measures. In addition, males sterilized by irradiation are used (sterile male technique) thereby ensuring that the females also remain infertile. The irradiation of female flies has proved to be even more effective, because their gonads are less likely to "recover" than those of the males (VAN STRYDONCK, 1982). Depending on the species, glossinas may frequent the savannah (e.g. *Glossina morsitans*) or the rain forests. They are frequently found along the banks of rivers and lakes. The growth of vegetation along the side of rivers is a typical feature of the landscape of tropical Africa. *G. palpalis* is found in these forest arcades. In West and Central Africa these "fly zones" along rivers and lakes are particularly high risk regions.
A biological method of controlling the tsetse fly consists of artificially altering the biotope, e.g. through the elimination of the vegetation along the banks of rivers. Another method of biological control consists of the use of small parasitic wasps of the genus *Mutilla* and Bombyliidae of the genus *Thyridanthrax*. Great importance is attached to this method for the future. The successful eradication of *Glossina* species will only be achieved by a combination of all the available mechanical, chemical and biological methods, and this is urgently required, not least in the interests of cattle breeding (LAIRD, 1977).
Repellents (preparations which are applied to the skin and which repel insects) offer people individual protection against infection. DDT and dieldrin are widely used as insecticides. It is worth noting that to date no signs of resistance to

insecticides have been noted in tsetse flies (the only important vector with no resistance).
Although man must be considered the main host Ⓐ, some mammals deserve attention as animal reservoirs for *T. b. rhodesiense* ©, e.g. antelopes and domestic pigs, and probably also goats, sheep and cattle. Locally, a direct relationship is seen, for example, between the extent to which sheep are bred and the incidence of trypanosomiasis.
T. b. gambiense, on the other hand, is primarily transmitted from man to man by *Glossina* vectors. According to ZILLMANN et al. (1983, 1984) however, dogs, pigs, goats and also antelopes must be seen as reservoir animals for *T. b. gambiense* as well. The dissemination of trypanosomes takes place through migrations of herds to new areas; dry periods lead to such migrations. As a result the boundaries of animal reservoirs can only be given imprecisely.
The difficulties of differentiating between *T. b. brucei* and *T. b. rhodesiense,* which is of epidemiological importance, have led to an in vitro test for distinguishing between the two subspecies. Morphological criteria are not sufficient if one is to distinguish between them in tsetse flies and potential parasite reservoirs in mammals (see Zymodemes, p. 11). Congenital transmission and acquisition of infection through sexual intercourse are also possible but probably rare.

Diagnosis by Microscopy. Trypanosomes can be detected by examining (1) the peripheral blood, (2) the cerebrospinal fluid, or (3) fluid expressed from enlarged lymph nodes. Besides GIEMSA-stained thin and thick blood films, preparations of fresh blood and cerebrospinal fluid should always be examined by microscopy as well; since parasites may be more easily detected by their movements in these than in a stained preparation (see p. 306). Suitable experimental animals for the demonstration of the pathogens are, in particular, mice, but also rats or guinea-pigs. SACHS et al. (1984) recommend a haematocrit centrifugation technique and also the so-called miniature anion exchange process which may detect even very light trypanosomal infections of the peripheral blood. According to MEHLHORN and PETERS (1983), the enrichment procedure with double centrifugation has proved to be particularly useful. (5 ml of venous blood is mixed with an identical amount of a sodium citrate solution, centrifuged for 10 min at 150 g, the sediment is discarded and the supernatant is then centrifuged again at 900 g for 10 min. In a positive sample the sediment will contain motile trypanosomes.)

Diagnosis by Immunobiological Methods. Of the serological methods available, besides the complement fixation test, the indirect immunofluorescence test, enzyme-linked immunosorbent assay (ELISA), the agglutination reaction and demonstration of a marked increase in the IgM fraction of the blood, have all proved useful in the diagnosis of trypanosomiasis. The IgM fraction may be 2–8 times above normal. The results are admittedly not species specific but in cases with negative findings with microscopy and where the clinical suspicion is high, this procedure can be an important diagnostic aid.

Chemotherapy and Chemoprophylaxis. Various effective preparations are available for both the treatment and prevention of trypanosomiasis. The choice of remedy is based on the stage of the disease. Suramin is still favored in the first stage of the disease (dosage, 20 mg/kg body weight, up to a maximum of 1 g; 3–5 doses administered with a 7-days interval between each dose). The preparation must be used cautiously because the side effects can lead to death. The first test dose (0.1–0.2 g intravenously) should be injected slowly over 3 min, keeping the patient under observation because of the danger of collapse. If fairly severe side effects occur, the treatment should be discontinued. The side effects of the drug are closely related to the state of health of the patient (side effects include albuminuria, exfoliative dermatitis, diarrhoea lasting more than 3 days, sometimes high fever, and bronchitis). VON KÖNIG (1984) recommends that the patient should be closely monitored for hypoalbuminaemia, which necessitates a reduction in the dose of suramin. Other useful drugs include the diamidines (pentamidine isethionate or Lomidine), which have a particularly prolonged prophylactic action.

Diminazene aceturate (Berenil) is considered an alternative drug to suramin for the treatment of stage I of African trypanosomiasis. Side effects must also be expected with Berenil (5 mg/kg body weight, 3 times at intervals of 1–2 days, intramuscularly or orally) but these are short-lived (ABARU et al., 1984).

Various arsenic and antimony preparations are used for treatment in the second stage. Tryparsamide is considered to be the drug of choice. In addition, melarsonyl potassium (Mel W), a derivative of melarsoprol (Mel B; danger of reactive encephalopathy), or the antimony compound MSbB are effective, so that basically the treatment of the second stage of sleeping sickness should not give rise to any difficulties. In early cases which are resistant to therapy nitrofurazone is recommended (1500 mg orally each day for 10 days; toxic side effects: WHO, 1986).

Both suramin and pentamidine are used for **chemoprophylaxis.** The mean duration of the prophylactic action of suramin (1–2 g for adults, 0.3–0.75 g for children), can be considered to be 3 months, and for pentamidine, 6 months.

Plate II ⇨

Trypanosoma cruzi
Trypanosoma rangeli

Plate II

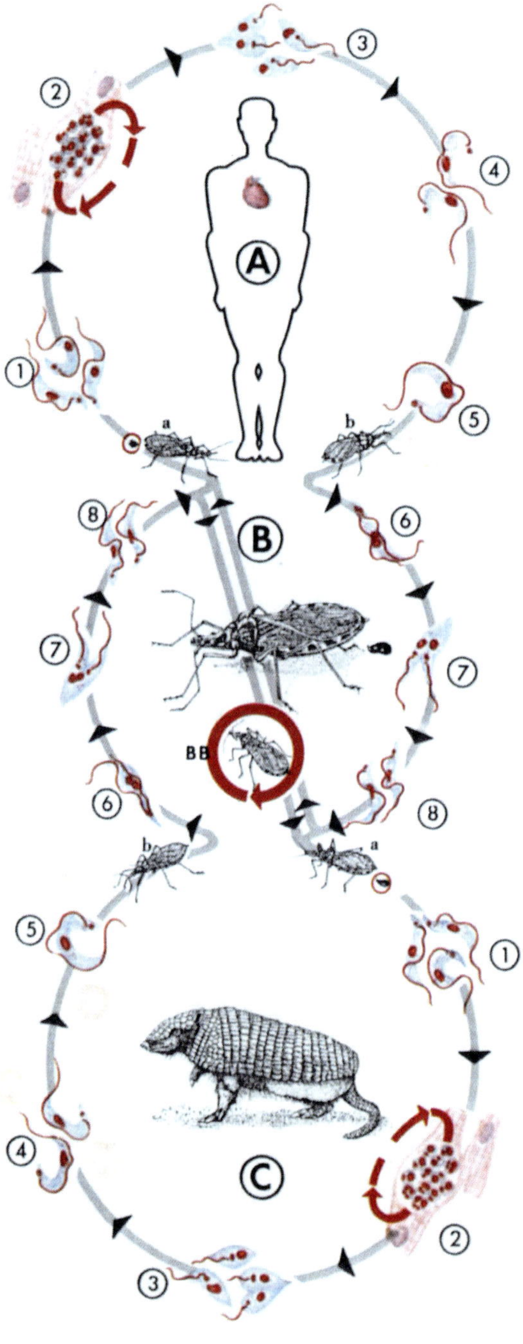

Trypanosoma cruzi CHAGAS, 1909

Pathogen of American trypanosomiasis (CHAGAS' disease)

Ⓐ Development in man (GIEMSA stain)

1 Metacyclic trypanosomes reach man in reduviid bug faeces (*a*)
2, 3 Intracellular transformation to amastigote stage in the "pseudocyst" with marked multiplication and development through epimastigote stage into trypomastigotes (see text p. 20)
4, 5 Trypomastigote forms from the peripheral blood

Ⓑ Development in the intestine of the reduviid bug vector (*Triatoma* sp.)

6 Freshly sucked up trypanosomes in the process of division
7 Transformation to the epimastigote form
8 Metacyclic trypomastigote form from bug faeces
BB Coprophagy and cannibalism lead to infection of the young reduviid bugs at the larval stage of development

Ⓒ Development similar to that in man Ⓐ occurs in the parasite reservoir (armadillo, opossum, dog and other animals).

Transmission by reduviid bugs can take place:

1. From man to man — Ⓐ → Ⓑ → Ⓐ
2. From animal to animal (parasite reservoir) — Ⓒ → Ⓑ → Ⓒ
3. From animal to man and — Ⓒ → Ⓑ → Ⓐ
   vice versa — Ⓐ → Ⓑ → Ⓒ
4. From bug to bug through coprophagy and cannibalism (*BB*)

CHAGAS' disease (American trypanosomiasis) develops from an infection with *Trypanosoma cruzi* and is indigenous to South and Central America. The main areas of distribution are Mexico, Guatemala, San Salvador, Costa Rica, Panama, Venezuela, Brazil, Argentina, Uruguay and northern Chile. The importance of this disease to Latin America becomes obvious if one considers that about 65 million people are exposed (over 6 million in Brazil alone) and 24 million are infected, with about 50% being asymptomatic (WHO, 1982). Approximately 10% die from the disease, which generally takes a chronic course. The pathogen is transmitted by bugs of the genus *Triatoma* (Reduviidae) and related species, which are also found outside of the regions mentioned (see dotted region of the map).

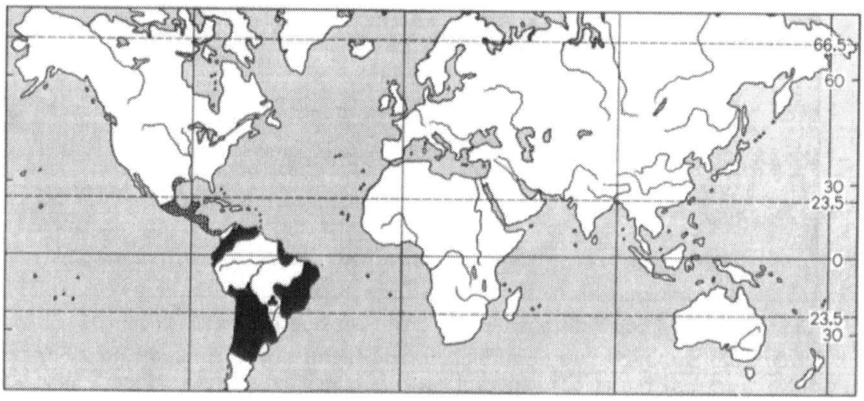

Regional distribution of *Trypanosoma cruzi*

Morphology and Development. *Trypanosoma cruzi* closely resembles the somewhat larger African *Trypanosoma* species (including the flagellum the organism is about 15–20 µm long) and has an undulating membrane. The basal granule from which the flagellum originates lies beside the kinetoplast, which is particularly prominent. In the centre of the cell is the nucleus which stains red with GIEMSA-stain (see Plate II, *1*). *T. cruzi* characteristically occurs in stained blood films as C-shaped trypomastigotes.

The trypomastigotes remain in human peripheral blood for a time (Plate II, Ⓐ *1, 4, 5*), but multiply very little. Multiplication takes place predominantly in endothelial cells of the lungs, liver, lymph nodes, gonads and other organs, and also in muscle tissue, particularly the heart. They develop to the amastigote stage intracellularly (Plate II, *2*), and then through an epimastigote stage to the trypomastigote stage (MEHLHORN et al., 1977). After several divisions a host cell contains about 500 parasites and eventually ruptures. Some of the amastigotes degenerate, the rest attack new host cells.

The reduviid bugs can take up the trypanosomes with their blood meal (Plate II, Ⓑ *6*). The parasites then develop into epimastigotes in the gastrointestinal tract of the bug (Plate II, Ⓑ *7*), i.e. the kinetoplast migrates anterior to the nucleus. This means that more stumpy forms develop, which then transform into metacyclic trypomastigotes in the insect's hindgut and at the entrance to the midgut (Plate II, Ⓑ *8*) (Böker and Schaub, 1984). These trypomastigotes are excreted with the faeces ("Stercoraria").

The surface coat already described for the *Trypanosoma brucei* group also occurs in the trypomastigote blood form of *T. cruzi*. It is absent from the epimastigote, however, but develops anew with the metacyclic trypomastigote stage in the hindgut of the cone-nosed bug.

There are variations between different strains of *T. cruzi* with regard to virulence and also tropism (myotropic, reticulotropic).

One possible way of characterizing different strains of *T. cruzi* is by separation of isoenzymes by electrophoresis. The resultant typical combinations of bands are called zymodemes (see *T. b. gambiense*, p. 11). A second method relates to the biochemistry of the parasite, including an analysis of the kinetoplast DNA. Other possibilities include analysis of the plasma membrane, which consists of a number of surface glycoproteins which are apparently of vital importance in the parasite–host relationship.

Besides *Trypanosoma cruzi*, reduviid bugs may be naturally infected with a morphologically similar flagellate, *Blastocrithidia triatomae* (Cerisola et al., 1971). This organism is also found in the bug's intestine but is non-infective for man. Transmission from bug to bug takes place via cysts which are excreted with the faeces and apparently taken in orally during the usual coprophagy. This parasite causes delayed development and increased mortality in the fourth and fifth stage larvae of the bug, so that the idea of using these flagellates for the biological control of *Triatoma* species has already been considered (Jensen, 1984; Schaub, 1986). *Blastocrithidia triatomae* can be easily grown in cell culture using tissue from the host (e.g. *Triatoma infestans*) and may also be cultured in mice (Mehlhorn, personal communication).

Clinical Symptoms. In man, particularly in children, there is an acute pyrexial stage of the disease following an incubation period of 10–20 days, when the parasites are found in the blood. This stage is characterized by oedema of the face, particularly the eyes (unilateral eyelid oedema is known as Romaña's sign), swelling of the lymph nodes, hepatosplenomegaly and anaemia. The acute phase generally lasts for 1–2 months. Death may occur within 2–4 weeks however, as a result of an acute myocarditis (up to 10% of cases depending on the region). In such cases there is lymphocytic infiltration and destruction of normal myocardial cells, sometimes even when no parasites have been demonstrated. The administration of corticosteroids in the acute phase of the infection leads to greater parasitaemia and cell parasitism. This effect of cortisone is not seen in the chronic phase. In the chronic stage the heart, central nervous system, thyroid gland, adrenal glands and also the gastrointestinal tract are affected. Denervation through *T. cruzi* gives rise to the "mega" syndrome – cardiomegaly, megaoesophagus, megastomach and megacolon (possibily involving autoimmunity) (see Köberle, 1968, 1972). Typical signs and symptoms include fullness, hypotonia of the small intestine, segmental dystonia, accelerated or slowed intestinal transit, a bloated abdomen, chronic constipation, and difficulties with defaecation. It is worth noting that certain manifestations of the disease are strain dependent and thus linked to a geographical region. For example, in Venezuela the mega-syndrome does not occur, whereas it is common in central and eastern Brazil (Miles 1983). These strain differences also become apparent in isoenzyme analysis (see above).

The number of asymptomatic individuals with chronic disease is greater than the number with manifest disease. The prognosis is often poor in children. In the endemic areas about

30% of those aged over 60 years have positive serology (even up to 90% locally in Brazil), and almost 50% of these have myocardial damage.

Transmission to man does not take place through the bite of the reduviid bug but through the contamination of bite wounds or of mucous membranes with insect faeces containing trypanosomes. The faeces are deposited on the host as the bug takes its blood meal (Plate II, Ⓐ *a*). The pathogen is then rubbed into the bite wound by the host in response to the itching. The most important vectors are *Triatoma infestans, T. brasiliensis, T. dimidiata, T. sordida, Panstrongylus megistus* and *Rhodnius prolixus*.

Congenital transmission of *Trypanosoma cruzi* can occur, particularly between the 26th and 37th week of pregnancy. Infection through breast milk is possible, but according to experimental results both routes of transmission are largely dependent on the strain of *T. cruzi* (ANDRADE, 1982). The transmission of the pathogen by blood transfusion from latent infected carriers has become of great importance. According to ROHWEDDER (1969) this route of infection is in fact the second most frequent. The trypanosomes survive in stored blood at +6 °C for about 14–21 days.

It is worth noting that a trypanosome infection within a reduviid population is maintained through coprophagy and cannibalism, so that it appears to be hereditary (Plate II, *BB*).

Epidemiology. Besides armadillos and opossums other possible animal reservoirs are dogs, cats, monkeys, bats(?) and presumably numerous species of rodents which often live in close association with the bugs (100 species of domestic and wild animals may be involved). CHAGAS' disease is thus a zoonosis. Man is nevertheless the main host, because he has the longest life expectancy and therefore remains a source of infection for a long time.

It should be borne in mind that many wild animals harbour species similar to *Trypanosoma cruzi,* including *T. rangeli* which is not pathogenic for man (see p. 25). Differentiating between *T. cruzi* and similar species can apparently be carried out with the aid of lectins (SCHOTTELIUS, 1982).

There are about 100 members of the genus *Triatoma* and related genera, most of which have been found to be naturally infected and once infected remain so for life. The *control* of reduviid bugs by using insecticides has generally proved to be very successful and is at present the only practicable control measure against CHAGAS' disease. However, insecticide resistance is occurring, as with many other arthropod hosts (see p. 102). In many parts of Venezuela signs of resistance of *Rhodnius prolixus* to dieldrin have occurred. Natural enemies of the bugs are to be found amongst the *Hymenoptera* (e.g. ichneumon flies, which oviposite the triatomid eggs), spiders and others.

Prophylactic Measures. In the main endemic regions prophylactic measures include the monitoring of blood donors, who, according to the investigations of SAGUA et al. (1982), are to be considered as suspect in about 12% of cases on the basis of serological results. In order to identify infected donors a micro-test is used to recognize parasitized blood. Health education is very important so that the indigenous population is informed about the routes of infection and can help in the control of the vectors. This includes improving living conditions so that hiding places are not offered to the bugs, e.g. in bamboo walls.

Diagnosis by Microscopy. In the acute stage of the disease the pathogen can be demonstrated by examining fresh blood or a stained thin or thick blood film or a muscle biopsy (see p. 306). Animal tests are also advisable (e.g. the injection of fairly large amounts of the patient's blood into guinea pigs or young dogs). After about 2 weeks one can then expect the trypomastigote form in the blood, but microscopic demonstration of the pathogen in a fresh blood preparation and also in the bone marrow and cerebrospinal fluid is difficult (care must be taken as double infections with *T. rangeli* are possible in animals).
In xenodiagnosis the pathogen is detected by means of the vector. Flagellate-free bugs are allowed to feed on the blood of a suspected case (up to 4 ml). Three weeks later trypomastigote forms may be found in the hindgut or faeces of the bugs. (*T. rangeli* in the salivary glands). The best results are said to be obtained by examining crushed bugs. Using this method parasites can be demonstrated in about 60% of patients in whom they could not be found by microscopic examination. The parasites can also sometimes be demonstrated by the blood culture method of NÖLLER (1923). WERNER and MERKS (1984) also recommend an artificial xenodiagnosis technique which involves feeding suspect blood samples to bugs using a plastic sheet as a membrane. The blood sample would have to be sent to a specialist laboratory for this. A positive result may still be obtained after 72 hrs. An improvement in microscopic techniques is the detection microhaematocrit method as used with *T. b. gambiense*. In this, blood is obtained from the finger tip, mixed with heparin in a haematocrit tube and centrifuged. Even a light infection can be detected (see pp. 14, 306; SACHS et al., 1984; LA FUENTE et al., 1984). It should be borne in mind that infections with the non-pathogenic species *T. rangeli* occur in man (see p. 25) and can lead to confusion with *T. cruzi* (see xenodiagnosis, above).

Diagnosis by Immunobiological Methods. In CHAGAS' disease, in addition to xenodiagnosis, the demonstration of antibodies is of great importance. It can be very difficult to detect trypomastigotes in the peripheral blood by microscopy, especially in patients with chronic disease and indigenous blood donors. The immunofluorescence test, the complement fixation test, ELISA and the indirect haemagglutination test (IHAT) have proved useful, making it possible to pick up all chronic cases even after infections of varying duration. Results of the complement fixation test generally do not become positive until the second month, and of the IHAT even later. To date however, none of the methods have been standardized. A network of cooperating laboratories has been formed in order to achieve standardized serological diagnosis. A particularly simple test, which is suitable for field investigations, is the staphylococcal-binding test. This is based on the fact that cultured epimastigotes form specific antibodies, and this complex will bind the bacterium *Staphylococcus aureus* (CAMARGO et al., 1983). Improved serological results are expected from initial attempts to detect circulating antigens with monoclonal antibodies and with a modified ELISA procedure. In serial investigations an intradermal test with an antigen extract called cruzin has proved of value. This extract has been produced from cultured trypanosomes. Cross-reactions may be expected only with leishmanial infections (CERISOLA et al., 1972). Experiments with vaccination against *Trypanosoma cruzi,* using live attenuated epimastigotes grown in culture, have given rise to the hope that this method of control might also be used in humans (ENDERS et al., 1984).

Chemotherapy. Specific effective remedies such as nifurtimox and benznidazole which eliminate both the blood forms and also the intracellular (amastigote) stages, are available (RAAFLAUB, 1980). Continuous treatment for 90 days is required with nifurtimox in children (15–20 mg/kg body weight per day) and young people (12.5–15 mg/kg body weight per day), and for 120 days in adults of 17 years or older (8–10 mg/kg body weight per day). This protocol is effective in both the acute and chronic stages (WEGNER and ROHWEDDER, 1972a, b; GÖNNERT, 1972; SCHENONE et al., 1981). There may be slight side effects, such as loss of appetite and headaches. Some strains of *Trypanosoma cruzi* are showing signs of becoming resistant to nifurtimox (CANCADO and BRENER, 1979). Benznidazole is given in a dosage of 4–5 mg/kg body weight per day for 30 days. However, the preparation is not devoid of side effects (e.g. polyneuropathies, exanthema).

Trypanosoma rangeli (Tejera, 1920) Hoare, 1972

Amongst the trypanosomes resembling *Trypanosoma cruzi*, *Trypanosoma rangeli* is of special importance because it develops in numerous mammals as well as in man but is non-pathogenic. The latter occurs in an area covering Central and South America, particularly Guatemala, El Salvador, Costa Rica, Panama, Columbia, Venezuela and Brazil (with a high frequency locally of over 30%). The carriers are reduviid bugs of the genus *Rhodnius*, which infect man with trypanosomes via their saliva during biting. The main carrier is *Rhodnius prolixus*, and where this occurs it is also likely that *T. rangeli* is present.

Morphology and Development. The appearance of *Trypanosoma rangeli* is substantially different from that of *T. cruzi*, because the blood form of *T. rangeli* is larger (about 30 µm vs. 20 µm). The nucleus lies in anterior half of the cell, and the kinetoplast is very small. The level of parasitaemia in man is generally low. Dividing forms are rare, while tissue forms apparently do not exist. There are also no pathological changes.

It is notable that *T. rangeli* is highly pathogenic for the reduviid bug. The parasites enter the haemocoelom and multiply very extensively in the haemolymph, where they produce epimastigote stages, then transform into trypomastigote or metatrypomastigote stages (size only about 10 µm) and enter the salivary glands (Evans and Ellis, 1983). The extent of the damage depends on the stage of development of the bug, the extent of the infection, and the strain of trypanosome. The early nymph stages are most severely affected, whereas primary infection in the adult bug does not produce any substantial injury. If the first nymph is infected, 34%–98% of the bugs die before reaching the adult stage, the proportion depending on the virulence of the trypanosome strain. All organs are affected. Ecdysis is delayed, and the haemolymph volume is increased. Particularly important is the destruction of symbionts, as these have metabolic and physiological functions (D'Alessandro, 1976). The bugs also become infected with *T. rangeli* by cannibalism, i.e., the bugs bite each other and thereby ingest trypanosomes with the haemolymph and blood (see also p. 22).

Xenodiagnosis is the best way of detecting *Trypanosoma rangeli*. In a positive case, the flagellates occur after 20–100 days in the salivary gland and/or the haemolymph. Trypanosomes are not found in the faeces. Serological methods are of little value because of cross-reactions with *T. cruzi*.

GUHL et al. (1986) found an antibody to a 90-kDa glycoprotein in patients with *T. cruzi* infections which was lacking in *T. rangeli* infections. It is possible also to distinguish between the two antibodies of the two species with an ELISA test. It is possible to distinguish between *T. cruzi* and *T. rangeli* by agglutination with lectins (SCHOTTELIUS, 1982) and also by means of the neuraminidase test (SCHOTTELIUS, 1986).

T. rangeli can be cultivated in vitro on NNN agar (NNN agar: 900 ml distilled water, 14 g agar, 6 g NaCl; cool to 45°–50 °C; then add 1.5 ml sterile defibrinated rabbit blood to each 5 ml and allow to solidity at a slant). It is of note that the cultured forms are infectious.

Plate III ⇨

Leishmania donovani, L. tropica,
L. braziliensis, L. mexicana

Plate III

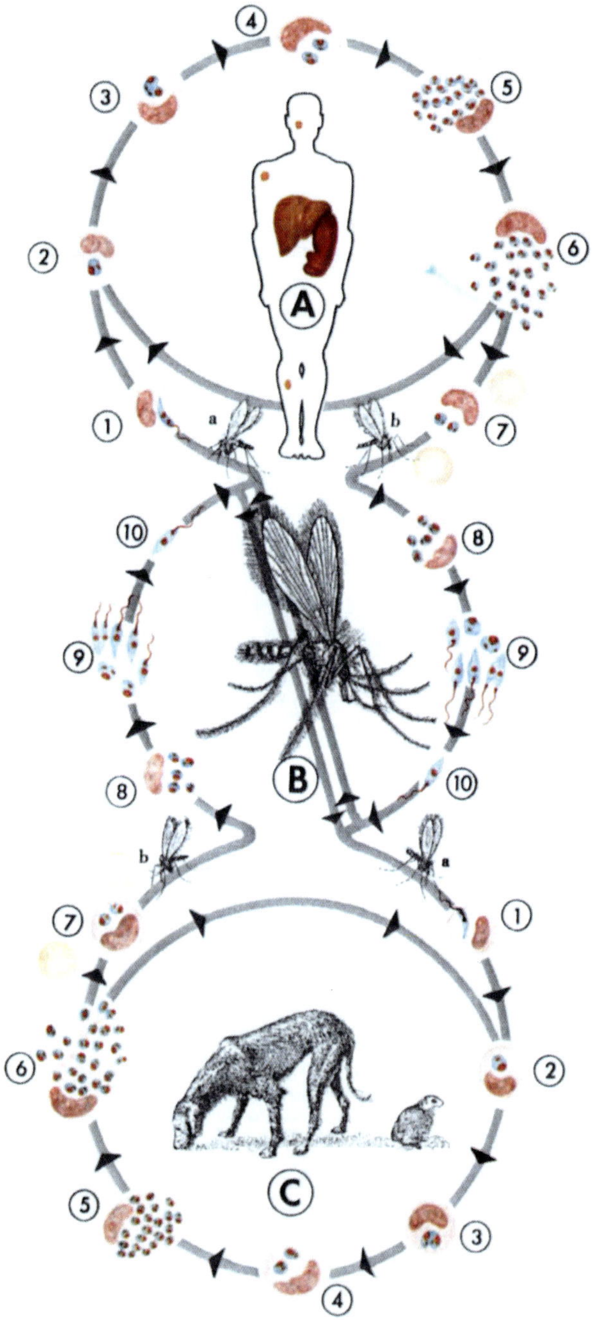

Leishmania donovani (LEVERAN and MESNIL, 1903) ROSS, 1903
L. tropica LÜHE, 1906, **L. braziliensis** VIANNA, 1911
L. mexicana (BIAGI, 1953) GARNHAM, 1962

Pathogens of leishmaniasis

Ⓐ Development in man

1 Flagellated *L. donovani* parasites carried by *Phlebotomus* species (promastigote stage) enter macrophages (GIEMSA stain)
2–6 Intracellular development in macrophages and later in endothelial cells

7 Macrophages in peripheral blood containing amastigote stages

Ⓑ Development in the sandfly vector

8 Amastigote stages, within host cells, in the fly's midgut
9 Growth and multiplication of the promastigote stage
10 Flagellate stage (metacyclic promastigote form) from fly's proboscis

Ⓒ Development as in man Ⓐ in animal reservoir (dogs, small rodents, etc.)

Transmission by the sandfly can occur as follows:

1. From man to man Ⓐ → Ⓑ → Ⓐ
2. From animal to animal Ⓒ → Ⓑ → Ⓒ
3. From animal to man and Ⓒ → Ⓑ → Ⓐ
   vice versa Ⓐ → Ⓑ → Ⓒ

All the parasites are contained in parasitophorous vacuoles.

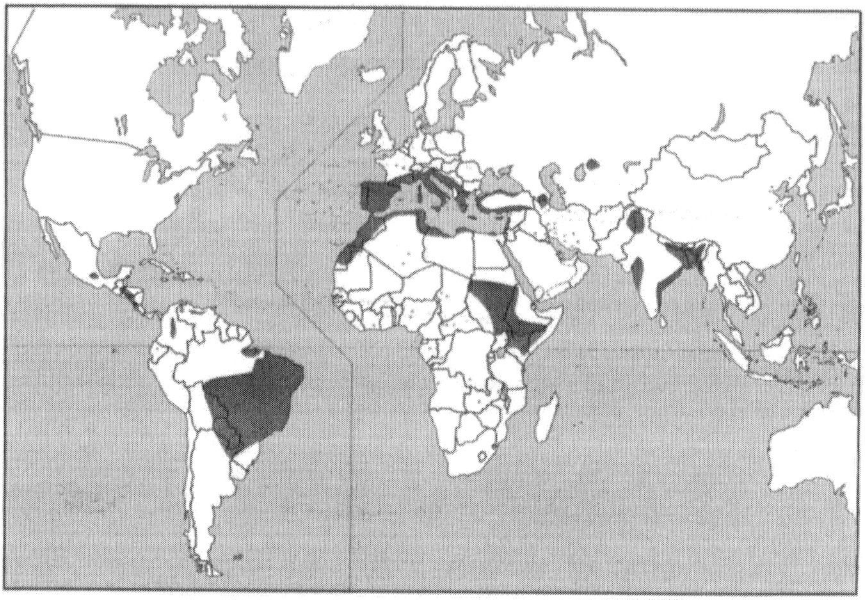

Distribution of visceral leishmaniasis (WHO, 1984)
grey sections endemic areas; *small shaded sections* indicate sporadic cases

The leishmanias belong to the family Trypanosomidae. At present, at least 13 different *Leishmania* species and subspecies are recognized from 80 different countries. The parasites are transferred to man by sandflies of the genus *Phlebotomus* and closely related genera.

Leishmaniasis in man occurs in three different forms: visceral, cutaneous and mucocutaneous. Each form shows local distinct features and manifestations depending on the species or subspecies of pathogen.

Visceral leishmaniasis or kala-azar is common in India, which is probably where the disease originated. (The causative organisms are *Leishmania donovani donovani* LAVERAN and MESNIL, 1903, ROSS, 1903; *L. d. infantum* NICOLLE, 1908.) In Europe, it occurs in central and southern parts of Portugal, and in Spain, France, Italy, the Balkans, and Turkey, as well as in the Soviet Union, where mainly children are affected. In addition, it occurs in the Near East and Egypt, in certain parts of East and Central Africa, but only occasionally in West Africa (Gambia). There are two large areas of occurrence in Brazil and Central America, although here the number of cases is generally small.

Cutaneous leishmaniasis (causative organisms *Leishmania tropica* WRIGHT, 1903, LÜHE, 1906; *L. major* YAKIMOW and SCHOCKOW, 1915 and *L. aethiopica* BRAY, ASHFORD and BRAY, 1973) occurs mainly in the Near East and throughout the Mediterranean, but also in Turkey, the Caucasus, the Sudan, Nigeria, and Senegal, as well as in Gambia and India. Special *Leishmania* strains are being

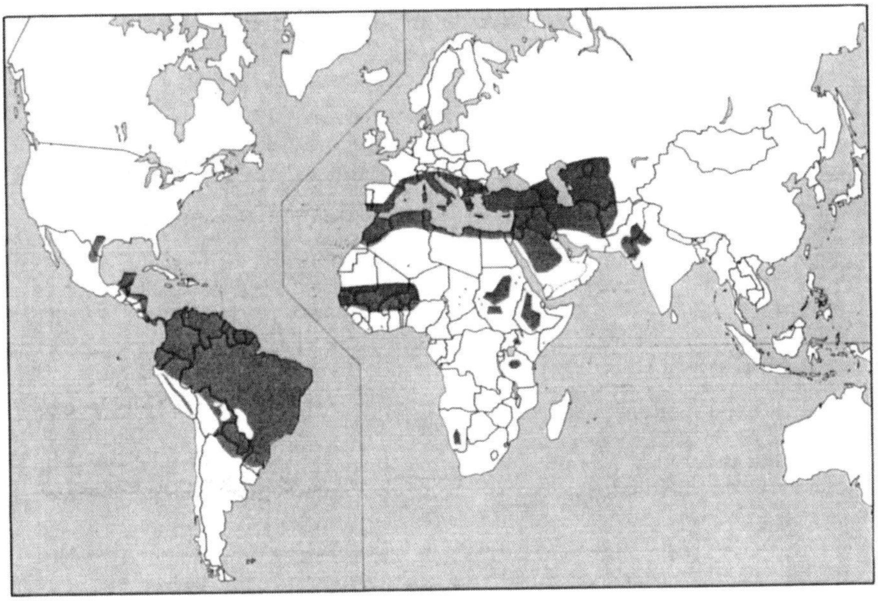

Old World, *Leishmania tropica;* New World, *Leishmania braziliensis*
grey sections endemic areas; *small shaded sections* indicate sporadic cases (WHO, 1984)

reported with increasing frequency, principally from Africa, which show a tendency to visceralization (*L. aethiopica* in Kenya and Ethiopia).

Mucocutaneous leishmaniasis (espundia, uta; causative organisms *L. braziliensis braziliensis* VIANNA, 1911, *L. mexicana mexicana,* numerous subspecies of *L. peruviana*) is at present restricted to South and Central America, but is on the increase. It has become a public health problem on account of its debilitating effect on the body and damaging effects on the face (MARINKELLE and RODRIGUEZ, 1981).

North and Central Europe are free from *Phlebotomus* species and *Leishmania* species, but cases of leishmaniasis are to be expected because of tourism and visits to various regions where the disease occurs (FALKNER VON SONNENBURG et al., 1979).

Morphology and Development. The intracellular stages (2–5 µm in size), which occur only in man Ⓐ, multiply by binary division (Plate III, *2–6*). In a GIEMSA-stained smear the plasma appears blue and the nucleus and kinetoplast red. Within the gut of the sandfly vector Ⓑ are found the flagellate promastigotes, which also divide (Plate III, *9*). The basal granule and kinetoplast lie anterior to the nucleus in the promastigote. These stages migrate into the proboscis of the sandfly and when the flies bite the parasites are inoculated into man or another suitable vertebrate Ⓒ (Plate III, *10, a*).

The various species are morphologically almost identical. However, they can be distinguished on the basis of their isoenzymes. This provides a form of biochemical taxonomy for the species and strains of *Leishmania* (CHANCE et al. 1974; GARDENER et al. 1974). Typical differences also occur in the buoyant density of kinetoplast DNA, which allows one to differentiate, for example, among the South American species or subspecies (BARKER and BUTCHER, 1983). There is further scope for distinction by lectin agglutination, although the interpretation of the results is not unambiguous (EBRAHIMZADEH and JONES, 1983). Specific monoclonal antibodies and DNA hybridization techniques can also be used in species identification.

Clinical Symptoms. The incubation period may vary from 10 days to more than a year, although infection may be completely asymptomatic. The disease picture in visceral leishmaniasis usually begins with vague general pains. Characteristically there is enlargement of the spleen and liver following invasion by the parasite. The rise and fall of fever can be documented by regular 4-hourly measurements, as is generally recommended by RODENWALDT for tropical diseases, and shows a distinct daily double peak. The blood picture consists of leucopenia, thrombocytopenia, a left shift, lymphocytosis, and monocytosis, as well as elevation of IgG. In general, one gets the following typical disease picture: anaemia, increasing weakness, notable enlargement of spleen and liver, dry and darkening skin, haemorrhages in skin and mucosa, hair loss, dysenteric diarrhoea, oedema, and effusions in the body cavities (DUARTE et al., 1983). During the course of a few months or years the disease steadily progresses such that untreated cases are generally fatal.
These general comments on visceral leishmaniasis do not cover the full variety of disease symptoms, as these depend on local strains of the parasite and the immune status of the patient. On account of the parasite's immunosuppressive effect, the immune status is unfavourably altered and the patient becomes more susceptible to bacterial infections. In the Mediterranean area, and in Southwest Asia, China, and Latin America, the disease mainly affects children aged 1–4; in East Africa and India, it mainly affects children aged 5–9 or older.
Subclinical infections exceed the number of clinical cases in some areas (Italy and Kenya); stress situations can lead to exacerbation (WHO, 1984). Consequently, no definite incubation time can be quoted although it is usually between 10 days and 1 year.
A specific post kala-azar leishmanoid is characteristic of India and Bangladesh. This appears some years after the healing of visceral leishmaniasis. The chronic lesions consist of multiple nodular infiltrations in the skin, usually without ulceration. Hypopigmented or erythematous patches on all parts of the body later become nodular, particularly on the face. In India, one frequently finds distinctive dark pigmented skin on the face, the hands and feet, and the abdomen (hence the name kala-azar = black disease). This disease picture also usually occurs in East Africa.

Travellers from the Mediterranean area usually carry visceral leishmaniasis. The disease picture is largely unknown in Northern and Central Europe and often causes substantial diagnostic difficulties. If there is a clinical suspicion of the disease, serology can prove valuable (see below). The clinical history may aid the diagnosis and also may the fact that there has been no response to antibiotics, which are often given in cases of pyrexia of unknown origin (KRAMPITZ, 1980).

The clinical symptoms in cutaneous leishmaniasis in the Old World differ with the geographic area and reflect the various *Leishmania* species and local variants. The incubation period is generally 2–8 months. The classical lesion of the so-called urban form (causative organism *Leishmania tropica minor*) begins with a small nodule at the point of puncture. The lesion develops a central depression, ulcerates and slowly heals, often leaving a permanent scar (oriental sore). Satellite nodules frequently occur at the margins of the lesion. Healing is usually spontaneous within a year. Natural healing is followed mostly by immunity, and consequently the indigenous population often infect others by direct transfer and decline treatment. Some strains of *L. tropica minor* tend towards visceralization. The spread to the viscera arises from simple and in part recurrent skin changes. Local manifestations of cutaneous leishmaniasis in specific geographical areas are known as Aleppo button or Baghdad boil.

The rural form of disease is produced by *L. tropica major* results in ulceration, often associated with severe inflammation, which heals, however, within a few months but may leave large scars. In addition to these localized lesions, lupoid or tuberculoid recurrent chronic forms may occur which persist for many years and are very resistant to treatment. Leishmaniasis due to *L. aethiopica* leads to typical oriental sores, to mucocutaneous leishmaniasis (in Kenya and Ethiopia), as well as to a diffuse form of cutaneous leishmaniasis. The first case of mucocutaneous leishmaniasis outside South America was observed in Senegal by LARRIVIÈRE and CAMERLYNCK (1978). These authors suspect that it is more common there than had formerly been supposed.

KRAMPITZ (1981) commented on the symptom pattern in "harara" (arabic: heat, sun's warmth), characterized by skin changes indicating an immediate allergic reaction to the salivary secretion of the sandflies. This involves an itchy papular urticaria (urticaria multiformis endemica) with flat, hard, bright red to haemorrhagic vesicles, whose maximum size may be that of a pea. These occur mainly on uncovered parts of the body, but also have a tendency to accumulate on the extremities, as the author has observed in Elba. These symptoms can also occur in tourists and have no relationship to cutaneous leishmaniasis, although they can be significant in indicating leishmaniasis if there is other appropriate symptomatology.

Cutaneous leishmaniasis in the New World is caused by various subspecies of *L. braziliensis*. The most important are *L. braziliensis braziliensis*, *L. b. guayanensis* and *L. b. panamensis*. *L. b. braziliensis* produces single or multiple ulcers, which rarely heal spontaneously. The lymph nodes are generally soon involved early. Typical mucocutaneous leishmaniasis or espundia shows a tendency to metastasize. Ulceration and erosion progressively destroy the soft tissues and cartilage in the oral, nasal, and pharyngeal cavities, whilst the lips and nose can become irregularly swollen (so-called tapir nose). Secondary infections are common. In contrast to typical oriental sores, the lesions do not heal spontaneously. The disease leads to severe suffering, mutilation and possibly death due to bronchopneumonia or malnutrition.

L. mexicana mexicana is responsible for the well known rubber-workers' disease, and chiclero ulcer. The lesions localized to the ears in 60% of cases are generally painless, and often heal spontaneously within a few months. Chronic progressive disease often ends with destruction of parts of the ear cartilage.

L. m. amazonensis produces individual or multiple skin lesions, which seldom heal spontaneously. The disease is relatively rare. About 30% of patients show diffuse cutaneous leishmaniasis. *L. peruviana* is responsible for the skin disease known as uta, which occurs mainly in children. The painless lesions usually heal spontaneously within 4 months.

Transmission. The various kinds of leishmaniasis are related to the occurrence of sandflies (genera *Phlebotomus*, *Sergentomyia* and *Lutzomyia*), which transmit leishmanias while sucking blood. In any given geographical area only certain species are responsible for this. Consequently, control measures can be restricted to these species (so-called species sanitation). Not all species prefer to bite man. The phlebotomid larvae live on organic detritus, often near poorly maintained buildings or in animal houses. For example, the carriers of *L. tropica* often live in rodent burrows, the rodents acting as reservoirs for the parasite. The main periods of activity are in the twilight and at night, but in areas where the days are warm and the nights are cool (for example, Egypt in April), the sandflies are also active during the day and enter houses. Protection from phlebotomid bites may be provided by using fine netting, which must be placed well away from the body in buildings. Repellents are often helpful (see malaria, p. 102 and 270). In addition, as a general health care measure organic rubbish that can act as a breeding site for the phlebotomids should be removed (MARINKELLE, 1980).

The following species of sandfly are the main carriers:
For visceral leishmaniasis (kala-azar):
Phlebotomus perniciosus, P. perfiliewi, P. major, P. chinensis, P. argentipes, P. papatasii.
For cutaneous leishmaniasis:
P. papatasii, P. sergenti, Sergentomyia garnhami, and others.
For American mucocutaneous leishmaniasis:
Lutzomyia longipalpis (also for *Leishmania aethiopica*), *L. intermedia, P. longipes,* and others.
The association of a species of *Leishmania* with the various phlebotomid species is not particularly strict, since overlaps occur.

Control of phlebotomids is based on the use of synthetic insecticides as the most important prophylactic measure. However, there is the possibility of insect resistance occurring (see p. 102). The considerable decrease in the control of malaria mosquitoes (*Anopheles* species) has meant that sandflies have also escaped and this has resulted in the unrestricted multiplication of sandflies in the Middle East, India and South America, which has thus led to an increase in leishmaniasis.
Leishmania species occur also in many vertebrates, which can act as significant reservoirs for the parasite (©, zoonoses). The dog family (fox, wolf and other wild predators?) plays an appreciable role in kala-azar. In endemic areas where these animals have been systematically eliminated, kala-azar has been almost completely eliminated. Thus far the rat (*Rattus rattus*) appears so far to have been underestimated as a reservoir for *Leishmania donovani infantum,* the cause of visceral leishmaniasis in man and dog in the Mediterranean area (Italy). In India, there appears to be no animal reservoir; only man is the source of parasites. Small wild rodents (ground squirrels, gerbils, etc.) can act as reserve hosts for the parasites that cause cutaneous leishmaniasis. The main reserve hosts for *L. aethiopica* are porcupines (*Procavia capensis, Heterohyrax brucei*).
The rodents and the phlebotomids continuously cross-infect one another because they live in close contact. Consequently, systematic extermination of rodents, as in Turkmenia, has produced a significant decrease in oriental sore. The relationships are similar for Mexican leishmaniasis.
The transmission of the parasites that cause mucocutaneous leishmaniasis is still being investigated. Transmission is by sandflies of several subgenera of the genus

Lutzomyia. Potential animal reservoirs are provided by several mammals living in forests, for example various rodents including rats, aguti, sloths, ant-eaters, and opossums, as well as monkeys (possibly also horses?).

Diagnosis by Microscopy. This is based on examining GIEMSA-stained smears. In kala-azar, one can use not only puncture material from the spleen, liver, lymph nodes, or bone marrow (particularly sternal puncture, in about 70% of cases) but also blood smears (spleen biopsy is more risky, but is up to 98% positive). In cutaneous leishmaniasis, material is collected from the margin and centre of the lesion. The diagnosis can similarly be confirmed in mucocutaneous leishmaniasis if material is taken from the margins of the lesions. In each case, amastigotes may be found (Plate III, Ⓐ *5, 6;* see also p. 306). The organism can be cultured relatively easily on blood agar (NNN medium, see p. 26), with the development of the flagellate promastigote stages (see Plate III, Ⓑ *9*).
PALOMINO et al. (1983) developed a selective fluid medium for the direct isolation of *Leishmania* species from mucocutaneous lesions. One can also inoculate suspect tissue from lymph nodes, liver, or spleen into the hamster to detect the parasite (found after 5–6 weeks in the liver and spleen of the test animal).

Diagnosis by Immunobiological Methods. In visceral leishmaniasis, the methods of indirect immunofluorescence, complement fixation, ELISA and agglutination can be used for indirect detection. Cultured promastigotes can act as a source of antigen, as they are readily produced on blood agar media. The indirect immunofluorescence test is the most reliable (significant titres between 1:64 and 1:4000), using cryo-sections of infected hamster liver. Antibodies can be demonstrated within 1 week, but more reliably after 3 weeks. An intradermal test (the Montenegro test) with an antigen prepared from cultured leishmanias is sometimes employed. However, the tests are group reactions without species specificity (note should be taken of travel history). Positive cross-reactions can occur with trypanosomes. In cutaneous leishmaniasis, the serum reactions are usually positive if the lymph nodes are involved (see GREENBLATT 1988).

Chemotherapy. The drugs of choice for treating kala-azar are the antimony preparations sodium stibogluconate (Pentostam) and meglumina antimonate (Glucantime). These are better tolerated than the preparations previously in use, but they can still give rise to side-effects such as loss of appetite, nausea, vomiting, headache, and pains in the limbs, while occasionally there may be circulatory or kidney damage. Consequently, it is necessary to perform careful individual and regional dose adjustment (Pentostam for example 10–15 mg/kg i.v., given slowly over 5 min, daily for 20 days; the course may be repeated after an interval of 10 days but there is the danger of accumulation and thrombosis; see HARTMANN, 1972). If the antimony preparations are not tolerated, alternatives are the aromatic diamidines (pentamidine isethionate 2–4 mg/kg i.v. or i.m., slowly, three times a week for 15–25 weeks). Allopurinol is also recommended (3–6 mg/kg three times daily for up to 6 weeks; the i.m. injection is painful). Allopurinol ribose can be given orally (SAENZ et al., 1986). Finally, amphotericin B should be mentioned. It is very active, but is also very toxic (liver and kidney damage), and should be used only in antimony-resistant cases (the dose slowly being increased up to a maximum of 0.5–1 mg/kg i.v. every other day). General treatment is also very important in kala-azar (nutritious food, parenteral nutrition, and in severe cases with pronounced anaemia also blood transfusion) (see also WHO, 1984).

Pentostam is suitable for treating cutaneous leishmaniasis. One or two injections (with a 1–2 day interval) of 1–3 ml are given near the skin lesions (concentration 100 mg antimony/ml). The possibility of secondary infection must be borne in mind and may require appropriate treatment. In addition, metronidazole has been recommended for both forms of cutaneous leishmaniasis. In mucocutaneous leishmaniasis, meglumine antimonate (Glucantime) can be used, the treatment (10–20 mg/kg) being continued for at least 3 weeks. Individual tolerance must be carefully observed during daily administration.

In general, chemotherapy of leishmaniasis is not very satisfactory, because all the drugs have various and sometimes pronounced side effects. They require great care on the part of the physician because of differing tolerance in individual patients.

Plate IV ⇨

Flagellates and Amoebae of the Gut and Genitalia

Plate IV

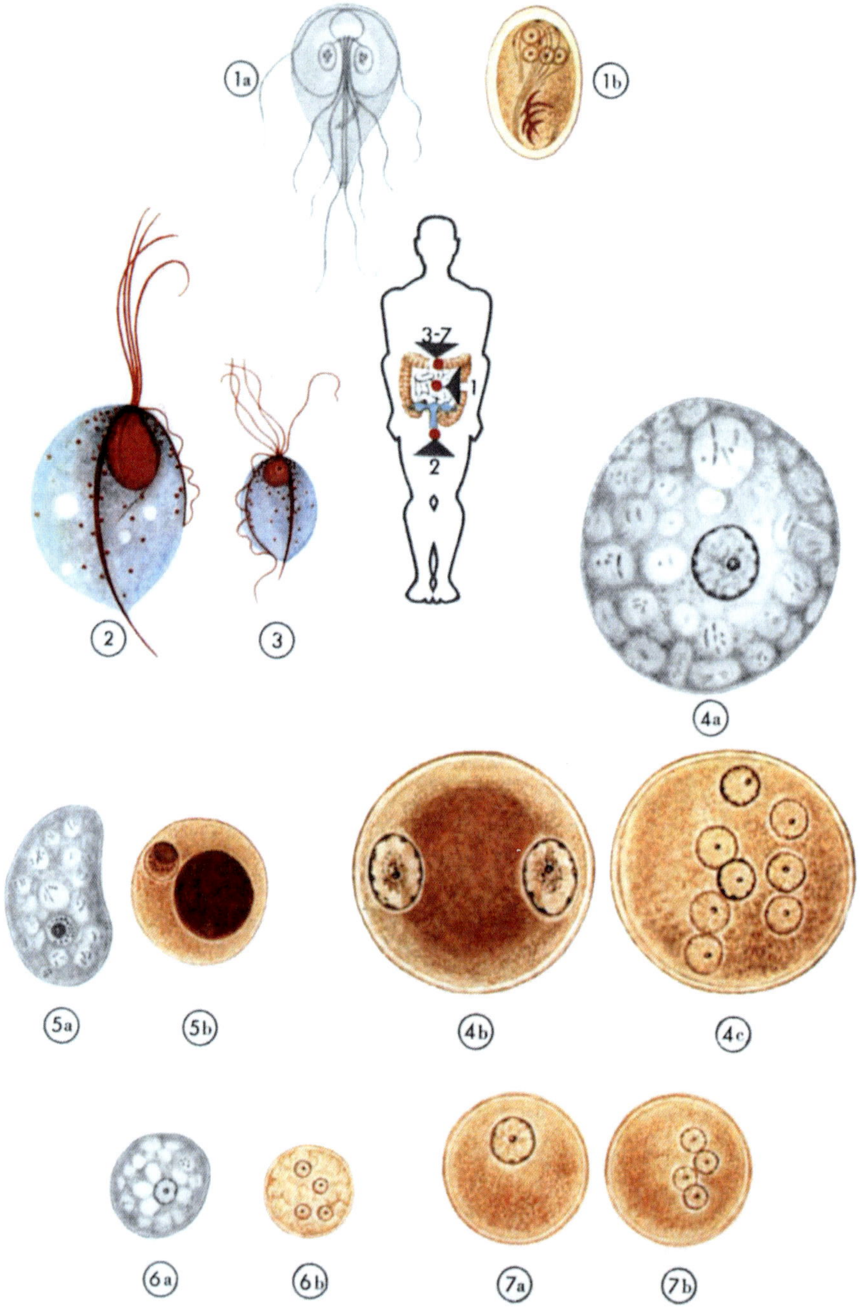

Flagellates and Amoebae

1 *Giardia lamblia* STILES, 1915

a Trophozoite HEIDENHAIN's iron-haematoxylin stain
b Cysts, iodine stain

2 *Trichomonas vaginalis* DONNÉ, 1937: GIEMSA stain

3 *Pentatrichomonas hominis* (DAVAINE, 1860) LEUCKART, 1879: GIEMSA stain (three, four, or five free flagella)

4 *Entamoeba coli* (GRASSI, 1879) SCHAUDINN, 1903

a Trophozoite, HEIDENHAIN's stain
b Binucleate cysts, iodine stain
c Eight-nucleate cysts, iodine stain

5 *Iodamoeba (Pseudolimax) bütschlii* (VON PROWAZEK, 1912) DOBELL, 1919

a Trophozoite, HEIDENHAIN's stain
b Cysts, iodine stain

6 *Entamoeba hartmanni* VON PROWAZEK, 1912

a Trophozoite, HEIDENHAIN's stain
b Four-nucleate cysts, iodine stain

7 *Entamoeba histolytica* SCHAUDINN, 1903

a Mononucleate cysts, iodine stain
b Four-nucleate cysts, iodine stain

Magnification in each case about × 1500 (this also applies to Plates V, and XXXII, *a–c*)

Flagellates of the Gut and Genitalia

Certain flagellates living in the oral cavity, gut and genital area are generally considered to be of only slight pathological significance for humans (see Plate IV). In some cases, however, they can act directly or indirectly as pathogens. Examples are *Giardia lamblia* Stiles, 1915 (= *Lamblia intestinalis* (Lambl, 1859) Blanchard, 1888; Plate IV, *1*) and *Trichomonas vaginalis* Donné, 1837 (Plate IV, *2*). The large gut has three species of flagellates, *Chilomastix mesnili, Pentatrichomonas hominis,* and *Retortamonas intestinalis,* and in the oral cavity one finds *Trichomonas tenax,* but these species have no clinical significance (commensals). As an example of a commensal *Pentatrichomonas hominis* is shown (Plate IV, *3*), since this species is of interest in differential diagnosis with microscopic examination.

Giardia lamblia Stiles, 1915

Pathogen of giardiasis

Giardia lamblia Stiles, 1915 (*Lamblia intestinalis*) (size 10–20 µm, Fig. IV 1), which is a parasite of the small gut, is found in all parts of the world especially the tropics, and is more common in children than in adults. As a trophozoite, the parasite is rounded anteriorly and pointed posteriorly, is highly motile and has a concave ventral surface which allows it to attach itself to the gut epithelium. There are four pairs of flagella arising from the eight basal granules. Two nuclei occur in the front part of the cell and give the flagellate its characteristic appearance in stained preparations (Plate IV, *1a*). In the cytoplasm, one often finds an arc-shaped parabasal body. There is no cytostome, nutrients being taken in by endocytosis (Meyer and Radulescu, 1979). The quadrinucleate cysts (cysts, 8–14 µm, Plate IV, *1b*) allow the transfer from host to host, e.g. with contaminated food. The cysts may be taken in by mouth, pass through the gut, and be excreted in the faeces. They can also be distributed in water (waterborne disease), and contaminated surface water drunk or used in agriculture may lead, for example, to the contamination of vegetables. Some observations suggest a short incubation period (about 7 days). The first cysts appear in the stools after 3–4 weeks. In the laboratory, various species of animals can be infected (e.g. rats, gerbils, guinea pigs and dogs, but not hamsters, rabbits or white mice). *Giardia lamblia* from naturally infected beavers can be transferred to man and animals (zoonosis?; BEMRICK and ERLANDSEN 1988).

The **pathogenic significance** of *Giardia lamblia* is in general very much underestimated. *Giardia lamblia* are quite frequently found, however, in those free from gut disease, but the infection can lead to persistent diarrhoea (danger of dehydration in children, see p. 87). In tropical areas, *Giardia lamblia* is frequently found in dysenteric stools and then quite often in large numbers. It can restrict the absorptive area of the gut epithelium and cause mechanical irritation and may lead to a malabsorption syndrome. The condition gives rise to abdominal pain, headache, a general feeling of illness, and irritability (AMENT and RUBIN, 1972). The occurrence of *Giardia lamblia* in the duodenum and upper jejunum is evidently due to the high bile concentration which favours its growth (see Culture methods below, and NAIK et al., 1979, and HOLLANDER et al. 1988 on immunity). In rare cases, *Giardia lamblia* may invade the gall bladder.

Diagnosis by Microscopy. This is based on examining a stool specimen, either: (a) a wet film preparation in which the mobile trophozoites with their characteristic appearance and erratic motility may be seen, (b) with iodine stain (addition of 4% aqueous iodine solution), or (c) with HEIDENHAIN's iron-haematoxylin stain, both stains making the cysts readily visible. *Giardia lamblia* may be found under some circumstances in the duodenal fluid (see p. 307 and Plate IV, *1a, b*) so a duodenal probe can be used if there is a strong clinical suspicion and results from a stool sample are negative. Concentration methods have been used to detect *Giardia lamblia* cysts (see p. 308). If results from the original sample are negative, it is necessary to repeat the stool examination because excretion of cysts is intermittent.

Indirect immunofluorescence testing is available this should not replace microscopic diagnosis.

In vitro culturing techniques involve the use of TYI-S-33-medium (DIAMOND et al., 1978). It is notable that the addition of bile substantially favours multiplication of the parasite (FARTHING et al., 1983; KEISTER, 1983; KASPRZAK and MAJEWSKA, 1983).

Prophylaxis. Protection from infection is provided by the consumption exclusively of clean or if necessary filtered or possibly chlorinated water. In any case it is recommended, that a known case of *Giardia lamblia* infection should always be treated with suitable drugs: The latter consist of the nitroimidazole group (see below). The frequency of *Giardia lamblia* infection varies very greatly with the region and is dependent on local living conditions. Local epidemics are mostly related to the quality of the water supply (contamination hazard). Boiling the water kills the cysts immediately.

Homosexual males are more frequently infected than the average adult male. A measure of this is provided by the incidence of mixed infection with *Entamoeba histolytica,* which on average occurs in 0.4% of European adults, whereas it is found in about 7.4% of homosexual males. In the latter it seems that immunosuppressive factors are involved (see p. 6, Mayers et al., 1977).

Chemotherapy. A *Giardia lamblia* attack can readily be treated with nitroimidazoles (ornidazole 25 mg/kg, 1–2 days; metronidazole 15 mg/kg for 5 days; tinidazole 40 mg/kg for 2 days; nimorazole 40 mg/kg for 5–10 days). The preparation may have to be changed with repeated treatment (a stool check should be made after 2 weeks, and use in pregnant patients requires caution). Chloroquine has also been found to be active.

Trichomonas vaginalis Donné, 1837

Pathogen of trichomoniasis

Trichomonas vaginalis (size about 10–30 µm; Plate IV, *2*) is a relatively common flagellate with world-wide distribution and is a parasite mainly of the female genital organs (about 10%–20% in the middle age range, but in up to 70% of patients with vaginal discharge). It also occurs in males but not so frequently. Its occurrence must, however, be related to individual and regional factors.

Morphology. The round to oval cell body contains a hyaline axial rod (the pelta axostyle), which stains readily in smears and which extends beyond the posterior cell margin. The nucleus lies at the upper pole. There is also a parabasal body together with a group of basal granules, from which four flagella extend forwards and a fifth extends backwards, forming the leading edge of the short undulating membrane running along the cell margin and gives the impression of a gearwheel in motion. Cysts are not produced, but the trichomonads usually round up, particularly shortly before dying, and can then readily be confused with cysts of other species. Another source of error in microscopic diagnosis is the presence of leucocytes which occur in increased numbers in inflammatory processes and can erroneously be taken for rounded up flagella-less *Trichomonas vaginalis* (see below). They are very sensitive outside the host; drying kills them at once, as does heating above 60 °C.

Clinical Symptoms. The clinical symptoms differ considerably and depend on the virulence and serotype of the parasite strain (Warton and Honigberg, 1983). The incubation period is about 4–24 days, after which discomfort occurs, frequently with complaints of irritation in the introitus and vagina. If there is extensive inflammation and swelling in the vagina, there is a greenish-yellow offensive discharge (in 40%–60% of all cases). In chronic cases, there is verrucose proliferation with massive leucoplakia, superimposed on a background of hyperkeratosis in the vagina and on the portio vaginalis (Rüttgers, 1982a, b).

KORTE (1973) deals with inflammatory processes in the adnexa and the consequent stages. It has even been suggested that sterility in the female and infertility in the male can be due to *Trichomonas vaginalis* infection, although detailed evidence is lacking. This suggestion should be supported, however, by detailed studies, particularly to establish the exact clinical significance of *Trichomonas vaginalis* infection, which at present remains unclear. It has also been suggested that *T. vaginalis* might lead to colpitis or urethritis (possibly peritonitis) by itself or in conjunction with associated bacterial flora, mostly species of coccus (ASAMI and NAKAMURA, 1955; SCHOLTYSECK et al., 1985). When the symptoms cited above occur in the genital region *Trichomonas vaginalis* should always be considered, since it is possible, that this parasite can cause prostatitis and urethritis, and particularly since latent infections are more common than symptomatic ones in women as well as men. In women, the evidence indicates that *Trichomonas vaginalis* infections persist much longer than in men, who appear to be infected mostly in the urethra, and who lose the parasites within a few weeks, although they can survive longer. RÜTTGERS (1982) has given a critical survey of the clinical features and epidemiology. *Trichomonas vaginalis* infections are rare in children of age 2–15 years. The parasite has, however, sometimes been observed in female infants, who have acquired it prenatally from the mother.

Transmission. *Trichomonas vaginalis* occurs not only in women but also in males, and is mainly transmitted by sexual intercourse. This explains the increasing occurrence of *Trichomonas vaginalis* infections everywhere; the rise is evidently parallel to the current increase in sexually transmitted diseases. The infection rate is particularly high in prostitutes (over 30%; in the Federal Republic of Germany prostitutes are required by law to be treated for syphilis and gonorrhoea, but not for *Trichomonas vaginalis;* TERAS et al., 1985). It has also been repeatedly suspected that a source of infection lies in public swimming pools and other baths. The parasites enter the water, but die in a few minutes, particularly with the high chlorination used in swimming pools (44 mg/l). Exchange of wet bathing suits should be discouraged, as these can infect the next wearer if they are contaminated by *Trichomonas vaginalis* discharge.

The parasites will multiply in the peritoneum of mice following intraperitoneal injection. Strains of low virulence invade the surface of the liver, but highly virulent strains enter it. While they retain their typical internal structure (axostyle), they lose their long flagella and move in an amoeboid fashion in the tissue and multiply there (LUDVIK and KULDA, 1982). This has clearly demonstrated that *Trichomonas vaginalis* species should not be viewed as a uniform group, just as with other pathogens, this parasite can produce latent or serious disease according to the virulence of the infecting strain. CHOROMANSKI et al. (1985) have identified differences in virulence by means of fluorescent lectins.

The clinical symptoms of trichomoniasis can also be produced by anaerobes from the *Bacterioides* group. These respond to the same drugs (metronidazole, see below), so correct diagnosis requires microscopic identification or culture; a decision cannot be made without one of these.

Diagnosis by Microscopy. This is based on examining a fresh vaginal or urethral smear in physiological saline, preferably by dark-field microscopy. The causative organism can also occur in the urine. In addition, it is recommended that vaginal and urethral smears stained with GIEMSA are examined. The cytoplasm stains blue while the nucleus, axostyle and flagella appear red. The nuclei of epithelial cells from the vagina appear violet and the cytoplasm reddish (staining as on p. 306). TERAS et al. (1985) recommend culturing the parasites from vaginal washings in order to give more reliable identification.

Trichomonas vaginalis can be cultured successfully only on fresh or well-preserved media (lyophilized or deep-frozen). One can also use a medium prepared by the method of JOHNSON et al. (1945) with ASAMI's modification (1952), which consists of two components. According to SAATHOFF (1985), it is also suitable for culturing *Entamoeba histolytica.*

I. 100 ml liver broth
   Preparation: 15 g of bovine or porcine liver is cut into small pieces and treated for 1 h with 100 ml 0.5% NaCl solution in a steam oven, after which the suspension is filtered. To this the following are added: 2 g peptone, 0.2 g cysteine hydrochloride, 0.5 g dextrose, and 0.2 ml 1% solution of methylene blue. The mixture is then autoclaved at 1.1 atmospheres for 20 min and cooled to room temperature.
II. 20 ml sterile inactivated human or horse serum I and II are mixed, and to each 1 ml of the finished medium, 1000 IU penicillin and 1000 IU streptomycin are added. The liver broth and the medium keep for about 1 week in the refrigerator, or for at least 3 months at -20 °C.

Diagnosis by Immunological Methods. *Trichomonas vaginalis* antibodies can be detected by complement fixation, indirect immunofluorescence and by indirect haemagglutination, and more recently also by ELISA (STREET et al., 1982, 1983). Experience indicates that the reactions are 86%–90% positive if the infection has been confirmed microscopically; 67% of microscopically positive women were positive in the ELISA test, and 18.5% also had IgM antibodies. These methods, however, at best give an indication of the diagnosis, which must be confirmed by microscopy or culture.

TERAS (1961, 1966) considered that there could be various serotypes of *Trichomonas vaginalis*, a finding which could influence the serological results, particularly using agglutination methods. Using homologous serotypes as antigens, optimal titres are possible, whereas heterologous ones sometimes give negative results. It is therefore necessary to establish which serotypes predominate in the region and to base the immunobiological reactions to them.

Chemotherapy of trichomoniasis has become very reliable following the use of nitroimidazoles. These compounds inhibit protein and RNA synthesis. They can be used orally or locally (metronidazole 250 mg, 2–3 times daily by mouth for 6–10 days, or 100 mg daily in the vagina for 6–10 days; nimorazole, ornidazole, or tinidazole, 500–1000 mg in 1–2 daily doses, if necessary combined with vaginal tablets). Nitroimidazoles should not be used during pregnancy [1]. The nitroimida-

[1] The treatment duration should not usually exceed 10 days because of the possibility of carcinogenic or genetic damage. Consequently, one should not prescribe, for example, metronidazole during pregnancy and breast feeding.

zoles are also active against trichomoniasis in the male. However, resistance has been reported but can mostly be overcome by increasing the dose, since the preparations are well tolerated. Any instances of therapeutic failure, though, must be considered critically, since reinfection or underdosage may often be confused with resistance. Favourable results have also been obtained with nifuratel (Inimur), a nitrofuran derivative, which is also active in thrush mycoses.
The principle of obligatory partner treatment applies in any genital infection (KORTE, 1973), and also for *Trichomonas vaginalis* infection. It is not possible to establish the success of treatment reliably without the aid of microscopic checks or culture, because the clinical symptoms can be produced by anaerobic organisms (*Bacteroides fragilis* etc.). These, however, can be treated with the same nitroimidazole preparations (WANDMACHER, 1979; WERNER et al., 1980; WERNER, 1981).

Non-specific **vaccination** against *Trichomonas vaginalis* can be provided by using an inactivated *Lactobacillus* vaccine (PAVIĆ and STOJKOVIĆ, 1982). This vaccine can produce persistent immunity against *Trichomonas vaginalis*. It probably acts indirectly by reducing the number of coccoid forms in the *Lactobacillus* flora, which predominate in the vagina in *Trichomonas vaginalis* attack. Consequently, this type of treatment provides protection not only in vaginal infection but probably helps prevent reinfection. These observations by RÜTTGERS (1982b, 1985) require further examination (see also BONILLA-MUSOLES, 1985; HARRIS, 1985).

Commensal Flagellates of the Large Intestine

(see identification table on p. 327)

The flagellates of the large gut, *Pentatrichomonas hominis* (Plate IV, *3*), *Chilomastix mesnili, Retortamonas* (= *Embadomonas*) *intestinalis,* and others are relatively rare, but sometimes occur in very larger numbers in people with diarrhoea and are seen by microscopy when examining a fresh stool. They are never the cause of such gut disease but have merely multiplied considerably on account of the favourable conditions.
It may be mentioned in passing that the view, which is sometimes expressed, that trichomonads in the gut and the genitalia merely represent local variants of the same species, is erroneous.
Pentatrichomonas hominis: this species does not produce cysts. The trophozoite (about 10 × 8 µm) has five flagella which arise from the anterior pole; a sixth flagellum accompanies the undulating membrane and extends well beyond the cell margin.
Chilomastix mesnili: the trophozoite is a pointed flagellate (about 10–15 µm), which produces uninucleate oval cysts (about 6 × 8.5 µm) resembling lemons. The vegetative form is distinguished by a short flagellum in the cytostomal area and three anterior ones about equal to the body length. The cyst wall is relatively thick and refractile and the nucleus is large. *Chilomastix mesnili* can also be found in the lower part of the small intestine (BARNHAM, 1977).

Enteromonas hominis: the trophozoite is round to oval in shape, about 5–7 × 3–4 μm in size, slightly refractile, with one nucleus and three free flagella; a fourth lies close to the body. The cysts are oval (6–8 μm), and contain up to four nuclei which are grouped in pairs at each end.

Retortamonas intestinalis: the trophozoite is elongate and oval in shape (about 4–10 μm) and has two flagella. It produces uninucleate cysts (about 6 × 4 μm), which resemble those of *Chilomastix mesnili* in being oval and having a protuberance at the anterior pole.

Plate V ⇨

Amoebae

Entamoeba histolytica

Plate V

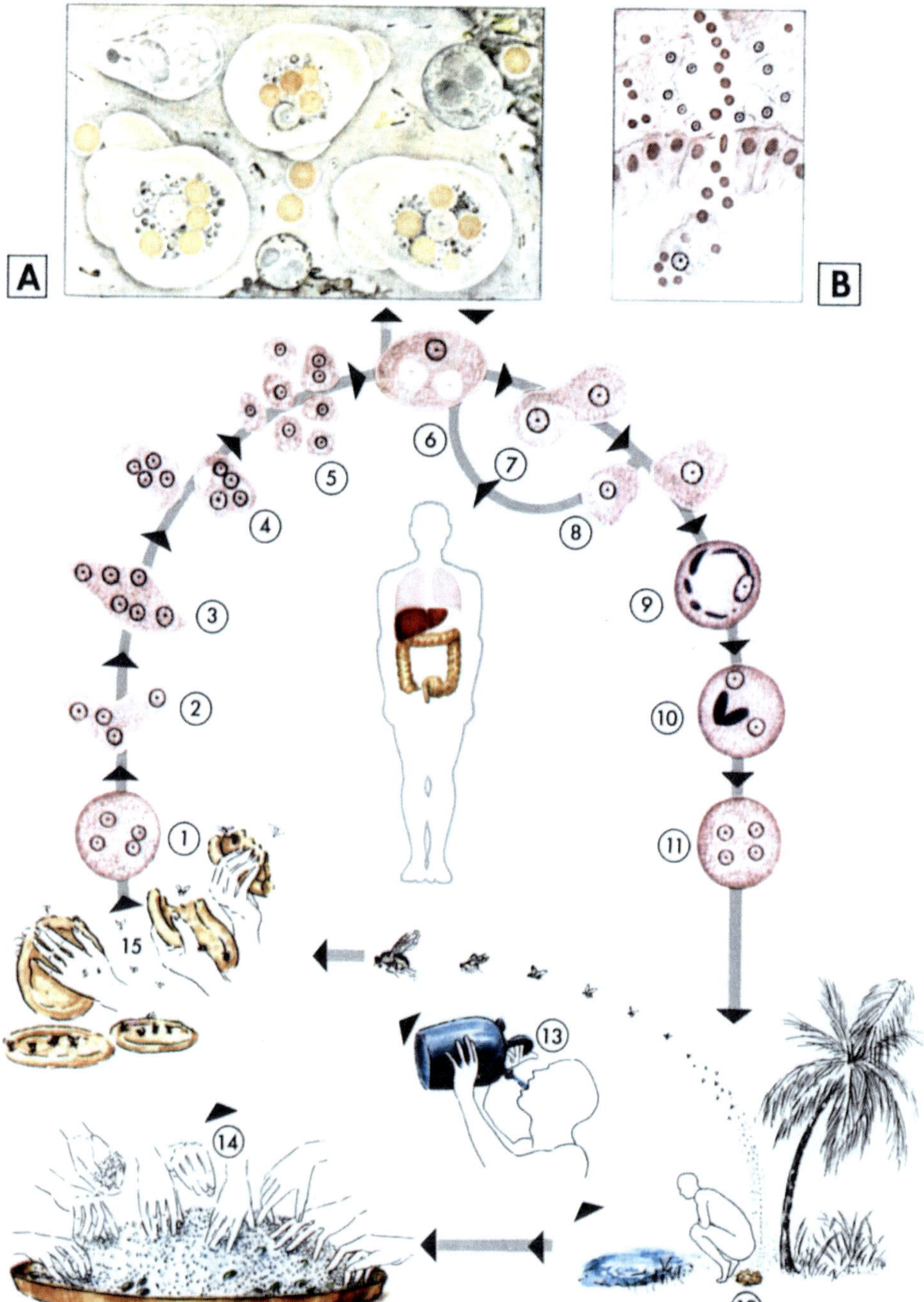

Entamoeba histolytica SCHAUDINN, 1903

Pathogen of amoebic dysentery (amoebiasis)

[A] Trophozoites from fresh stools in acute amoebic dysentery (so-called tissue or magna form), wet-film preparation

[B] Schematic diagram of a section through the wall of the large intestine containing entamoebae (abscess formation, ulceration); in addition to the amoebae, some of which contain phagocytosed erythrocytes, there are some free red blood cells (HEIDENHAIN's iron-haematoxylin stain)

Biology of *Entamoeba histolytica* (HEIDENHAIN's iron-haematoxylin stain; magnification about ×900)

1 Mature four-nucleate cyst in the gut
2–5 A four-nucleate amoeba after release from the cyst starts to divide; with further nuclear divisions, eight individual amoebae arise
6–8 Multiplication of the gut lumen form (so-called minuta form), from which (*6*) the tissue form [A] and [B] can arise
9–11 Cyst production
9 Typical uninucleate stage with marginal chromatoid bodies
10 Binucleate cyst containing chromatoid bodies (whetstone form)
11 Mature four-nucleate cyst

Mode of Infection

12 Flies transfer cysts from faeces to food (see *15*)
13 Drinking water contaminated by faeces
14 Contaminated food

(See also Plates IV, *4–7* and XXXII, *c*)

Amoebic dysentery (cause: *Entamoeba histolytica*) is a typical protozoan gut disease of warm countries. It is frequent in the region between the 25 °C July isotherm and the 20 °C summer isotherm, although this region does not correspond with the distribution of *Entamoeba histolytica*, which is much wider.
The prevalence of infection is 1% – 5% in Southern Europa, the USA and Canada, whereas it is 30% or more in Asia and Africa. The pathogenicity is particularly great in the region between 10 °N and 10 °S, the countries most affected being Bangladesh, Burma, China, Korea, India and Iraq, as well as Mexico and other Latin American states (50% – 72%). The pathogenic significance of the parasite has been greatly exaggerated in warm countries because the region of amoebic dysentery is also always one where bacterial dysentery occurs. Detection of amoebic cysts in the stool of someone with diarrhoea, therefore, does not necessarily mean that the amoebae are responsible for the diarrhoea. To confirm the diagnosis of amoebic diarrhoea, entamoebae (magna or tissue form) must be seen under the microscope in the stool (see below).
It is estimated that there are annually 30,000 deaths amongst the 400 million amoeba carriers.

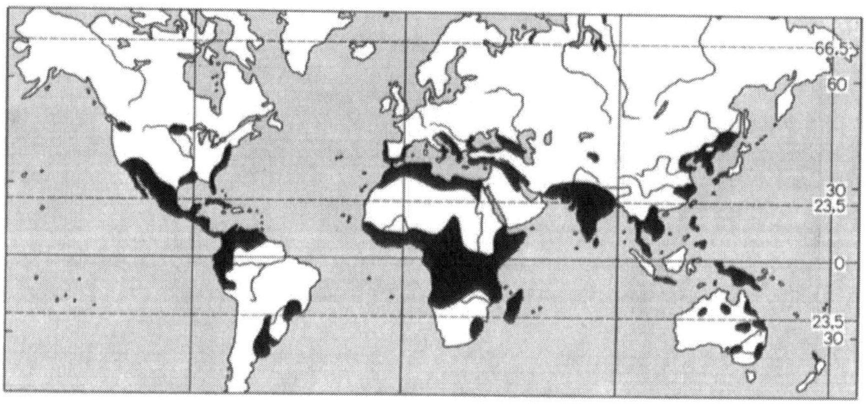

Distribution of amoebic dysentery

VON PROWAZEK (1912) found *E. polecki* in a child in Saipan as well as *E. histolytica*. In 1949, KESSEL and JOHNSTONE found it in California in two patients and in an Indian monkey. Subsequently, several authors reported the occurrence of *E. polecki* in human faeces as well as in pigs and monkeys (= *E. suis?*). It is commonly found in man in a central area of New Guinea, where people live in close contact with their pigs. So far, not more than 300 cases have been identified in man, although it must be assumed that there is world-wide and much more frequent infection, as well as confusion with *E. histolytica* cysts (see pp. 55, 56; KESSEL and JOHNSTONE, 1949).

Morphology and Biology. *Entamoeba histolytica* occurs mainly in the large intestine, where it has three characteristic stages (Plate V): (1) in acute amoebic dysentery, the haematophagous tissue form (magna form, about 20–30 μm, Plate V, [A], [B]); (2) in latent infection, as the gut lumen form (minuta form, about 10–20 μm, Plate V, *5–8*); and (3) as the persistent form or cyst (about 10–15 μm, Plate V, *1, 9–11*). The amoeboid forms (trophozoites, Plate V, [A], [B], *2–8*) are motile and continually change in appearance because of the pseudopodia. The magna forms use pseudopodia to engulf solid food, such as erythrocytes or cell debris, by flowing around them. At the same time they take up dissolved nutrients. The typical hyaline ectoplasmic pseudopodia may be thrust out explosively, which is a mode of movement (Plate V, [A]) that does not occur in any other gut amoeba. A feature characteristic of all *Entamoeba* species is the structure of the nucleus. This is particularly well seen in stained stool preparations, as a ring with a central pin-point nucleolus (Plate V, *1–11*). In microscopic diagnosis, it must be remembered that the human gut contains not only *E. histolytica* but also possibly three other non-pathogenic species of *Entamoeba: E. coli, E. polecki* and *E. hartmanni* (see pp. 55, 56, and Plate IV).

In acute amoebic dysentery, only the magna forms are found and these mostly contain erythrocytes but not bacteria (Plate V [A], haematophagous amoebae). These can enter the gut tissue (submucosa) and then pass via the blood vessels to the liver, lungs, brain, skin and other parts of the body, where they can lead to abscesses. As the acute amoebic dysentery settles, the magna form gives rise to the somewhat smaller minuta form, which alone is capable of producing cysts. In latent amoebic infection (amoebae carriers), one finds the minuta form and cysts (see also Plate IV).

It is possible to store *E. histolytica* in a viable form by adding glycerol or dimethyl sulphoxide (DMSO) and keeping the samples at very low temperatures (for example, in liquid nitrogen).

Strain differences determine virulence, which is related to two major factors: the capacity for phagocytosis of erythrocytes and the production of a cytopathic effect in epithelial cell culture when a cell-free extract of *E. histolytica* is added. The cytotoxic action of *E. histolytica* arises from direct contact between the amoebae and the tissue cells as well as from the action of a protease. An enterotoxic effect is demonstrable in rabbit ileal loop. According to Reed et al. (1983), virulent strains show resistance to lysis or killing by human complement-containing serum, whereas avirulent strains do not. Zymodemes (compare p. 11, 21) can also be used to characterize strains of *Entamoeba*. Sargeaunt et al. (1984) found that of 54 individual isolates from India, 28 pathogenic ones had identical zymodemes.

According to Mattern et al. (1977), virus-like elements can be detected in *E. histolytica* cells; these occur regularly in amoebal strains from patients with invasive amoebiasis but they are lacking in non-invasive strains.

Clinical Symptoms. The clinical picture in acute amoebic dysentery is typically characterized by pronounced abdominal pain and severe diarrhoea. Soft liquid stools, containing bloody mucus, are produced many times a day. In mild cases, diarrhoea and constipation alternate, and the abdominal symptoms can be quite uncharacteristic. Following a temporary cessation of pain, there are frequently recurrences and development of chronic disease. Leucocytes are not present in the stool, whereas in dysentery of bacterial origin they are regularly found. In severe forms the colon is ulcerated, and the ulcers can perforate and lead to peritonitis. Such cases usually result in death within a few days, although the patient may be saved by prompt chemotherapy. Amoebic dysentery is particularly serious if a *Shigella* infection is also present.

Various factors may favour the penetration of amoebae through the gut wall and affect the incidence of amoebic dysentery (reduction in gut wall resistance due to damaging environmental factors, bacteria, etc.). A change in bacterial flora can cause conversion of a non-pathogenic strain to a pathogenic one, which can be recognized by an altered zymodeme pattern (MIRELMAN et al., 1986). Chronic amoebic dysentery often leads to extensive persistent damage to the wall of the large intestine (resulting in scarring for example), which can mean that consequently there is substantial loss of the haustra. On X-radiographs the large intestine then appears as a smooth tube.

A complication of the disease is the production of liver abscesses, mostly in the right lobe (common in Southeast Asia, India, North Africa and Mexico), or less often lung or brain abscesses. It appears that this is due to invasive highly virulent strains of *E. histolytica* or to immunosuppression due to drug treatment (steroids), or immunosuppression as a consequence of viral infections (AIDS, see p. 6, and also 91). In these cases previously asymptomatic carriers can develop severe clinical symptoms, particularly during pregnancy or childbirth. The liver disease can occur without preceding colitis. Chronic alcoholism predisposes to liver abscess.

Computed tomography and sonography give reliable identification of liver abscess (computed tomography from diameters of 1.5 cm).

In women with bloody vaginal discharge, which does not respond to the usual treatments, *E. histolytica* infection should be suspected as the cause if there is a corresponding history (occurrence of amoebic dysentery, residence in hot countries, particularly India). Microscopic examination of the discharge enables the cause to be rapidly identified. Frequently, abdominal pain in patients returning from hot countries are erroneously assigned to *E. histolytica* infections, although other infections (e.g. *P. falciparum*; see p. 100) are much more often responsible for such pains, as STEFFEN (1984) has pointed out. The cause may be dietary indiscretion in the visited country, particularly the consumption of raw food of plant or animal origin, as pointed out by KOZICKI and STEFFEN (1983).

Transmission of entamoebae from man to man is orally via the cysts. The cysts are taken in with contaminated food and reach the gastrointestinal tract. House flies (*Musca domestica*) are frequently the agents that contaminate food (although cockroaches have also been suspected) (Plate V, *12–15*). They take up the cysts from human faeces and can later excrete the cysts in an infectious state. One thus gets new infections by the contamination of food.

Contaminated drinking water can also be responsible (waterborne disease), as can general insanitary conditions where cysts are allowed to come into contact with food (Fig. V, *13*). In hot countries, there may be carriers of amoebae working as food handlers in restaurants, and so on, who should be checked. These carriers can be the starting point of epidemics. Household pets, domesticated animals and wild animals are hardly ever sources of infection for man, although rats, guinea pigs, cats and dogs have been infected in the laboratory. (The cysts can be killed, for example, with hot water or with 5% acetic acid solution for 15 min at 30 °C.) Fresh vegetables and fresh fruit should be avoided in areas where there is a risk of amoebic dysentery or should be eaten only after proper cleaning.

Microscopic Identification of entamoebae is essential for confirming the diagnosis. However, inexperienced operators may have considerable difficulty in identifying entamoebae in fresh preparations. There may be confusion with leucocytes (bacterial dysentery!), plant structures, moulds, and artefacts, so it is essential to know the characteristic features of *E. histolytica,* as otherwise there is a risk of wrong interpretation. If the typical features are not present, a doubtful case should be considered negative (see ELSTON-DEW, 1968 as well as p. 304). Trophozoites and cysts can be found in carriers of entamoebae after treatment with saline purgatives.

In acute amoebic dysentery, the magna form as shown in Plate V, [A] is found in the bloody mucus, and is recognizable from the orange-coloured erythrocytes in the cytoplasm and the typical pseudopodia. Such examinations must be made on fresh material, if possible at the sick-bed and within 10–20 min of the sample being passed; the dispatch of unfixed stool samples to a remote laboratory is unsuitable because of the disintegration of the magna forms. HÖFLER (1980b) recommends the preservation of stool samples with diluted formalin (35% formalin diluted 1:10) or with sublimate solution (1 part 96% alcohol +2 parts saturated aqueous sublimate (mercuric chloride) solution = 5.7%). A fresh stool sample the size of a walnut or a piece of haemorrhagic mucosa obtained by proctoscopy is mixed with four times the volume of either of these solutions and packed securely for dispatch.
In latent asymptomatic amoebal infection, amoebic cysts, containing from one to four nuclei, can be detected by adding 4% iodine solution to a fresh stool smear. The nuclei are then clearly visible and the glycogen in the cysts is stained brown (Plate IV, *7*). The merthiolate-iodine-formaldehyde concentration (MIFC) method (see p. 310) has definite advantages over microscopic examination of fresh stool preparations (for details of the technique, see MEHLHORN and PETERS, 1983; PIEKARSKI and SEITZ, 1987).
Plate IV shows the non-pathogenic species *Entamoeba coli* (Plate IV, *4*) and *E. hartmanni* (Plate IV, *6*). These must be clearly identified in examination by microscopy of stool samples in order to distinguish them from *E. histolytica* (compare Plate IV, *7a, b* and Plate V). Both of these species have typical entamoeba nuclei with small karyosomes, which are easily identified by iodine staining (Plate IV, *4b, c, 6b, 7a, b*); the karyosome is eccentric in *E. coli* (see p. 56). One can prepare permanent specimens by staining with HEIDENHAIN's iron-haematoxylin (see p. 308). Plate V, *1–11,* [B] shows the stained amoebal stages. *E. histolytica* is also quite readily cultured in vitro (monophasic liquid media, e.g. for producing antigens or testing drugs; DIAMOND et al., 1978). According to MICHEL (personal communication), culturing is reliable only with the trophozoites or four-nucleate cysts.

Diagnosis by Immunobiological Methods. Serological tests are necessary only if extraintestinal disease (liver, lung or brain abscess) is suspected. Complement fixation and indirect immunofluorescence tests (IIFT) are mainly used. The reactions are positive in about 60%–70% of cases in which the disease is confined to the intestines but nearly 100% in extraintestinal diseases. This applies also for the enzyme immunotest (EIT), the indirect haemagglutination test (IHAT), and the latex agglutination test (LA). The reactions give maximal titres after varying periods, namely 2 weeks for LA, 3 weeks for EIT, 4 weeks for IHAT, and 2 months for complement (MANNWEILER and KNOBLOCH, 1983). DISKO and SCHINKEL, (1979) have recommended countercurrent immunoelectrophoresis as a rapid, simple, sensitive and economical method. ELISA is a method suitable for epidemiological studies. However, serology cannot replace stool examination by microscopy in acute dysentery (see DENIS and CHADEE 1988).

The following rule applies: symptomless, non-invasive amoebic infections do not lead to measurable antibody titres. If there is invasion of the gut epithelium or extraintestinal disease with abscess formation, antibody levels are usually elevated. The titre can also sometimes rise considerably during drug treatment. On the other hand, the persistence of antibodies for some years after successful drug treatment can make it difficult to decide whether there is a persisting abscess. Clinical examination is then required for confirmation (sonography or computed tomography). The various classes of immunoglobulin are not always found simultaneously as they have a tendency to appear in a recognized sequence. For example, one finds IgG and IgM antibodies in primary infection, but the IgM class disappears quite rapidly, whereas IgG antibodies persist longer. According to JACKSON et al. (1984), negative IgM findings contraindicate acute amoebiasis.

The following notes may be of assistance in interpreting the results of the routine methods. In the *indirect immunofluorescence* test (IIFT), one gets antibody titres of up to 1:16,000, but low values (up to 1:64) persist after successful chemotherapy for quite a long time (90%–100% probability of acute disease for a titre of 1:320).

The *complement fixation test* is positive only in the acute stage of the disease, with titres of up to 1:320. Following drug treatment, the titre falls more rapidly than with IIFT.

The *latex agglutination test* (one common commercial product is Serameba) is relatively easy and quite sensitive. However, it remains positive for quite a long time after the disease is cured and should be used only in conjunction with one of the other two tests (IIFT, complement fixation test).

Chemotherapy. In the treatment of amoebic dysentery, it is necessary to determine whether there is *intestinal* or *extraintestinal* disease, quite apart from the important symptomatic characteristics. The nitroimidazoles are very effective against both forms of amoebiasis (HÖFLER, 1980 recommends: metronidazole, 500 mg three times daily for 5 days; ornidazole 500 mg twice daily for 5 days; or tinidazole 1 g twice daily for 5 days). However, these drugs can pass through the placenta and

therefore should not be given before the fourth month of pregnancy. Side effects are restricted mainly to gastrointestinal pains. Headache and dizziness are rare. In the extraintestinal case (e.g. liver abscess), dehydroemetine or chloroquine are particularly active.
A new drug with activity against the gut form is diloxanide furoate. Cyst carriers travelling in hot countries where there is a risk of amoebiasis, or upon returning from such areas can be treated reliably (dosage, about 20 mg/kg in children, 1.5 g in adults daily in divided doses for 10 days).

Experimental work has provided some interesting facts on vaccination. Monkeys (*Ateles* and *Cercopithecus*) produce a lysosomal antigen-specific humoral antibody following inoculation. This immunization has been found to be protective against intrahepatic infections and virulent entamoebae (Sepulveda et al., 1980). This is evidently a cell-mediated immunity.

Non-pathogenic Amoebae of the Large Intestine Important in Differential Diagnosis
(see differentiation table on p. 327)

During the microscopic examination for amoebae, six other species may be encountered but they are non-pathogenic: *Entamoeba polecki, E. hartmanni, E. coli, Endolimax nana, Iodamoeba (= Pseudolimax) bütschlii* and *Dientamoeba fragilis* [1]. The last two species have been considered by some authors as potentially pathogenic (see p. 56 and 57).
Reliable distinction of *E. histolytica* from the non-pathogenic species must be based on some morphological features. The similarity to the dysentery amoeba is greatest in the species *E. hartmanni, E. coli* and *E. polecki. E. hartmanni* is so similar to the dysentery amoeba that for a long time it was termed a "small race" of *E. histolytica.*
All the *Entamoeba* species have the same typical nuclear structure (Plate IV); all the species produce persistent forms (cysts). The infective cysts of *E. hartmanni* contain four nuclei which resemble those of *E. histolytica* each with a central karyosome, but are less than 10 µm in size. On the other hand, the cysts of *E. coli* contain eight nuclei each with an eccentric nucleolus while the cell may be up to 25 µm in size. *E. polecki* has mainly uninucleate cysts, but occasionally binucleate ones.
The microscopic differences are clear in a good preparation, but it is recommended that species differentiation should be undertaken only after training in a special laboratory. It is known that wrong diagnoses are very frequent without such training (see identification table on p. 327).
Entamoeba hartmanni (Plate IV, *6*): Trophozoite (a), size 6–8 µm, single nucleus, vacuolated cytoplasm containing isolated bacteria. Cysts (b), 5–8 µm,

[1] *D. fragilis* is no longer assigned to the Amoebae in the revised classification of the protozoa given by Levine et al. (1980) but instead is assigned to the aflagellate mastigophorans on account of the pelta-axostyle complex (Camp et al., 1974) which are demonstrable by electron microscopical investigation.

one, two or four nuclei. There is a characteristic spoke-like structure in the cytoplasm. Considerable morphological similarity exists between these and *E. histolytica* cysts (see Plate V).

Entamoeba coli (Plate IV, *4*): Trophozoite (a), size 20–30 μm, single nucleus; numerous vacuoles in cytoplasm, mostly containing bacteria. The cysts can have one nucleus or two (b), or four or eight (c); size between 15 and 20 μm; cysts with two or eight nuclei are common. The karyosome is mostly eccentric.

Entamoeba gingivalis (GROS, 1849) BRUMPT, 1913: This is a relatively common organism found in the oral cavity, which can be identified in plaque, particularly if there is poor oral hygiene. The organism occurs world-wide.

The trophozoites in the dysentery amoeba (p. 48) are morphologically very similar. They are between 10 and 30 μm in size. The granular endoplasm gives rise to distinctly separate ectoplasmatic pseudopodia, while the nuclei are typical of *Entamoeba* species (2–4 μm). Bacteria from plaque and cell detritus act as food. Multiplication is by binary fission. Cysts have not so far been identified. *E. gingivalis* is non-pathogenic and therefore clinically is hardly ever observed. Careful cleaning of the teeth and gums protects from amoebal invasion, which some authors consider leads to gum abscesses. The transmission from man to man is mainly by oral contact (kissing), or by shared toilet articles (cups and toothbrushes). The organism can be *demonstrated* by microscopy in fresh plaque. Chemotherapeutic measures are not necessary.

Entamoeba polecki VON PROWAZEK, 1912: This species occurs mainly in pigs and monkeys, although it is also found in man (in Pakistan, India, Taiwan, New Guinea and Venezuela). Morphologically, the cysts of *E. polecki* (about 12 μm) mostly have a single nucleus, although isolated ones with two nuclei occur (5–10 μm); some of them have dark-stained inclusions (after iron-haematoxylin staining), as well as abundant polymorphic chromatoid material, and a polymorphic karyosome. *E. polecki* does not cause diarrhoea or pain. So far, of the drugs active against amoebae only metronidazole has been found to be effective (750 mg orally, three times daily for 5 days; CHACIN-BONILLA, 1980).

Endolimax nana: The trophozoite has a vesicular nucleus and irregularly shaped nucleolus; the cytoplasm in the trophozoite (10–15 μm) contains bacteria in the vacuoles. The four-nucleate cysts (about 5–10 μm) are oviform and often contain glycogen bodies.

Iodamoeba (Pseudolimax) bütschlii (Plate IV, *5*): This is another non-pathogenic species of amoeba, which is distinguished by its size (about 10–20 μm). Other characteristic features are the glycogen bodies in the cysts. Unlike in *Entamoeba* species, the addition of iodine produces distinctive brown staining and enhances the regular outlines of these glycogen bodies. (This feature determined the choice of name of the genus.) These bodies fill more than half of the cysts, which are always mononucleate (about 8–15 μm), and which have the nucleus displaced towards the cyst wall (Plate IV, *5b*). The nucleus in this species has a relatively large karyosone, which is particularly well seen in HEIDENHAIN's iron-haematoxylin staining (see p. 308 and Plate IV, *5a*).

Dientamoeba fragilis (see p. 55, footnote): This is another species of non-pathogenic protozoa found in man, although some authors have considered it potentially pathogenic (about 6–22 µm). It is suspected that this species, which occurs world-wide, is more frequent than stool examination would indicate. Children are infected more frequently than adults. The nucleus, often double, contains four to six chromatin granules (arrested telophase), which distinguishes it from all other species of amoebae. Cysts have so far not been recognized. A correlation has been demonstrated between *Dientamoeba fragilis* infection and abdominal symptoms, including intermittent diarrhoea, meteorism and spasmodic pains. With such symptoms, treatment should always be given. Specific treatment with nitroimidazole preparations stops the pains (see *Entamoeba histolytica*, p. 54). Some authors consider that the eggs of the pinworm *Enterobius vermicularis* (see p. 217) can act as vectors (OCKERT, 1972). This view however requires confirmation.

We may also mention here *Blastocystis hominis*, which is the most important source of error in stool examination for protozoa according to REICHENOW (1953). These are gut inhabitants, so far classified as moulds, but which have been assigned to the amoebae by VON ZIERDT and SWAN (1981) and ZIERDT (1988). The spherical single cells vary greatly in size (5–120 µm), while growth is by a form of budding. These are features indicative of a mould rather than an amoeba (see also MATSUMOTO et al., 1987). *Blastocystis* cells occur frequently in stools, but no chemotherapy is required. ZIERDT (1988) now drews attention to its unusual biochemistry and supposed an agent of intestinal disease in primates. Systematic clinical studies on this gut inhabitant are urgently required.

Plate VI ⇨

Acanthamoeba castellanii
Naegleria fowleri

Plate VI

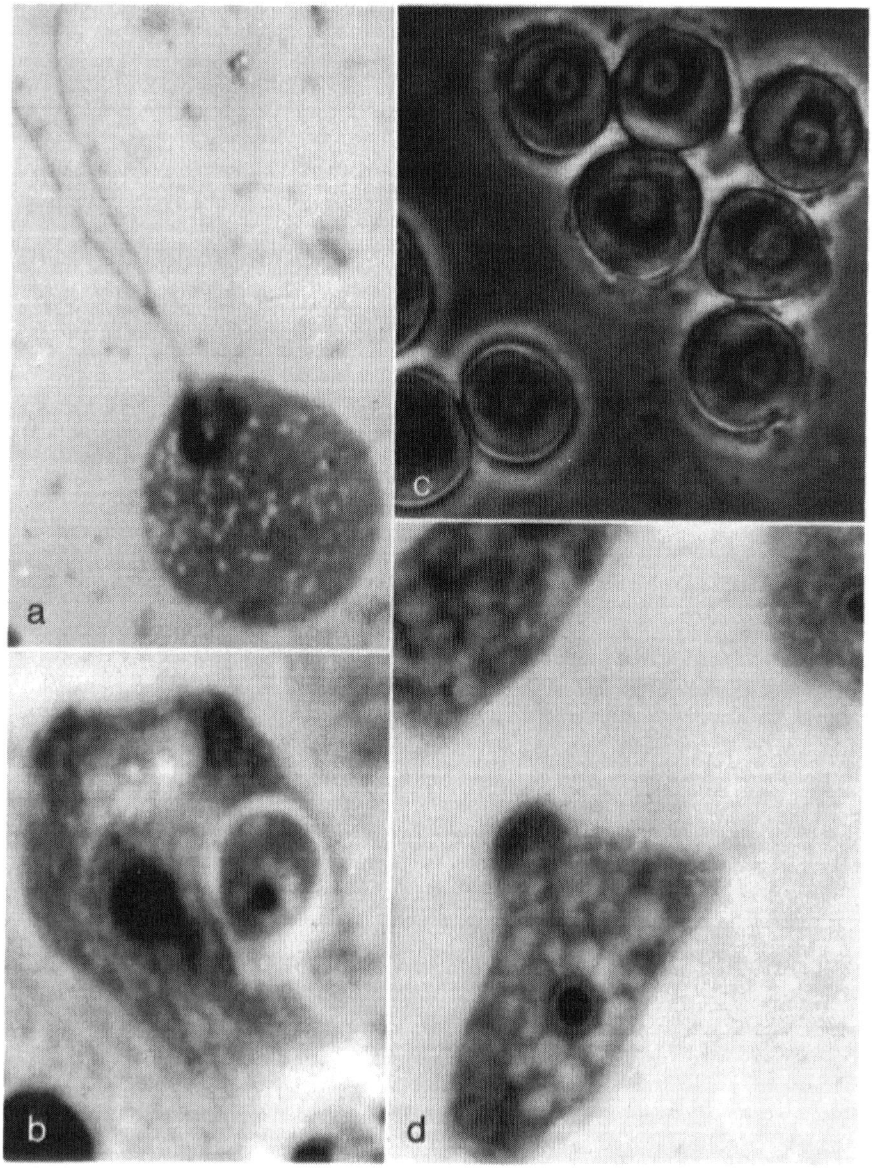

Naegleria fowleri

a Flagellate form, Vítek strain, GIEMSA stain, ×2300

b Affected PURKINJE cell, Usti strain, ×2300

c Cysts from agar culture, Vítek strain, ×1500

d Culture form under axenic conditions, ×1500 (after ČERVA, 1970).

Primary Amoebic Meningo-encephalitis (PAME) and Granulomatous Amoebic Encephalitis (GAE)

Amoebae of the genera *Naegleria* and *Acanthamoeba*

Amongst the free-living amoebae, which mainly live in damp soil and fresh water, representatives occur which are able to colonize man and can lead to severe diseases of the central nervous system, to meningitis or meningo-encephalitis. The best known species belong to the genera *Naegleria* and *Acanthamoeba*. The number of clinical cases recorded in the world literature is not very large, but increases constantly and was about 120 in 1984. The disease picture was observed for instance in Czechoslovakia, Belgium, the United Kingdom (Manchester), different regions of the USA (Texas, Virginia, Florida) as well as in South Australia and New Zealand. In the Federal Republic of Germany, so far two cases of amoebic keratitis due to acanthamoebae have been noted (LUND et al., 1978; WITSCHEL et al., 1984).

Morphology and Development. The morphological characteristics of the different species are related essentially to the size of the individual cells (between 10 and 40 μm), the form of the pseudopodia and also the structure and mode of division of the cell nuclei (mainly with large karyosomes).
The formation of cysts is common to all species (persistent stages), which makes possible their wide dissemination. For example, one finds them in the soil and in dust; they are carried by the air and thus get into the inspired air. As a result the amoebae can be isolated from the nasal mucous membrane and also from throat smears (e.g. ČERVA et al., 1973; OCKERT, 1974; MICHEL et al., 1982). They can easily be cultured in the presence of bacteria (the bacteria act as a food source). *Naegleria* species are smaller (about 22 × 7 μm) than *Acanthamoeba* species (15–45 μm), and can also be deeply stained, which makes them more difficult to recognize in histological preparations, and they can also easily be confused with degenerate neutrophil cells. Trophozoites from the cerebrospinal fluid measure about 8 × 10 μm, and are actively motile. The persistant forms (cyst, about 9 μm) possess a tough outer coat (with acanthamoebae the cyst is stellate or polyhedral; 15–20 μm), and contain one nucleus. On plate culture *Naegleria* species develop two flagella and thereafter assume a typical, longish flagellate form within 2–12 hrs, following the addition of warm water (see Plate VI, *a*).
The temperature tolerance of all pathogenic strains of these amoeba is ranging remarkable, between 30 °C and 48 °C. Many pathogenic *Naegleria* species can still multiply in vitro at 42 °C (*N. australiensis*) or 45 °C (*N. fowleri*), and pathogenic *Acanthamoeba* species at 40 °C. However, as individual non-pathogenic species

which multiply at these elevated temperatures (e.g. *N. lovaniensis*), there remains only the mouse inoculation test (MIT) is necessary to establish whether a thermo-tolerant strain is pathogenic. In this test virulent strains cause an infection of the brain and also of the lung (especially after infection with acanthamoebae) following intranasal administration (see the following table).

The differential diagnosis of the most important *Naegleria* and *Acanthamoeba* species (after MICHEL, unpublished)

		Remarks
Vahlkampfiidae:	*Naegleria gruberi*	Non-pathogenic
	Naegleria fowleri (= *Naegleria invadens*)	Pathogenic in human and mouse Multiplies at 45 °C
	Naegleria australiensis	Multiplies at 42 °C; average mouse pathogenicity
	Naegleria jadini	Multiplies at 43 °C; non-pathogenic for mouse
	Naegleria lovaniensis	Multiplies at 45 °C; non-pathogenic for mouse
Acanthamoebidae:	*Acanthamoeba castellanii*	Individual strains multiplying at 40 °C are pathogenic for mouse
	Acanthamoeba polyphaga group (PAGE, 1967)	*A. polyphaga, A. lugdunensis* pathogenic; *A. quina* non-pathogenic
	Acanthamoeba culbertsoni	Multiplies at 40 °C; obligate mouse pathogen
	Acanthamoeba lenticulata	Multiplies at 40 °C; obligate mouse pathogen; very similar to *A. culbertsoni* and *A. palestinensis*; differentiation only possible by zymogram

Of special importance with regard to pathogenicity are the species *Naegleria fowleri, N. australiensis, Acanthamoeba culbertsoni* and *A. castellanii. Naegleria* species cause primary amoebic meningo-encephalitis (PAME), pathogenic *Acanthamoeba* species on the other hand cause chronic granulomatous amoebic encephalitis (GAE), in which the meninges can also be affected.

Naegleria species infections have occurred with increased frequency, especially in connection with visits to contaminated swimming pools. When water containing *Naegleria* species gets into the nose, there is initially an acute mucous membrane inflammation, which extends to an inflammation of the olfactory mucous membrane. The amoebae penetrate via the submucous nerve plexus into the axon cylinders and intermediate spaces of the olfactory nerves. They then migrate centripetally through the cribriform plate of the ethmoid into the subarachnoid space and from there into the brain (PHILLIPS, 1974; JANITSCHKE, 1982).

Initial signs and symptoms following infection include headache, fever, neck stiffness, vomiting, confusion, delirium and rhinitis. The protein content of the spinal fluid tends to be

elevated leucocytosis of 380-7,300 leucocytes/mm^3). Finally, an acute fulminating disease develops, with a fatal outcome, which affects both children and adults (BRASS, 1973; KADLEC et al., 1980).
According to the observations of BUTT et al. (1968), in one patient the symptoms began with increasing frontal headache, which by the next day had led to nausea, vomiting and fever. On the fourth day after the occurrence of the first symptoms he became delirious and talked unintelligibly. On admission to hospital, first bilateral parotitis (with considerable right-sided enlargement of the lymph nodes) was diagnosed, with a suspicion of a mumps encephalitis. One day later death occurred – at autopsy all that was found were pathological changes in the brain and respiratory tract with signs of bronchopneumonia. The lungs showed no granulomatous reaction or parasite infection. The oedematous brain showed flattened gyri with a yellowish, fibrinous exudate, which covered the medulla and cerebrum. In the thick exudate of neutrophil plasma cells and eosinophil leucocytes few amoebae were detectable. The exudate filled the arachnoidal, subarachnoidal and VIRCHOW-ROBIN space.

Of the first cases reported from Czechoslovakia during 1962–1965, it is established that the 16 youths from Northern Bohemia had all been infected in the same swimming pool (ČERVA and NOWĂK, 1968). After about 12 years several species, including *N. gruberi* (?), were isolated from the water and damaged parts of the pool wall, despite a temporary embargo on its use. Of course, no more cases occurred.
The risk of acquiring a *Naegleria* species infection by swimming in affected waters is estimated from American studies as 1 per 2.5 million exposures. It is further suspected that many cases of fatal aseptic meningitis in actual fact represent undiagnosed cases of *Naegleria* species infection. When strongly suspected a careful examination of the spinal fluid for trophozoites (size about 8–15 µm) should be carried out at the beginning of the illness.
Acanthamoebae cause a granulomatous encephalitis in chronically ill or immunosuppressed patients, in which, generally, no contact with swimming bath water can be demonstrated. The central nervous system is probably secondarily affected from another active focus, for example in the lungs or skin, and death occurs following a very chronic disease course (there may be opportunistic parasites, for instance in AIDS cases; MARTINEZ and JANITSCHKE, 1985).

Acanthamoeba polyphaga, A. culbertsoni and *A. castellanii* cause, in particular, severe eye diseases (keratitis, uveitis; see p. 61). They are in certain cases amenable to drug therapy, whereas the encephalitis cases nearly always end fatally. In addition to the involvement of the central nervous system, under certain conditions amoebic hepatitis can occur. Cases with lung involvement (other organs affected include kidneys, pancreas, lymph nodes and myometrium) have been described (DUMA et al., 1969; see also MARTINEZ and JANITSCHKE, 1985).
Experimentally, severe rhinitis and encephalitis occur in mice, rabbits and apes after intravenous, intraperitoneal or intranasal inoculation of *Naegleria* species. With acanthamoebae pneumonia also occurs. In monkeys a typical meningo-encephalitis occurs after intrathecal inoculation, which results in death 6–10 days later.

Epidemiology. All free-living species of amoeba form cysts, which occur ubiquitously and are dispersed with dust. Probably only the flagellate and amoeboid stages in water lead to disease. The nasal mucous membrane is in all likelihood the point of entry.

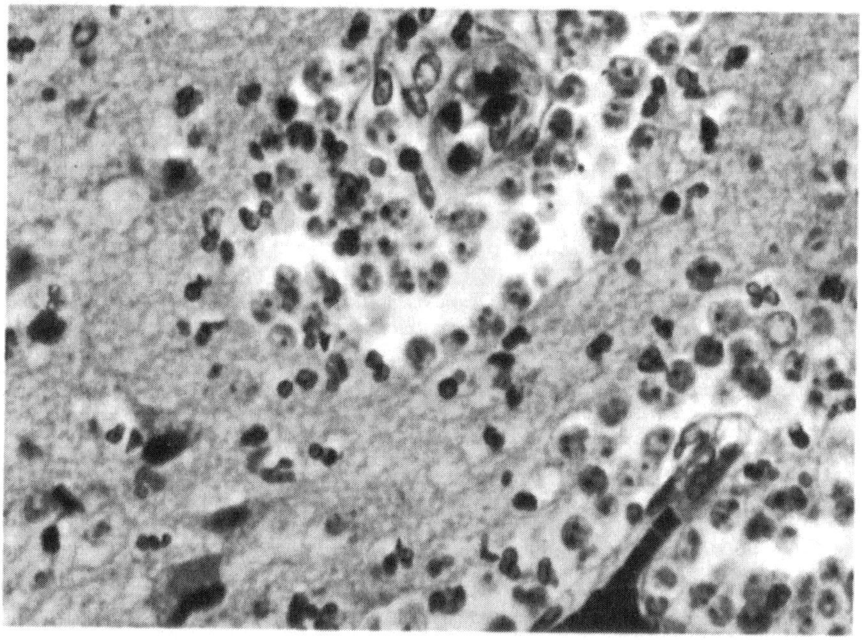

Naegleria fowleri: Brain section of a patient who died of primary amoebic meningoencephalitis; perivascular infiltration of amoebae, × 500 (from ČERVA, 1968)

KINGSTON and WARHURST (1969) could detect by culture methods, both inside and outside their laboratory in London, 12 strains of *Acanthamoeba castellanii,* 13 strains of other species, and further, three *Naegleria* species. All *Acanthamoeba castellanii* strains are pathogenic in vitro in HeLa cells. Thirty-eight *Hartmannella-*(*Acanthamoeba?*)-positive throat swabs, which WANG and FELDMAN (1967) found in routine investigations of 2289 healthy persons, testify to the occurrence of an airborne infection pathway. They isolated the amoebae predominantly from children about 1 year old, who had probably picked up the infection by crawling on the ground, but who remained asymptomatic (for details, see CARTER, 1972). Also, MICHEL et al. (1982) were able to identify 13 positives in 140 healthy national service recruits in the Federal Republic of Germany, in which four different species of amoebae (*A. mauretaniensis, A. lenticulata, A. quina* and *A. rhysodes*) were demonstrable in culture (according to isoenzyme analysis; MICHEL, personal communication). Amoebae of these species were also found in neglected tubing of dialysis units, of dental treatment units and similar fixed water conducting equipment (see CASEMORE, 1977; MICHEL and JUST, 1984). According to all previous experience these findings posed no danger for the patients.

Diagnostic Methods

Amoebae can be cultured at 37 °C on NNN agar plates (according to PAGE), which have been previously seeded, for example, with *Escherichia coli*. Suitable specimens include both fluid obtained by puncture and small tissue specimens of brain obtained at autopsy. After 48–72 h amoebae from the inoculum multiply on the bacterial culture and encyst after consumption of the bacteria.
A satisfactory culture medium for *Naegleria* species is a 2% agar (Bacto-Agar; Difco) with heat-killed bacteria of the species *Aerobacter aerogenes* (20 g Bacto-Casitone with 100 ml fresh sterile horse serum to 1 l distilled water; ČERVA et al., 1969). MUNGELLUZZI and BIANCHINI (1969) have reported the successful axenic culture of amoebae. WANG and FELDMAN (1967) used tissue culture (and the cytopathic effect) for the routine detection of acanthamoebae from throat swabs. For the immunological detection of an experimental *Acanthamoeba* infection in research animals, ČERVA used the immunofluorescence technique. This has also proved valuable in serological investigations on man. However, as yet there is no commercially available product for this, and so this kind of research is confined to specialist research laboratories. ELDRIDGE (1967) recommends the complement fixation test, which he used both in patients and in healthy controls. Cultured *Acanthamoeba* species were used as the antigen.

Chemotherapy. Only high doses of amphotericin B in combination with high doses of sulphadiazine have proved experimentally successful. Antibiotics and sulfonamides alone show no activity. Of many other drugs tested in vitro so far, only clotrimazole and acriflavine have shown themselves to be active in *Acanthamoeba* eye infections. However, differences exist among species of amoebae. In immunodeficient patients reduced activity of the drugs must be anticipated. Reliable statements on therapy cannot be made to the small number of known cases (STEVENS and WILLEART, 1980).

Plate VII ⇨

Sporozoa, Coccidia

Sarcocystis suihominis, S. bovihominis
and Related Species
Isospora belli

Plate VII

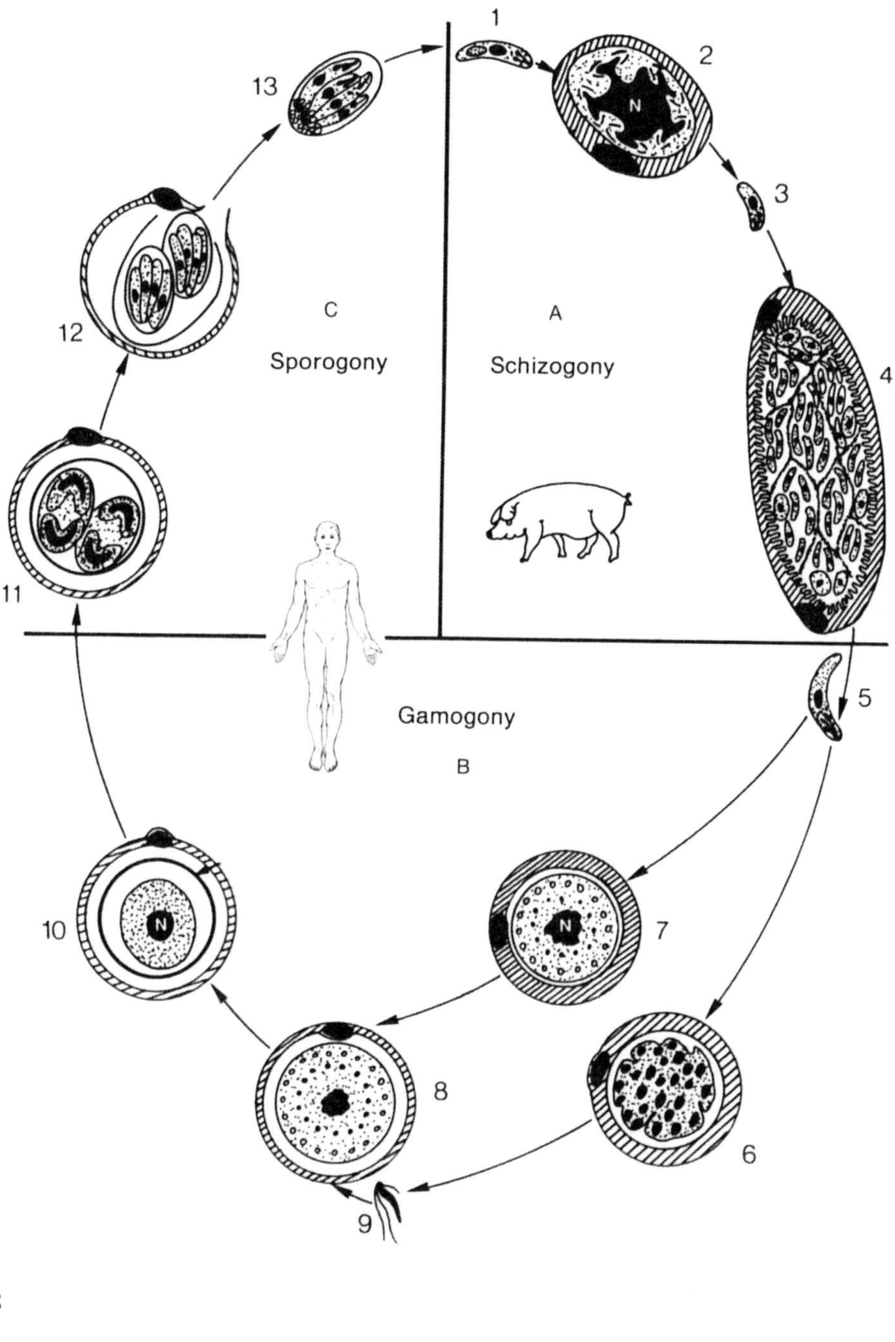
1
2
N
3
4
5
6
7
N
8
9
10
N
11
12
13
A
Schizogony
B
Gamogony
C
Sporogony

Sarcocystis suihominis

Developmental cycle (after MEHLHORN, 1980)

Ⓐ Schizogony (asexual reproduction)

Fully developed sporocysts (*13*) are ingested orally and reach the gastro-intestinal canal; from them sporozoites are released (*1*), which multiply asexually in the endothelial cells of the liver, kidneys, lungs and other internal organs and form merozoites (*2–5*). This intracellular reproduction can repeat itself many times (*5*→*2*)

2 Schizont
3 Merozoite
4 Cysts in the musculature
} Stages of development in pigs
5 Merozoites from one cyst after consumption of infected muscle

Ⓑ Gamogony (sexual development)

Merozoites (*5*) develop in the lamina propria to macrogamontes (*7*) or microgamontes (*6*) and finally to macrogametes (*8*) or microgametes (*9*)

Ⓒ Sporogony

10 The oocyst develops from the zygote
11 Beginning of sporulation (still within the host cell)
12 Oocysts with two sporocysts; the oocyst wall splits open in the intestinal lumen
13 Free sporocysts containing four sporozoites (capable of infection)

Amongst the sporozoa which can attack man, only the malaria parasites (see p. 95) and *Toxoplasma gondii* are of great medical interest. The coccidian species occurring in man, known under the names *Isospora hominis* and *Isospora belli,* had so far required no special consideration; they were considered to be relatively harmless small intestinal parasites without serious clinical significance.

Intracellular development is typical of all Coccidia (see p. 68; Plate VII), ending with the formation of oocysts (Plate VII, *10*) after asexual reproduction (schizogony) and a sexual cycle (gamogony). The oocysts are generally very resistant stages that reach the intestinal epithelium as living forms and are excreted out of the host and develop into sporocysts (sporogony). This stage of development may occur, however, in the intestine. From the fully developed sporocysts (Plate VII, *12*), taken in orally by the host, sporozoites (Plate VII, *1*) are released in the gastrointestinal tract. These then penetrate, depending on the parasite and host species, into certain organs (intestine, liver, kidney) in order to begin their asexual development and multiplication anew. FAYER (1970, 1972) succeeded in demonstrating the complete development of sarcosporidia in cell culture.

Of great significance in the investigation of coccidia of man was the recognition that the pathogen of toxoplasmosis *Toxoplasma gondii* belongs to this protozoan group (HUTCHINSON, 1965), (see p. 78). The cyst stages (bradyzoites) for example, from the brain of a mouse, develop through their sexual stage in the intestine of cats (Felidae) and form the highly resistant oocysts after schizogony and gamogony in cells of the small intestinal epithelium. Outside the host the oocysts sporulate so that they contain two sporocysts each with four sporozoites. This knowledge lead to the recognition that MIESCHER's tubules, well-known since 1843, from the muscle of pigs and cattle, are cysts of members of the genus *Sarcocystis* and are thus also coccidia (like the *Toxoplasma* cysts). The parasites included in the sarcocysts continue their development in the small intestinal tissues of meat eaters (e.g. dogs, cats), and two species also develop in man. It was thereby demonstrated that the sarcosporidias of domestic animals, until then thought to be a single species, represent not one species, but numerous morphologically and biologically differentiable species. All undergo an obligatory change of specific hosts (see pp. 71 ff.).

Members of the genus *Sarcocystis* are characterized by the fact that in the final host, before gamogony, no schizogony occurs as it does in *Toxoplasma gondii.* The parasites released from the cysts of the intermediate hosts immediately develop into gamontes and gametes (see MEHLHORN and HEYDORN, 1978). Two *Sarcocystis* species can develop in man, one acquired from the cow and one from the pig. From the sexually differentiated stages in the small intestinal epithelium in man typical coccidian oocysts form. These sporulate in the intestine and are excreted in the stools. In this the sporocysts of the former species *Isospora hominis* are identical with oocysts or sporocysts of the following newly named *Sarcocystis* species, *S. bovihominis* and *S. suihominis*. Therefore, in man at least four different coccidian species must be considered: *S. bovihominis* and *S. suihominis* (see pp. 71 ff), *Isospora belli* (see. p. 73) and *Toxoplasma gondii* (see p. 77). A fifth species which belongs to the genus *Cryptosporidium* has in recent years gained a certain prominence,

representatives of this genus are primarily parasites of domestic animals, but can also colonize man and lead to intestinal disease especially in an immunocompromised host. Transmission to man takes place directly through ingestion of sporulated oocysts (see also p. 85). The systematic position of the species *Isospora belli* remains untouched by these new developments (see p. 73) and becomes more and more important as pathogenic agent in AIDS patients.
Of the four different coccidian species of man the pathogen that causes toxoplasmosis has the greatest clinical and pathogenic significance (see p. 77).

Sarcocystis bovihominis (Heydorn et al., 1975) **S. suihominis** (Tadros and Laarman, 1976) Heydorn, 1977

Sarcocystis bovihominis and *S. suihominis* are to be found wherever raw or insufficiently cooked beef or pork is consumed. Both species are therefore very frequent in North and Central Europe. The colourless very fragile sporocysts are not easily detectable in the faeces of man and so they may remain unnoticed during a routine stool investigation, as carried out for example in examination for worms (see below p. 308).

Morphology and Development. The individual banana-shaped intracellular parasite (merozoite) from pre-cystic schizont (Plate VII; *3*) is relatively large (about 10–14 µm long), possesses one nucleus and all the organelles of a protozoan cell (mitochondria, Golgi apparatus, etc.). Division takes place by endodyogeny (i.e. binary fission within the mother cell, Plate VII, *4*). The merozoites of the first generation measure only about 5–6 µm. After ingestion of the muscle cysts the merozoites are released into the gastrointestinal tract, penetrate the cells of the lamina propria of the small intestine, round up and develop within about 14 h. to micro- or macrogamonts. Both *Sarcocystis* species of man probably affect the small intestine, this being inferred from the behaviour of the related species (*S. bovicanis, S. suicanis, S. bovifelis*) which form sexually mature stages in the dog and cat. The gametes arising from these fuse to form zygotes. About 5–10 days after eating parasite-containing raw meat, the first oocysts (about 22 × 12 µm in *S. bovihominis,* and about 20 × 10 µm in *S. suihominis*) are excreted with the faeces (prepatent period). Because the oocyst wall is very thin, it is frequently torn, releasing the sporocysts (about 14 × 8 µm) into the intestinal lumen. Usually the oocysts and sporocysts are excreted over more than 6 weeks (Rommel et al., 1972; Rommel and Heydorn, 1972; Mehlhorn and Heydorn, 1979).

When pigs ingest mature sporocysts with their food, about 6 days after infection the first schizonts appear in the cells of the endothelium of the liver, and later also in other organs,

especially the kidneys and the brain. After asexual reproduction the merozoites penetrate into other epithelial cells and grow into a new schizont generation. During this phase infected animals develop a high fever and may die of severe internal haemorrhage. If the animals survive, then the merozoites penetrate the muscle bundles from the 20th day after infection and form the tissue cysts.
After repeated endodyogeny, the cysts contain the banana-shaped cysto-merozoites, which are then also capable of spreading the infection. These cysts can attain a considerable size (several millimetres) in the musculature of cows and pigs. Even mild infections in meat animals lead to reduced growth rates, so that high economic losses can occur because of sarcosporidia.

Clinical Symptoms. These do not usually occur in man infected with *S. bovihominis*. The symptoms described by some authors as occurring after the consumption of steak tartare, are, if aetiologically correctly interpreted, very mild. On the other hand, with meat infected with *S. suihominis* heavily violent intestinal disorders can occur. These begin about 6–10 h after eating the affected meat and continue with diarrhoea, vomiting and severe water loss. However, the symptoms, though often very dramatic, pass relatively rapidly, and the patient is usually well again after 24–48 h. As a rule the cause can only be detected from the clinical history, because the diagnosis of the parasite type is possible by stool microscopy only after the prepatent period of 5–10 days.

Diagnosis by Microscopy. Identification of the sporocysts is through concentration procedures (see p. 308).

Chemotherapeutic Measures. These are generally not required, and efficacious drugs are so far unknown. With persistent symptoms, sulphonamides, as used in coccidiosis of animals are recommended.

Sarcocystis lindemanni

The species *Sarcocystis lindemanni* is still considered a species specific for man. It is surprising that the parasite has very seldom been found. Only about 47 cases have been published. In the light of the more recent information on the genus *Sarcocystis*, Beaver et al. (1979) followed up all known cases and came to the conclusion that in seven cases there was a misinterpretation of non-parasitic objects. Of the 40 *Sarcocystis*-positive samples, seven were found to be morphologically different *Sarcocystis* forms. The isolates could be identified collectively as zoonoses, some coming from the skeletal musculature of monkeys, some from cattle. Thirty cases were probably native to Southeast Asia, eight to India, five to Central or South America, four each to Africa and Europe, three to the USA, and one to China. Two were of unknown origin. No indication of pathogenicity was available.
From the above findings, it seems highly probable that a *Sarcocystis* species specific to man does not exist, and that the species *S. lindemanni* lacks all validity.

Isospora belli WENYON 1923

The coccidian species *Isospora belli* is rarely found in Northern and Central Europe, and more frequently in the Mediterranean regions; it is widespread in Asia and South America, but according to the investigations of STAHEL (1984) is apparently rarely carried by European travellers (0.7% of the patients investigated by a tropical medicine practice; for comparison *Giardia lamblia* is 10 times more frequent). *I. belli* requires no intermediate host for transfer to man (cf. *Sarcocystis* species, p. 71).

Morphology and Development. Intracellular intestinal forms of *Isospora belli* in man can not as yet be studied. However, on the basis of experience with *Isospora* species of dogs and cats it must be accepted that intracellular development takes place in the epithelium of the small intestine, and ends with the formation of unsporulated occysts which can be detected in the stools. Only the oocysts of *Isospora belli* can be recognized. They have a characteristic oval form (about 30×20 µm) and according to MEHLHORN and PETERS (1983) the unsporulated oocysts are frequently slightly pointed at one pole, with a "neck-like constriction'. The cytoplasm is widely separated from the cyst wall. Within a few days two sporocysts each containing four sporozoites are formed outside the body. Thereafter the oocysts are capable of infection.
According to the observations of LAARMAN and VAN DER SLIK-VAN DER VEEN (1961) "unsporulated oocysts" can be excreted from man over a period of more than 18 months. These findings obviously relate to *I. belli*.

Clinical Symptoms. The pathogenic significance of *Isospora belli* is generally regarded as minor; in many cases an infection remains symptomless (in 61% of *I. belli* carriers according to STAHEL, 1984). However, intestinal inflammation (enterocolitis) can occur through infection of the intestinal epithelium. Symptoms may then include long episodes of diarrhoea, with pain, sensitivity in the ileocaecal region, colic, loss of appetite, and vomiting. These problems may continue for a considerable time and result, amongst other things, in weight loss. Therefore, with persistent and repeated diarrhoea following a sojourn in the tropics, an infection with *I. belli* should be suspected. Lasting damage is unlikely. In immunodeficient patients a severe colitis occurs, and electrolyte replacement must also be considered in therapy (here it is acting as an opportunistic parasite).

Epidemiology. The transmission of *I. belli* to man frequently takes place exclusively via the very resistant oocyst or sporocysts. Auto-infections do not occur as sporocysts, only sporulated (in about 3 days) when they have access to oxygen in the open air. The question of a parasite reservoir cannot be answered at present, but the host spectrum is certainly very small. Experimentally, the gibbon can be infected with *I. belli*.

Diagnosis by Microscopy. With the help of concentration techniques (the zinc sulphate method of FAUST et al., 1939 and the use of a saturated sugar solution

(= 37%; specific gravity 1.14–1.15, recommended by FRENKEL) it is relatively easy to establish whether a stool sample contains unsporulated oocysts; but three different samples taken at different times should be tested. The oocysts (about 30 × 18 μm) are still uninucleate in the fresh stool (see also p. 71) and are differentiated from the sporocysts of sarcosporidia by their characteristic form.

Chemotherapy. Therapeutic measures are, in general, not necessary in a human infection, for spontaneous cure usually takes place after a very short time. STAHEL (1984) has found all patients to be parasite free after the administration of trimethoprim and sulphamethoxazole (1 : 5), or sulphadoxine with pyrimethamine in the acute stage of the disease. However, some symptoms remain even after parasites can no longer be detected.

Plate VIII–X ⇨

Toxoplasma gondii
Cryptosporidium species
Pneumocystis carinii

Plate VIII

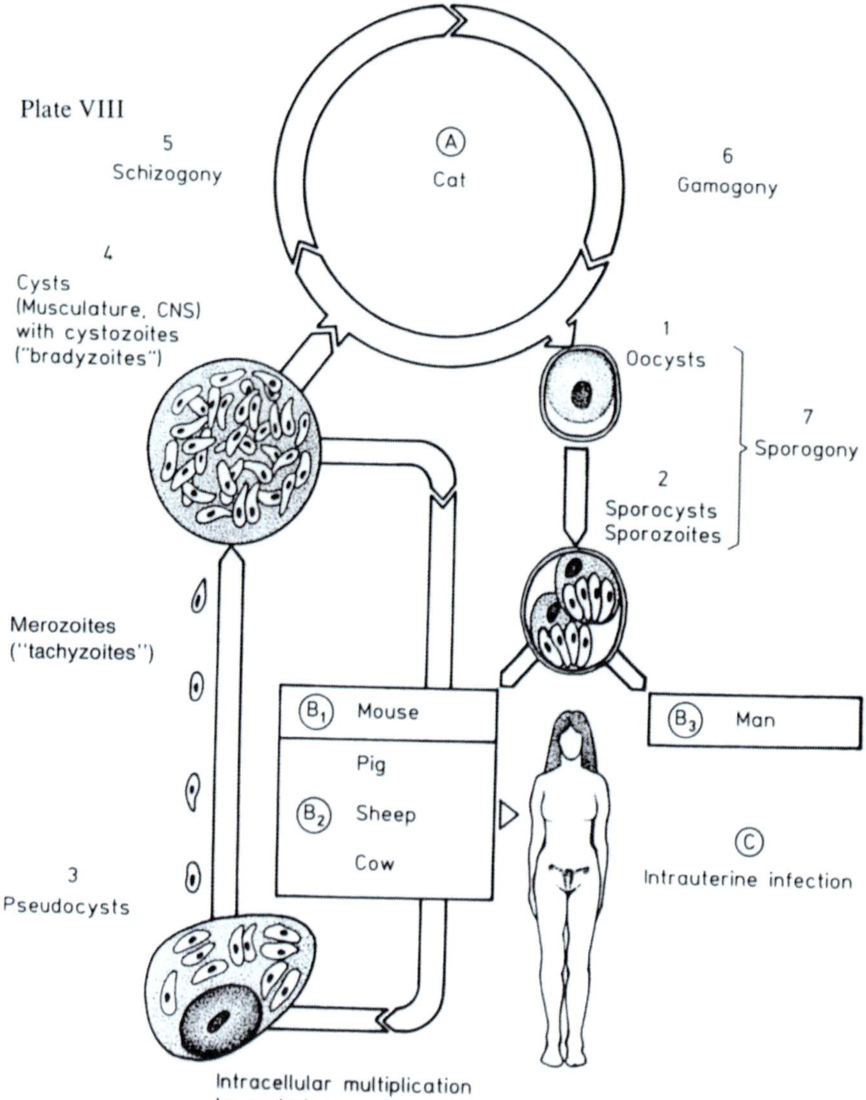

I

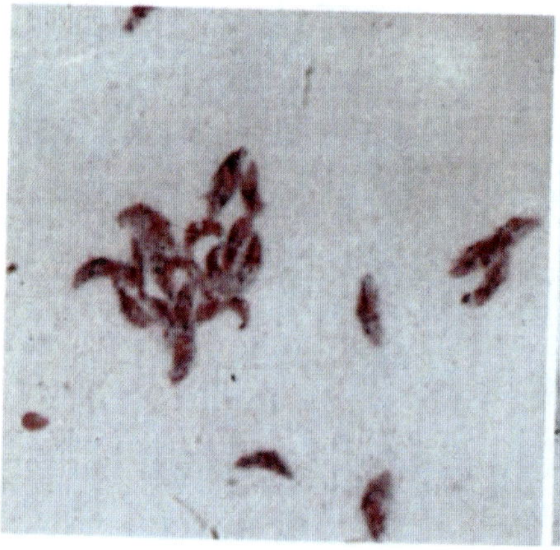

II

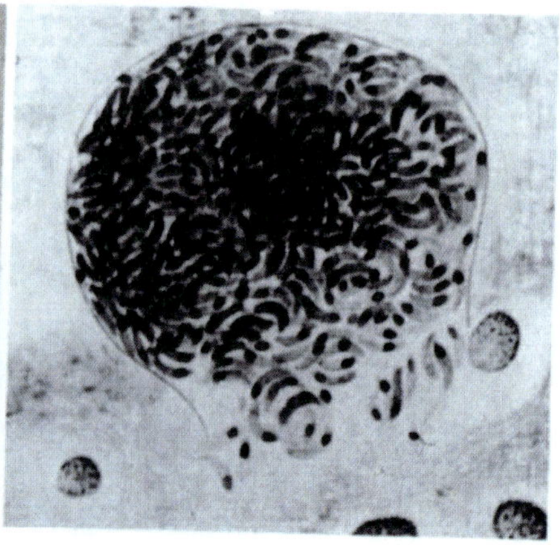

Toxoplasma gondii NICOLLE and MANCEAUX, 1908

Ⓐ Cats, the specific definitive hosts, excrete *Toxoplasma gondii* oocysts in their faeces (*1*). Almost all mammals (non-specific intermediate hosts) can be infected Ⓑ by the sporocysts (each with four sporozoites, *2*).

Ⓑ For example, mouse (B_1), pig, sheep, cow (B_2), and man (B_3).

The parasites multiply intracellularly asexually (*acute phase*) (*3*) and form cysts (*chronic phase*) (*4*). These lead to renewed infection in meat eaters. After a cat is infected, the organisms first multiply asexually in the small intestinal epithelium (schizogony *5*). Thereafter they form gamontes and gametes (gamogony, *6*). Development of oocysts and sporocysts each containing four sporozoites (sporogony, *7*) occurs after fertilization.

Ⓒ Intrauterine infection; congenital toxoplasmosis

Infection pathway in man (B_3):

a Through oral ingestion of sporulated oocysts (*2*)
b Consumption of raw, cyst-containing meat from sheep, pig or cow
c Intrauterine transmission (see p. 79)

I Toxoplasmas in a smear, individual cells stained with GIEMSA (×1000)
II Cysts from the brain of a child who died of toxoplasmosis (×500) (haematoxylin stain)

The causative agent of toxoplasmosis (*Toxoplasma gondii*) is pathogenically the most important coccidian species of man. Owing to its high host specificity and its capacity to be transmitted *in utero* both in man and in some mammals, it has aroused particular interest. *Toxoplasma gondii* is distributed world-wide and in cases of congenital toxoplasmosis a notifiable disease in some European countries.

Morphology and Development. *Toxoplasma gondii* lives strictly intracellularly and multiplies asexually in man, in almost all mammals, and in many birds (non-specific intermediate hosts). Development can take place in the cells of all organs and ends with the formation of cysts (resting stages). The muscles and CNS are favoured sites.

Sexual development (gamogony) occurs only in cats and other felids (the specific and definitive host) which infect themselves, for instance, by eating cyst-containing mice. Gamogony in the cat precedes an asexual multiplication (schizogony). Sexual reproduction ends with the formation of relatively small oocysts (about 12 × 10 μm), which reach the open air with the faeces while still uninucleate. In the oocyst, two sporocysts (about 8 × 6 μm) each containing four sporozoites (Plate VIII, *2*) develop during the course of the sporogony (2–4 days depending on the temperature). The prepatent period in cats after oral administration of tissue cysts is 6 days, and after administration of oocysts is 25–27 days. Cats are thus the sole excreters of potentially infectious stages of *Toxoplasma gondii*. Extraintestinal development occurs in the final host (= felid host) just as it does in the intermediate host, i.e. *T. gondii* has intra- and extraintestinal development in cat hosts, but only extraintestinal in intermediate hosts.

The oocysts are highly resistant and under some circumstances remain viable for longer than a year. The sporozoites are released from the sporocyst in the next host (which may be man) and further development is intracellular. The sporozoites enter cells of various organs and develop rapidly into pre-cystic merozoites (tachyzoites, endozoites) which multiply by endodyogeny (i.e. formation of two daughter-cells within the mother-cell) to form pseudocystes (*3*), and cysts (cystozoites, bradyzoites, *4*; cysts can be up to about 300 μm long; see Plate VIII, *1–4*). After this the cycle of development is complete (Scholtyseck, 1973, 1979).

Within *Toxoplasma gondii* there are many strains of varying virulence. Virulence is usually measured by the reaction of white mice. Depending on the strain and route of application, the mice may survive with little systemic upset or may die within 10–14 days of infection (Johnson, 1984).

Pre-cystic merozoites (also called tachyzoites), cystozoites (also called bradyzoites), and sporozoites are of approximately similar size (about 5–7 × 2–3 μm) and appear sickle-shaped (Plate VIII). However, the pre-cystic merozoites have a higher RNA and lower glycogen content than the cystozoites. The nuclei of the pre-cystic merozoites lie approximately centrally, whilst those of the cystozoites are almost terminal. All stages have at their anterior pole a conoid, an organelle which is typical of coccidia and which allows the parasite to penetrate into a host cell. The cystozoites have one characteristic in particular that differentiates them

from pre-cystic merozoites: they are resistant to gastrointestinal passage when taken orally so that they may penetrate into the host tissues (see above). In a first infection, during this phase there is haematogenous dissemination by which a foetus may also be infected through the placenta.
Toxoplasma gondii can be preserved for many months in a viable condition, both as pre-cystic merozoites and as cystozoites, by adding glycerol or dimethyl sulphoxide (DMSO) and then deep freezing in liquid nitrogen.

Clinical Symptoms. Clinical diagnosis is only confirmed by the demonstration of the parasite or specific antibody (see pp. 81 ff.), as the disease manifestations of toxoplasmosis are not pathognomonic. Moreover, latent symptomless infections are far more frequent than disease. The number of latent infections increases with age and in general correlates with age in Europe. Thus, approximately 20% of 20-year-olds and about 40% of 40-year-olds have latent infections; evidently individuals remain with latent infections for life. This fact must be borne in mind in every differential diagnosis and results of tests critically evaluated.
The clinical picture of toxoplasmosis changes with the age of the patient.
Infant toxoplasmosis (congenital toxoplasmosis) is obvious. A typical case has the triad hydrocephalus, chorioretinitis and calcification in the brain. This symptomatology is only confirmed as that of toxoplasmosis following the demonstration of the parasite or specific antibodies in the mother and child. In general, it appears that only with a first infection of the mother during pregnancy does an intrauterine infection occur (see below). However, this is not necessarily life threatening to the child. With opportune drug therapy the danger can be so minimized that the child may be healthy. However, latent infection of the newborn may occur, possibly with late manifestations (ocular toxoplasmosis, mental retardation) after many years, and so such cases must also be treated with drugs (HAY et al., 1984).
An infection in a woman that occurred before the beginning of pregnancy, i.e. a positive serological finding (see above), is of no danger to the child. A latent *Toxoplasma gondii* infection in the mother can be regarded as equivalent to vaccination, as it protects against parasitaemia arising from new infections and requires no drug treatment (THALHAMMER, 1981).
Toxoplasmosis in adults occurs mainly between the ages of 18 and 30 years and usually appears as a lymph node disorder, (predominantly nuchal) with a typical pathohistological picture (lymphadenitis according to PIRINGER-KUCHINKA, 1952). Lymphogranulomatosis and infectious mononucleosis should be considered in the differential diagnosis. High *Toxoplasma gondii* antibody titres in all serological tests always accompany the condition (p. 82). Rarely, there may be abdominal symptoms due to involvement of the liver and spleen, as well as lung and myocardial disorders. The incubation period is between 2 and 3 weeks. Juveniles may become ill with *Toxoplasma*-encephalitis and must be regarded as especially at risk. In older age groups ocular toxoplasmosis can occur.
In immunodeficient persons (e.g. induced by high doses of corticosteroids and other immunosuppressive drugs, or through infection with the AIDS virus) latent *Toxoplasma gondii* infections can be reactivated, leading to acute toxoplasmosis with a fatal outcome under certain circumstances (*Toxoplasma* is an opportunistic parasite) (REMINGTON, 1982b; DATRY et al., 1984; SEITZ and KERSTING, 1985). Some immunosuppression may also arise from latent *Toxoplasma* infection and under certain conditions may favour subsequent acquisition of other pathogens.
Experimental investigations in mice have shown that latent affection of the central nervous system can lead to behavioural changes (see WITTING, 1969; HUTCHISON et al., 1980). Laboratory animals react very differently to *Toxoplasma gondii* infection. Whereas, for example, mice and guinea pigs can succumb within a few days to infection with a virulent strain, rats and older dogs, though indeed susceptible to *Toxoplasma gondii*, are rarely ill and do not die of the infection. Avirulent and low-virulent strains are widely distributed in laboratory and

working animals (draught animals) and animals for slaughter. Congenital toxoplasmosis frequently occurs under natural conditions in sheep and goats. Considerable economic loss occurs through abortion and malformation. According to the findings of BEVERLEY and WATSON (1959) the degree of damage to the foetus depends on the timing of the intrauterine infection (see FRENKEL 1988).

Epidemiology. The transmission of toxoplasmas to man (see Plate VIII) may occur:

1. through the consumption of raw cyst-containing muscle of slaughtered animals (soft sausage meat, steak tatare),
2. through oral ingestion of sporozoite-containing oocysts with contaminated food,
3. by the intrauterine route (congenitally, see p. 79, 82 ff.).

1. Cysts in Raw Meat. The main route of infection for man consists of the consumption of raw flesh of infected meat animals (pigs, sheep, cattle). Deep-frozen meat (preserved at –30 °C for 24 h.), as well as meat heated to 50 °C for at least 20 min, can be eaten with safety. During the course of acute generalized infection in man and animals the organisms are present predominantly in the blood but also in secretions (e.g. saliva). This dissemination phase is relatively short and remains, contrary to many statements, epidemiologically without practical significance, although it may occur in blood transfusion(!) and tissue transplantation. It has been shown experimentally that cockroaches can ingest the highly resistant oocysts, excrete them again, and thereby theoretically spread the infection. However, this is also of doubtful significance epidemiologically.

2. Oocysts in Contaminated Food. House cats which excrete *Toxoplasma gondii* oocysts in their faeces are, under some circumstances, a dangerous source of infection for man (B_3) and animals (B_1), (B_2). Some wild species of Felidae must also be taken into consideration. Contact with fresh cat faeces, which only contain immature oocysts, does not lead to human infection. The oocysts become mature and can be considered infectious 48–72 h. after being excreted by the cat. Under favourable climatic conditions, i.e. with the correct humidity and warmth, they remain infectious for up to 18 months. The significance of the cat as a source of infection for man depends largely on their habits. In Europe and North America cats are usually kept very clean, whereas, for example, in Costa Rica cats are kept less clean, live very closely with man, and in this way play an essential part in the spread of *Toxoplasma gondii* (FRENKEL and RUIZ 1981; RUIZ and FRENKEL 1980). All other domestic and wild animals and birds are only potential *Toxoplasma gondii* carriers – there are no excreters like cats! Contact with infected domestic or wild mammals or birds thus does not lead to *Toxoplasma gondii* infection in man.

3. Congenitally. See the discussion on infant toxoplasmosis (p. 79) and immunological detection (p. 81).

Prophylaxis. Small children and pregnant women must be warned about close contact with cats which roam freely and catch mice, because under some circumstances such cats are infected and excrete oocysts. Cat faeces must be removed daily from the house and are best burnt. Likewise, pregnant women who are serologically negative for toxoplasmas must carefully avoid the consumption of raw meat; this also extends to tasting in the kitchen. Moreover, serological monitoring of pregnant women should be considered and is obligatory in some countries (see Aspöck, 1983).

Diagnosis by Microscopy. In practice the microscopic detection of toxoplasmas is limited to examination of the liquor cerebrospinalis sediment of a newborn child with appropriate symptoms (see above). Tissue samples suspected of containing parasites, e.g. obtained by biopsy from lymph nodes can be inoculated intraperitoneally into specially sensitive *Toxoplasma*-free white mice. This method is enhanced by the inoculation of artificially digested tissue (2% trypsin, 1:250 in 0.9% NaCl solution; after Sharma and Dubey, 1981). If a latent infection develops, the mice seroconvert within 3–4 weeks. They may die within 5–14 days of acute toxoplasmosis if the strain is highly virulent.

Besides the white mouse, *Mastomys natalensis* (the multi-mammate rat), a frequently used laboratory animal, was shown to be highly susceptible and sensitive, and may be used for the detection of toxoplasmas. Material suspected of containing toxoplasmas may be stained (blood and tissue impressions or sections) with fluorescent antitoxoplasma γ-globulin (Werner and Voss, 1970).

Diagnosis by Immunobiological Methods. Immunobiological detection methods are of essential value in the recognition of a *Toxoplasma gondii* infection, because the intracellular parasite is only diagnosed by microscopy, tissue culture or animal investigation with difficulty, and the symptoms of the disease are not specific in character. In order to decide the date of infection (e.g. in pregnancy), sera must be tested for the presence of both antibody classes IgM and IgG. The presence of IgM antibody, which appears first after infection, indicates an acute infection. IgM antibody may remain detectable, depending on the test used, for some months and under some circumstances for longer than a year. If IgM antibodies are absent, then an old latent infection probably exists. A first infection in a pregnant woman is recognized by seroconversion i.e. by a change from a negative serological result at the time of conception or at the beginning of pregnancy, to a positive serological result in the course of pregnancy up until the time of parturition. Seroconversion should in every case be occasion for careful drug treatment of the pregnant woman. Experience has shown that treatment will prevent infection in the fetus, and the omission of treatment at this period must be regarded as a professional mistake. Serological confirmation of *Toxoplasma gondii* infection is, therefore, not an indication for termination of pregnancy (see Thalhammer, 1981).

Frenkel (1981) indicated that serological detection methods are reliable for use in mammals, but can lead to false negative results in birds. Therefore, he recommends the inoculation of suspect tissue extract into laboratory mice for the detection of toxoplasmas in birds.

The detection of *Toxoplasma gondii* antibodies in immunosuppressed patients is either negative or at best unreliable.
The most important methods for the detection of IgG antibody are:

a) Complement fixation test (CFT)
b) Dye test according to the method of Sabin and Feldman (DT)
c) Indirect immunofluorescence test (IIFT)
d) Indirect haemagglutination test (IHAT)
e) Direct agglutination test (DAT)
f) ELISA technique (enzyme-linked immunosorbent assay)

Of these methods, a, b and c are used predominantly in practice (for ELISA see Dall and Johnson, 1984).

IgM antibodies at present may be detected by the following methods:
a) Indirect immunofluorescence test (IIFT; Remington, 1969) positive from the 5th day up to the 4th month after infection but with little specificity)
b) Double sandwich IgM ELISA (DSIgM-ELISA; Remington, 1982a)
c) Reverse enzyme immunoassay (REIA; Walls and Franco, 1982)
d) Immunosorbent agglutination assay (IgM-ISAgA; Desmonts, 1982)
e) Enzyme immunoassay (Wielaard et al., 1983)

(b–e: positive up to about 12 months after infection)

DSIgM-ELISA, REIA and the IgM-ISAgA test are clearly superior to the hitherto much used IgM-IIFT, with regard to sensitivity and especially specificity (72%–81% positive), and remain positive for up to about 12 months after infection (see Naot et al., 1981).
The good results obtained with the IgM-ISAgA test have induced Saathoff and Seitz (1985) to use this method in conjunction with the IIFT because by using both tests the time of onset of infection can be established relatively accurately. These tests can at present only be carried out in special laboratories (for monoclonal antibodies see p. 311).
For the assessment of antibody detection in man, the following guidelines with reference to the two most frequently used methods (IIFT and CFT, if necessary also including an IgM determination) may help:
Toxoplasma gondii antibodies can be first detected about 11 days after infection, and generally reaches a maximum titre 3–4 weeks after infection (IIFT ~ DT).

1. **Results of all serological investigations negative:** A toxoplasma infection very probably does not exist.
Prophylaxis for all: Eating of raw meat and close contact with cats is to be avoided. For pregnant women danger to the foetus only exists with the first infection, therefore prophylaxis is especially important. If necessary seroconversion should be checked for through repeated tests after 2–3 months.

2. **IIFT or DT up to 1:256, CFT 1:5:** Probably a latent, symptomless *Toxoplasma gondii* infection or the beginning of a first infection! Under certain conditions there may be no clinical symptoms. If there is lymphadenopathy, pyrexia, malaise, or pregnancy, immediate repetition of the test is necessary with measurement of the IgM level. If there is an increase in titre and a positive IgM value (ISAgA > 4000), see under 3.

3. **IIFT or DT 1:1000–1:64,000 and CFT 1:10–1:320:** Acute toxoplasmosis very likely. Therapy is in general only necessary if clinical symptoms are present. IgM-ISAgA determination should be carried out. In pregnant women only acute first infections require therapy (seroconversion).

If the values at the time of conception are not known, an IgM determination should be carried out immediately. If the results on repeated investigation remain unchanged or are higher, one should proceed as with a first infection requiring treatment. Positive serological results in no way imply fatal consequences for the foetus.

4. A two- to fourfold variation in titre, e.g. 1:64–1:256 can be due to physiological factors or to technical factors related to the test. It must not be interpreted directly as a decrease or increase in the antibody titre in the sense of an improvement or worsening of the disease condition. Controls are necessary.

5. In the newborn, passive transfer of antibodies across the placenta may be anticipated. The levels of these fall to zero within about 6–8 months (seldom longer) when no congenital toxoplasma infection exists. If there is an infection, then IgM antibody may also be present and specific therapy becomes necessary. Even when there appears to be no disease in the infant, drug therapy must be undertaken in order to prevent late manifestations (e.g. toxoplasmosis of the eye).

6. With the use of all available laboratory technical methods the serological results permit a relatively reliable estimate of the time of infection in a pregnant woman. Then, through opportune chemotherapeutic measures if necessary, the mother may expect a healthy child.

Chemotherapy. For the treatment of adult toxoplasmosis, sulphonamides alone (sulphadiazine 4 × 500 mg daily) or in combination with pyrimethamine (25 mg daily) are given for 2–3 weeks. Amongst the antibiotics only spiramycin has been found to be active (adults 6 million IU daily in two separate doses, infants 0.15–0.3 million IU/kg body weight). Pyrimethamine should not be used in the treatment of pregnant women before the 3rd month of pregnancy (not more than 25 mg daily), and should only be given with careful haematological and serological monitoring.
FRENKEL (1971) recommends administration of supplementary folinic acid (Leucovorin, in adults 2–10 mg daily, or baker's yeast 5–10 g; with children, 1 mg folinic acid daily or 100 mg baker's yeast) to act as antagonist to sulphonamide.

Cryptosporidium species

The cryptosporidia have aroused special interest in recent years, because they can cause a diarrhoeal disease in man. They are sporozoan species (Apicomplexa, Eimeriina) which can be transferred from man to man without an intermediate host but are predominantly passed from different domestic animals to man. The oocysts are ingested by man by the faecal-oral route. Immunodeficient persons are at high risk for opportunistic infection as *Cryptosporidium* is an opportunistic parasite.

Morphology and Development. The very small banana-shaped sporozoites (about 5–6 µm) initially attach themselves after ingestion to the surface of the stomach or the intestinal epithelial cells – probably in the lower part of the jejunum or ileum. An adhesive zone forms between the parasite and host cell whereby the microvilli (cytoplasmic processes) of the intestinal cells surround the cryptosporidia (Plate IX). The parasite lies quasi-intracellularly and not intracytoplasmically (an unusual position for coccidia), at first giving the impression of an extracellular parasite (see REICHENOW, 1953; PELLERDY, 1974). Moreover, the microvilli can be destroyed by the parasite. As a repair effect, the epithelial cell forms a microplica, which delimits the parasitophorous vacuole (GÖBEL, 1983). The young parasites develop to schizonts, each of which forms eight banana-shaped merozoites (about 5 µm; see Plate IX, *d, e, f*). One or more further cycles of schizogony are possible. Finally, the merozoites develop into macro- and (here without flagella) microgamontes, and finally into gametes. After fertilization and fusion of the cell nuclei, the zygote develops and from this the oocyst (3–6 µm). Within the oocyst the sporocyst develops, surrounded by a single coat, containing four sporozoites (GÖBEL, 1984).

According to CURRENT (1985, 1986) this developmental pathway has been extended, in the sense that after the formation of the zygote two forms of oocyst, a thin-walled and a thick-walled type, have been described. The thin-walled oocyst releases eight sporozoites into the small intestine which attack further intestinal cells (auto-infection). The thick-walled oocysts with four sporozoites reach the outside environment with the faeces; they serve to spread the infection and transmit it to a new host. Further investigations are necessary to explain the development of cryptosporidia, although CURRENT and HAYNES (1984) were able to propagate cryptosporidia in human foetal lung cells in vitro and observe the complete intracellular development up to the formation of the four sporozoites.

Plate IX

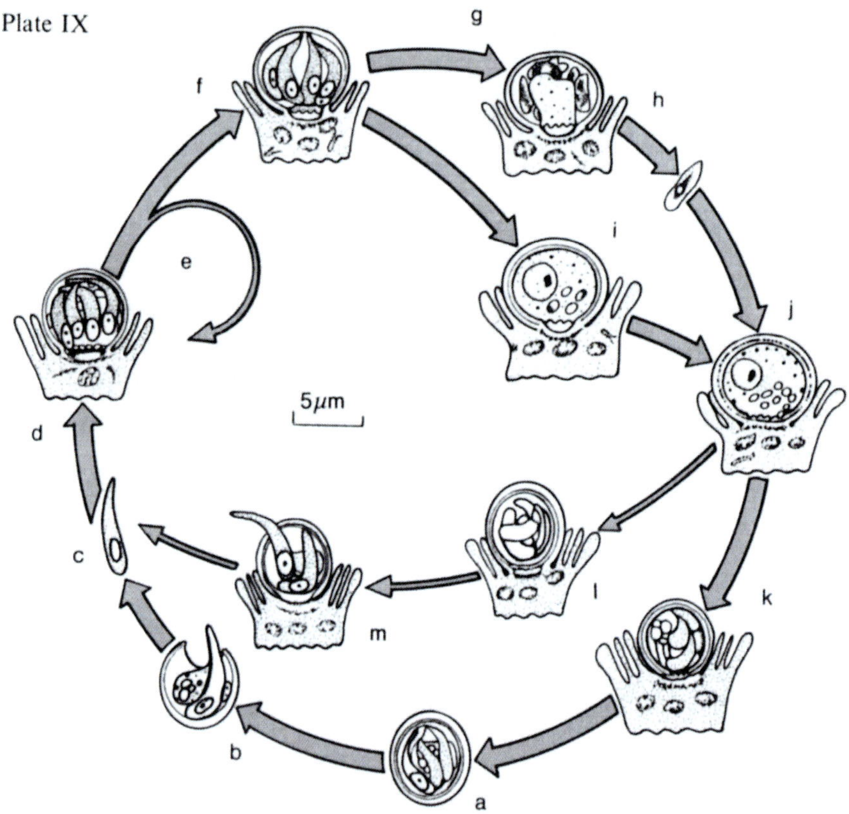

I

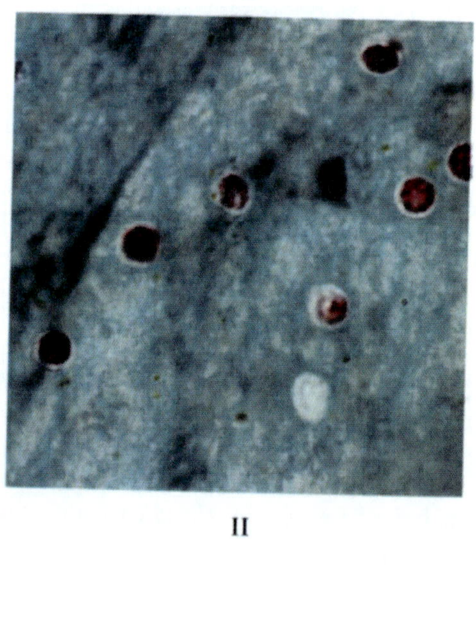

II

Clinical Symptoms. Clinical symptoms consist mainly of diarrhoea and watery stools, which can continue for 3–14 days. There may be considerable water loss, and cramp-like abdominal pains, and under some circumstances followed by constipation. Weight loss, vomiting and moderate fever are associated with these symptoms. There is also leucopenia and lymphopenia. The total picture resembles that of gastroenteritis (TZIPORI et al., 1983). After about 3 weeks the symptoms either disappear or the condition becomes chronic (WERK and KNOTHE, 1983). With immunosuppressed patients especially in case of AIDS all symptoms are frequently more severe and can lead to death (BURCHARD et al., 1985). Lung and biliary infections can also occur (CURRENT and HAYNES, 1984).

The incubation time is 3–12 days, the prepatent period about 1 week, and excretion of oocysts occurs over 2–3 weeks. The first oocysts can be found in stool preparations as early as 2–3 days after the onset of symptoms. In diarrhoea of unexplained origin, especially in children, cryptosporidiosis should always be considered.

Epidemiology. With the world-wide distribution of the parasite there is a high risk of infection for people working on land, because pets and domestic animals (cattle, pigs, sheep and goats) excrete the oocysts and constitute a source of infection. The infectious stages are taken in with contaminated food, water and also with dust (guinea pigs, mice, rats, and even birds, snakes and fish can be carriers of cryptosporidia).

Diagnosis by Microscopy. This can be performed by the examination of a methanol-fixed stool smear, stained with GIEMSA. The oocysts from fresh stool samples already contain sporozoites. In diarrhoeic stools oocysts with one to four nuclei can occur, these develop into sporozoites within a few days at 20 °C. The small size of the oocysts (4–7 µm) makes detection of the parasites difficult and so a concentration method should always be used (e.g. sugar solution; cf. p. 309). Differential staining with ZIEHL-NEELSEN carbol-fuchsin stain (after KINYOUN), by which the oocysts are clearly recognizable, is also recommended (see GÖBEL, 1983).

Therapy. Sulphadiazine and pyrimethamine do not act reliably. However, symptomatic treatment is important and the frequently high amount of water lost (dehydration!) must be replaced. Electrolyte replacement should also be considered (according to the WHO recommendations: to 1 l water, add 20 g glucose, 3.5 g sodium chloride, 3.0 g sodium citrate, 1.5 g potassium chloride).

◀ *Cryptosporidium* species.
I Developmental cycle (according to CURRENT, 1985). *a* Sporozoite-containing oocysts from the faeces; *b* Excysting sporozoites; *c* Free sporozoite; *d* Schizont with 6–8 merozoites; *e* New infection of intestinal cells; *f* Schizogony, which leads to micro- and macrogamontes (*g, h, i*) and a zygote (*j*) with a thick-walled oocyst (*k*); Thin-walled oocysts, which cause auto-infection can also be formed (*l, m*).
II Stool preparation stained with KINYOUN's carbolfuchsin solution (1985); oocysts partly red, partly unstained (preparation and photograph from EZZELDIN, Cryptosporidiosis, J Am Ved Med Assoc 187:1334–1335).

Plate X

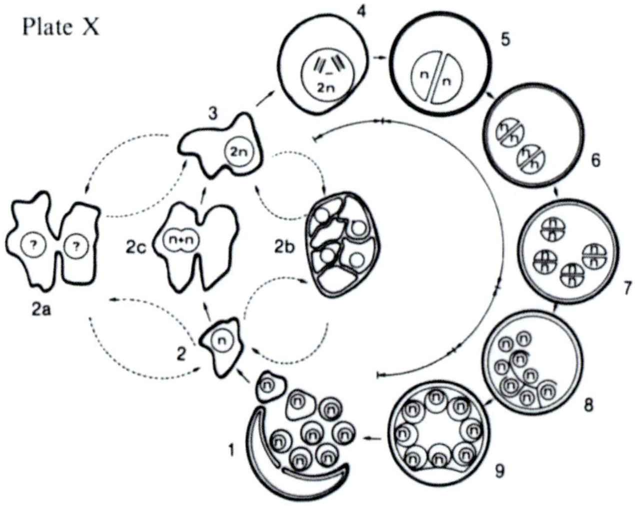

I

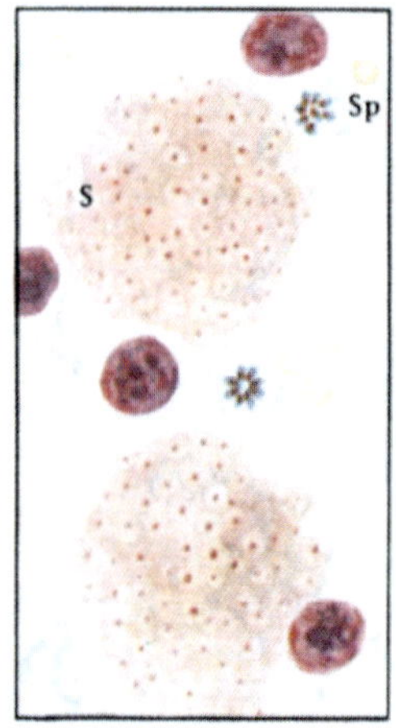

II

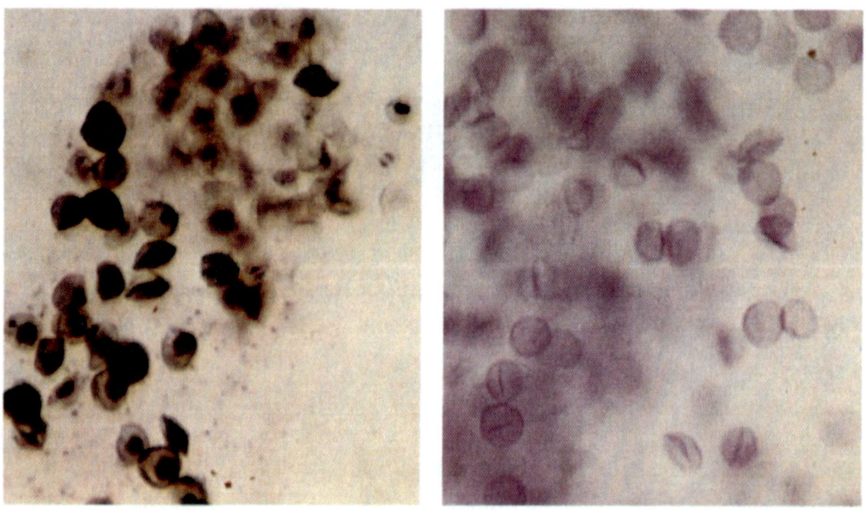

III

Pneumocystis carinii (CHAGAS, 1909) DELANÖE and DELANÖE, 1912

Pathogen of *Pneumocystis* pneumonia

Pneumocystis carinii belongs, judging by appearance, to the Protozoa, but in the new classifications of the Protozoa by LEVINE et al. (1980; see p. 5 footnote) and by CORLISS (1984) it is not mentioned. Therefore no one opinion prevails as to the nature of *P. carinii*.

Pneumocystis carinii was described for the first time in 1909 by CHAGAS in lung smears of guinea pigs, and in 1911 also in man, but was first connected with the disease picture of interstitial plasma cell pneumonia of infants by VANEK and JIROVEC (1952). Since then numerous cases of this, in particular in premature and undernourished infants, have been observed. There is often increased incidence locally and the disease usually has a fatal outcome. It was soon established than adults receiving immunosuppressive therapy or in whom other factors might limit immune potency could be severely ill with *Pneumocystis* pneumonia, although not necessarily with the involvement of plasma cells. Now, owing to the action of the human immunodeficiency virus (HIV) which leads to immunosuppression (AIDS) this problem has become particularly acute (see p. 6). *P. carinii* must therefore be considered an opportunistic parasite.

Present day information shows that latent infections with *Pneumocystis carinii* are frequent. There are reports from almost every country on the occurrence of this parasite, both in man (more than 50%) and in animals (e.g. rats, mice, rabbits, dogs, cats, cattle and sheep).

I Schematic representation of the developmental cycle of *Pneumocystis carinii* (after MATSUMOTO and YOSHIDA, 1984)
1 Ruptured cyst discharging eight "spores"; *2* Amoeboid trophozoites (haploid), *2a* Binary fission; *2b* Sexual fusion; formation of diploid trophozoites; *2c* Endogeny (thin-walled cysts, intracellular?); *3* Diploid trophozoite; *4, 5* First meiotic division; *6* Second meiotic division; formation of cyst membrane; *7* Mitosis; formation of intracystic bodies; *8* Segmentation into eight "spores"; *9* Thick-walled cyst with eight "spores" (sporozoites?)

II Lung swab preparation: two cysts each with eight sporozoites. Alveolar contents with numerous dividing stages (GIEMSA stain)

III Cysts from bronchoalveolar lavage; (a) Silver stain after GROCOTT: walls of viable and empty cysts black; (b) Toluidine blue O stain: the cysts stain metachromatically bluish-red (×1000; preparation and photograph by ELIAS)

Frenkel (1976) has expressed the opinion that the species of *Pneumocystis* occurring in man is different from that which occurs in the rat, called *Pneumocystis jiroveci*. He bases his view on the immunity occurring in man, which must account for the small number of parasite carriers. Furthermore, serum from rats infected with *Pneumocystis* does not react with parasites from man. On the other hand, no morphological differences can be established. This species differentiation has, of course, not so far been generally recognized (Meuwissen, 1976).

Morphology and Development. Our knowledge of the biology of *Pneumocystis carinii* is still very imperfect (see Plate X). Certainly, the characteristically round, relatively thick-walled cystic stage (about 3–8 μm) containing eight oval "sporozoites" (about 1–2 μm) is known. Yoshida et al. (1981) called these, neutrally, "intracystic bodies." Following infection by the inhalation of cyst-containing dust or through droplet infection the eight cystic bodies (sporozoites?) become free in the alveoli. They take on an amoeboid form and become trophozoites and these multiply through binary fission. According to Matsumoto and Yoshida (1984), after pairwise union the haploid trophozoites become diploid. In the precystic stage, meiotic division leads to the development of further haploid forms, a process which the above-mentioned authors now describe as sporogony (Frenkel, 1976; Yoshida et al., 1981). In the cysts, "cystozoites" lie together in rosettes and in old mature cysts exhibit rapid and active movements. These cysts could be completely inhibited by the administration of cotrimoxazole (Szabados et al., 1986). The cystozoites can apparently multiply by endogeny (endodyogeny?) within a thin-walled cyst in the alveolar epithelium (Matsumoto and Yoshida, 1984; Plate X). The characteristic thick-walled cyst can still be stained with a silver stain (methenamine silver) when the cyst is empty, which makes the histological diagnosis of the parasite easier (see below). The walls of empty cysts collapse and appear, therefore, to be of unusual shape, sometimes crescent shaped, or like coffee beans (Plate X).

Pneumocystis carinii can apparently multiply in vitro in chicken embryonic lung culture (Pifer et al., 1977; Murphy et al., 1977). An unequivocal decision as to which protozoan group the parasite belongs cannot be made from this. Yet from reproducible and continuous culture in vitro reliable information can be obtained on the nature and mode of multiplication of *P. carinii*.

Electron microscopic studies have indicated that the parasites in the alveoli are connected with the host cell by short, slender pseudopodial processes (Yoshida et al., 1984). The parasite attaches itself to the host cell, enlarges to the precystic stage, and then becomes quiescent and matures into a cyst (Hughes, 1982). The results of Vossen et al. (1978), who report intracellular multiplication, are at variance with these findings. Intracellular multiplication would explain the frequent large multiplications of the parasite and its high pathogenicity.
Extrapulmonary infections are rare, yet in a few cases the parasites have been found in lymph nodes, spleen, liver and peripheral blood. However, morphological similarities, e.g. with *Candida* cells, have been pointed out, and there is danger of confusion.

Clinical Symptoms. *Pneumocystis carinii* causes pathological changes as follows. In the first stage the parasites lie thinly scattered on the alveolar wall. During this period no clinical symptoms occur. In the second stage, the affected cells are desquamated into the alveolar lumen; the parasites multiply. There is little, if any, inflammatory reaction of the alveolar septa at this point. Also, at this stage there are still no clear observable disease symptoms. In the third stage the desquamation of the alveoli spreads, with numerous organisms in the alveolar cell debris. The consequences of this are thickened alveolar septa with mononucleate inflammatory cells. In this stage patients show a clinical pneumonia (WALZER et al., 1976). In typical cases, *Pneumocystis carinii* fills the alveoli and bronchioles so thickly that finally scarcely any of the respiratory surface remains, as a consequence of which the patient practically suffocates. *Pneumocystis carinii* infection thus acts primarily in a mechanical fashion (JIROVEC, 1954) in obstructing the alveoli and bronchioles. The patients are hypoxaemic and cyanotic. Fever does not generally occur. Radiologically there is no characteristic picture; the patients exhibit a diffuse interstitial infiltration. Untreated, almost all patients die and so in immunosuppressed cases prophylactic therapy should be considered (JIROVEC and ŸANĚK, 1954).

The number of adults who become ill with *Pneumocystis* pneumonia increases constantly. This is partly as a consequence of immunosuppressive treatment (given for many different conditions) which, by the exacerbation of latent infection, leads to specific disease. Other immunosuppressed patients are those with primary immunodeficiency (premature births) or persons damaged by defective nutrition. A remarkably large number of homosexuals has been found to be immunosuppressed, as with many AIDS cases (see p. 6). The virus infection has spread widely in this group of people (see ZIEFER et al., 1986).

Transmission. JIROVEC (1954) had already surmised that the pathogen was exhaled in its cystic form and was inhaled by healthy people, to then undergo multiplication in the lung alveoli as trophozoites (Plate X). Frequently, an increase in cases within a clinic is observed and this may even be linked to certain sick rooms. Whether, as is supposed, the infective stages are spread from nursing staff with latent infections, either by droplets or inhaled with dust, is still not clear. Air borne spread of cysts from animals is also suspected, but with no firm evidence for it. Rats, dogs and guinea pigs must be regarded as animal reservoirs, even if transmission to man cannot yet be proven. Transplacental transmission is also suspected (PAVLICA, 1966).

It is probable that the proportion of persons with latent infections increases with age. According to a study in the Netherlands on 120 randomly selected normal children (HUGHES, 1982), the frequency of seropositive patients increases constantly after the first year of life and attains 83% by the age of 3–4 years.

Diagnosis by Microscopy. The parasite is rarely found in the sputum or bronchial secretions. Therefore, for the diagnosis of *Pneumocystis carinii* infection in vivo when there is a strong clinical suspicion a lung washout (bronchoalveolar lavage) is necessary. In this way material for examination is obtained directly from the alveoli. For microscopic preparation, a smear of the lung fluid is made immediately upon obtaining the sample, or else the fluid is carefully centrifuged and the sediment used and stained. In urgent cases with negative results of lavage, biopsy material may be obtained for examination by transbronchial endoscopy or transcutaneously. In such an examination a spot preparation is first prepared, and afterwards a histological preparation. A suitable staining method is crucial for a good result.

The usual GIEMSA staining method (see p. 306) is not effective. Therefore, it is recommended that the specimen be stained overnight intensively with GIEMSA (stock solution diluted 1:20). With this method the cytoplasma stains blue and the nuclei red; the mucous coat assumes a reddish-violet tone. With the methenamine silver method (after GOMORI/GROCOTT) both the empty and the sporozoite-containing cysts appear black, and with toluidine stain, blue to blue-red (Plate X, *III a, b*) (JACOBS et al., 1984). Cyst wall thickenings are characteristic. These staining procedures are also suitable for pathological investigation after an autopsy.

In the absence of suitable expertise a specialist laboratory should be consulted, especially when other organic structures, e.g. fungal spores and erythrocytes, are possible contaminants.

Diagnosis by Immunobiological Methods. The methods available are the complement fixation and indirect immunofluorescence (IIFT) tests. Antigen is obtained from lung sections of rats. Acute pneumocystosis is produced experimentally in these animals by treating them with immunosuppressive drugs or irradiating them. Serological investigations with IIFT gave positive results in only about 40% of patients. The IIFT is rated in this case as the most sensitive test (1:16–1:64), but permits the recognition of only one-third of all infections (see below; KAGAN and NORMAN, 1976; MEUWISSEN, 1976). Moreover, the results have only very limited significance, as *Pneumocystis carinii* infections are common and in immunosuppressed patients there may be no antibody response.

Chemotherapeutic Measures. Treatment of *Pneumocystis carinii* infection consists of the use of either trimethoprim (20 mg/kg body weight) plus sulphamethoxazole (100 mg/kg body weight, daily, for 14 days), or pentamidine isethionate (3–4 mg/kg body weight, i.m., daily, for 14 days) either alone or in combination with pyrimethamine and hydroxystilbamidine (4 mg/kg body weight i.m. daily over 10–14 days). Of course, amongst other side effects, there is the risk of nephrotoxicity and therefore the patient must always be admitted to hospital. With the occurrence of thrombocytopenia or leucopenia replacement therapy with folic acid (Leucovorin, oral or i.m.) must be considered (YOUNG and DEVITA, 1976; ZIEFER et al., 1986). WHISNANT and BUCKLEY (1976) reported the successful use of pyrimethamine (1 mg/kg body weight, daily) and sulphadiazine (250 mg every 8 h over 8–12 days) plus cholic acid (once weekly, 6 mg) in an immuno-deficient child. Symptomatic therapy is particularly important (e.g. oxygen administration) in severe respiratory insufficiency in pneumocystosis.

Plate XI ⇨

Plasmodium falciparum, P. vivax
P. ovale, P. malariae
Babesia species

Plate XI

Life Cycle of Malaria Parasites
Malaria-Cyclus

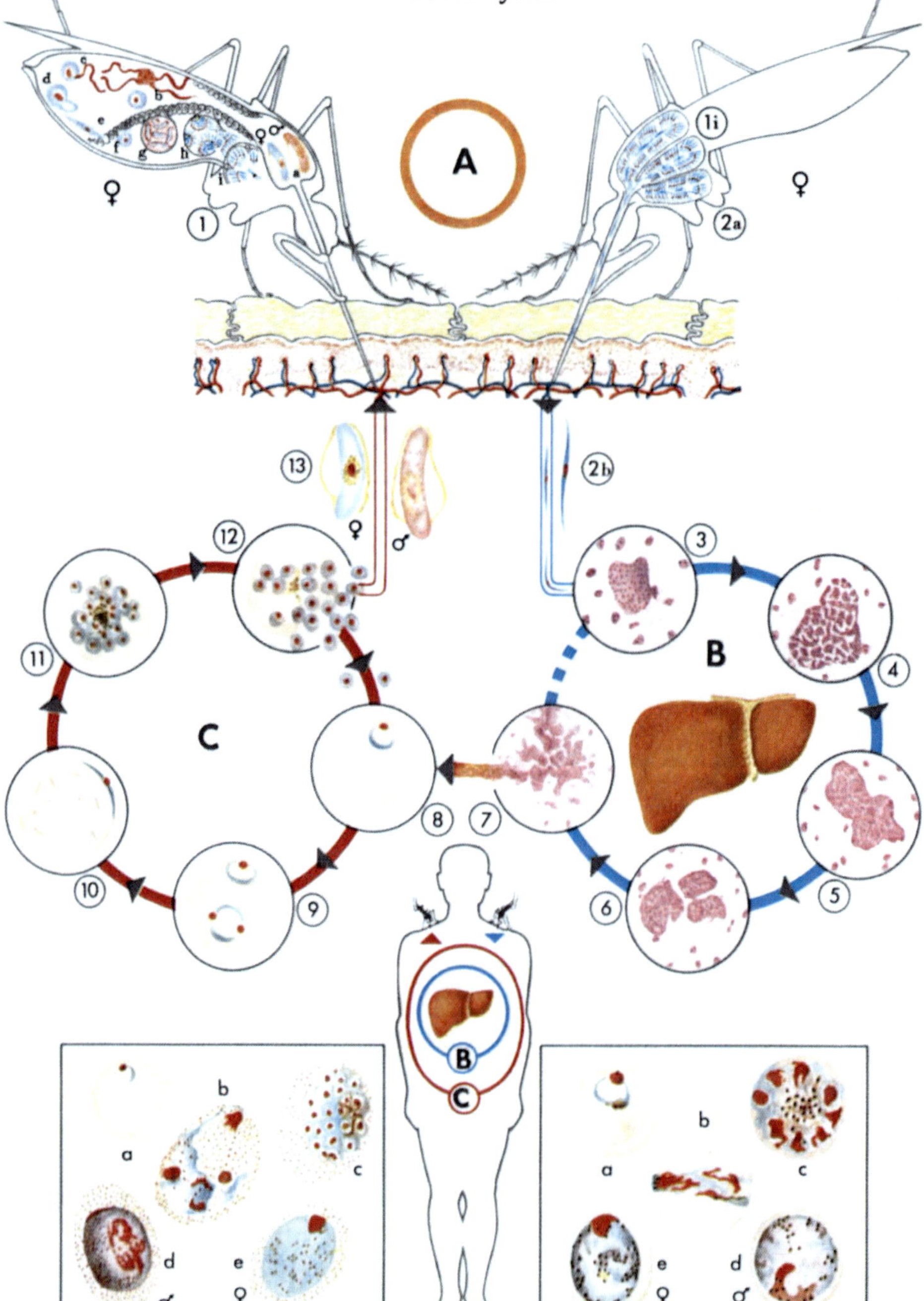

I Plasmodium vivax (Malaria tertiana)

II Plasmodium malariae (Malaria quartana)

Plasmodium falciparum (WELCH, 1897)
P. vivax (GRASSI and FELETTI, 1890) LABBE, 1899
P. ovale STEPHENS, 1922
P. malariae (LAVERAN, 1881) GRASSI and FELETTI, 1890

Pathogens of malaria

Life cycle of the malaria pathogens using *P. falciparum* as example (GIEMSA stain)

Ⓐ Sexual development (gamogony) in the female *Anopheles* mosquito:

1 a Gamontes taken with human blood
b Gametes
blue, female cells (macrogametes)
red, formation of microgametes (exflagellation)
c Fertilization, *d* Zygote (retort form), *e* Ookinete
f Oocyst, *g* Oocysts, *h* Mature, ruptured oocysts
i Sporozoites in the cells of the salivary gland
2 a Biting *Anopheles* mosquito transmits sporozoites
b Individual sporozoite

Ⓑ Pre-erythrocytic development in the liver cells of man (blue circle):

3–7 Different stages of the so-called endothelial forms from the liver (exo-erythrocytic stages); in *P. vivax* infections, possible development of hypnozoites (see p. 98)

Ⓒ Erythrocytic development (schizogony) in man (red circle):

8, 9 Young schizonts (trophozoites, ring forms) of pathogen of malignant tertian malaria
10 Marginal form of young schizont; erythrocyte with MAURER's clefts. (*8–10*) Different positions of young schizonts of pathogen of malignant tertian malaria in erythrocytes
11 Morula stage with about 20 merozoites
12 Liberated merozoites invade further erythrocytes, some developing into schizonts, some into gamontes
13 Gamontes (in *P. falciparum* crescent shaped) (see *I* and *II, d, e*)

Typical stages from the peripheral blood:

I *Plasmodium vivax*
II *P. malariae*
a Trophozoite, ring form *d* Male gamonte
b Multinucleate schizont *e* Female gamonte
c End-stage of schizogony – Morula (*I*) or "daisy head" (*II*)

In *P. vivax* infections there are characteristic changes in the erythrocytes; they are enlarged, become pale, and frequently appear stippled (SCHÜFFNER's dots).

Malaria

The name malaria (mal aria = bad air) points to the relationship between the "air" and the disease, suspected even before the discovery of the pathogen. This lies in the fact that mosquitoes, in this case species of the genus *Anopheles*, have a decisive role to play as vectors of the malaria pathogens, blood parasites of the genus *Plasmodium*. Mosquitoes "infect" the air as it were (GARNHAM, 1984).

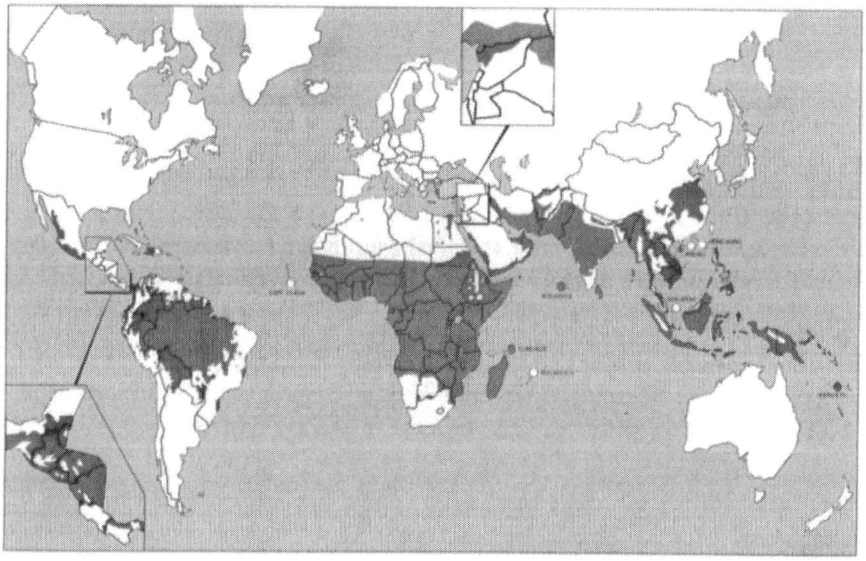

Distribution of malaria (WHO, 1986)

Even today malaria (intermittent fever, swamp fever) is one of the most widespread diseases of hot countries. It is found everywhere between latitudes 40° north and 30° south, and also sometimes outside this area. A global increase has been seen in the incidence of malaria since 1979, especially in East Asia and Central America. About 200 million people suffer from malaria and about one million children die annually from this disease in Africa alone. More than 1.62 billion people live in the 91 countries with a moderate to high risk of malaria (WHO, 1983).

It is not surprising, therefore, that visitors return from these regions with malaria. In the Federal Republic of Germany alone 2519 cases of malaria with 56 deaths

were recorded in the years 1978–1983 (BRUCE-CHWATT, 1970; WEISE, 1984). The risk of malaria has also increased because of the development of drug resistance, whereby active substances generally used until now have lost their value in some regions. New forms of chemotherapy are therefore being sought (see below).
The parasites are dependent on particular climatic conditions for their further development in the vector, which introduces the pathogen to man when feeding on blood.

The malaria pathogens are protozoa of the systematic group Haemosporidia. The most important species are:

1. *Plasmodium falciparum,* pathogen of malignant tertian malaria (falciparum malaria)
2. *Plasmodium vivax,* pathogen of benign tertian malaria (vivax malaria)
3. *Plasmodium ovale,* malaria pathogen, the febrile cycle of which corresponds to that of benign tertian malaria
4. *Plasmodium malariae,* pathogen of quartan malaria

The map shows the approximate geographical distribution of malaria (as in 1986). Benign tertian malaria occurs the furthest north there, whilst the dangerous malignant tertian malaria predominates in the tropical regions. For a long time *Plasmodium ovale* was thought to be confined to Africa, but it has increasingly been found in places outside Africa (e.g. Indonesia, Burma, Philippines, Thailand, Vietnam).

Humans are infected in natural ways through *Anopheles* mosquitoes, which develop in water. The female mosquitoes take in the parasites with the blood of infected humans. Development is restricted to two hosts, man and *Anopheles* mosquitoes, and comprises three stages:

A Sexual development and multiplication in the female *Anopheles* mosquito, in which sporozoites develop, through which humans are infected when bitten by mosquitoes (Plate XI, *1, 2*).
B Asexual multiplication in the hepatic cells (pre-erythrocytic stage; Plate XI, *3–7*): dormant sporozoites may be retained in the liver as hypnozoites in *P. vivax.*
C Asexual multiplication in the erythrocytes (erythrocytic stage; Plate XI, *8–12*); these stages cause the clinical picture characteristic of malaria in man, the intermittent fever.

Development. When sporozoites enter man with the mosquito bite (Plate XI, Ⓐ *1, 2*), they are carried to the liver via the bloodstream and penetrate into hepatic cells through the endothelium. There they multiply asexually (pre-erythrocytic schizogony; Plate XI, Ⓑ *3–7*). This continues for a minimum of about 6–9 days depending on the species. The merozoites are then released and some of them enter red blood cells through invagination of the erythrocyte membrane (Plate XI, Ⓒ *8*). In the erythrocytes the parasites, which on a stained smear initially appear to be ring-shaped, grow and multiply again (Plate XI, *8–12;*

erythrocytic schizogony); this cycle takes place within a certain period of time which differs depending on the type of malaria (48 h for benign and malignant tertian malaria; these may also show the quotidian-type, 24-h rhythm; 72 h for quartan malaria). The parasites escape from the erythrocytes again (merozoites; Plate XI, *12*), infect other erythrocytes and either repeat the same development several times (Plate XI, *8–12*) or become sexually differentiated cells, male and female gamontes (Plate XI, *13*). These can continue their development only in mosquitoes of the genus *Anopheles*.

The transformation of the gamontes into gametes takes place in the stomach of the mosquito (Plate XI, *1*), and here eight microgametes develop and are released by exflagellation. The fusion of microgamete and macrogamete (*c*) leads to a motile zygote, which as the ookinete (*d, e*) penetrates through the intestinal epithelium of the mosquito and develops into the oocyst between the basal lamina and the epithelium (*f, g*). The sporozoites develop within the oocyst (*h, i*), being released when mature and migrating into the salivary gland cells. They enter humans again with the saliva when the mosquito bites (Plate XI, *2a*). The development of the parasites in the mosquito lasts for about 15 days at 25 °C.

In addition to this "normal" cycle, in *Plasmodium vivax* and *P. ovale* resting stages, the hypnozoites, also develop. These develop directly from sporozoites, which in this case do not commence their pre-erythrocytic development immediately. They may remain in the liver for one year or more before continuing the usual developmental pathway. These stages are apparently considered responsible for the relapses of the disease (see below), infection taking place in late summer or autumn and malaria developing in the spring. These forms of disease with their extremely long incubation periods are now attributed to special strains of *Plasmodium vivax* in which hypnozoites hibernate, hence they are also called *P. vivax hibernans*. With *P. malariae* small populations of blood parasites survive and even after years may lead to new fevers (recrudescences).

Typical blood forms (see Plate XI) are as follows:

1. *Plasmodium falciparum:* Trophozoites (signet ring-shaped), very small (2–2.5 μm; Plate XI, *8–10*) and delicate; sickle-shaped gamontes (Plate XI, *13*). Segmenting forms are very often absent from the peripheral blood; as a rule they first occur there in moribund patients. Affected erythrocytes often show MAURER's clefts (Plate XI, *10*), which are not to be confused with the fine granular stippling, SCHÜFFNER's dots, seen in *Plasmodium vivax*.

The sexually differentiated early stages of *P. falciparum* disappear temporarily from the peripheral blood; they undergo maturation in the deep tissues of the spleen and bone marrow. Ten to twelve days later they reappear in the peripheral blood as crescent-shaped stages, becoming fully mature after another 1–2 days. The in vitro multiplication of *P. falciparum* as a source of antigen has become relatively simple and is of great experimental value (see p. 104); it is also useful for study of drug resistance (TRAGER and JENSEN, 1976).

2. *Plasmodium vivax:* Trophozoites are ring-shaped (*a*); multinucleate amoeboid segmenting forms with fine granular pigmentation (*b*) finally form the morula stage (*c*), from which about 12–16 merozoites develop. Gamontes also develop (*d, e*). The erythrocytes become distinctly enlarged because of the parasite infection; they often show the characteristic red stippling of SCHÜFFNER's dots.

3. *Plasmodium malariae:* In a blood smear trophozoites appear ring-shaped (*a*), but they mostly already contain rather coarse-grained golden-yellow pigment which is characteristic of all quartan parasites. The multinucleate schizonts initially assume a band form (*b*); finally six to twelfe merozoites group around the pigment ("daisy head") (*c*). The gamontes are similar (*d, e*) to those of the benign tertian pathogen. The affected erythrocytes do not exhibit any particular changes, however; there is no stippling of any kind.

4. *Plasmodium ovale:* The pre-erythrocytic phase with *P. ovale* lasts for about 9 days, the erythrocytic cycle for 48 h. The trophozoites are ring-shaped; the half-grown parasites mostly contain a brown-black, glistening pigment. The erythrocytic schizonts develop 6–12 merozoites. The schizonts and gamontes distinctly enlarge the erythrocytes. Besides the stippling of the erythrocytes (JAMES' dots, similar to the stippling of SCHÜFFNER's dots in *Plasmodium vivax*), an irregular, corrugated border develops in the erythrocytes. The course of the fever follows that of benign tertian malaria.

All the malaria parasites can be preserved in a viable state for months by adding glycerin or dimethylsulphoxide (DMSO) and then deep-freezing (e.g. in liquid nitrogen).

Clinical Symptoms. Malaria is characterized by intermittent fever, with one or two fever-free days between attacks of fever, the periodicity depending on the species of parasite present. These are preceded by an uncharacteristic fever (prodromal stage) in cases of "first fever" (not before relapses). During this period the malaria parasite often cannot be demonstrated with microscopy. The typical attack of fever starts often with violent shivering, rapidly followed by a febrile stage and, after a few hours, by complete defeverescence with profuse sweating. This is then followed by another attack after 48 h (benign and malignant tertian malaria) or 72 h (quartan malaria). The classical onset with shivering and fever-free intervals often does not occur in malignant tertian malaria however. The body temperature fluctuates in a similar way to the benign tertian type (only recordable if the temperature is measured every two hours), but the temperature remains high, sometimes at 38°–39 °C. Small children are protected up to the 3rd month of life (milk factor), but after that they are particularly at risk from *P. falciparum* which has high infant mortality.
Obscure pictures of fever are seen in mixed or double infections; attacks of fever then often occur daily (e.g. tertiana duplicata or quotidian malaria). Malarial recurrences can still occur after more than 2 years in untreated infections with *P. falciparum*, *P. vivax* and *P. ovale;* with *P. malariae* the parasites may sometimes persist for over 30 years. Some level of immunity is always associated with the presence of plasmodia but is strain specific.
In a typical case of malaria the development of parasites in the erythrocytes takes place in parallel with the fever, i.e. during the fever-free period the schizonts grow and at the time of

the attack of fever the merozoites are released. Anaemia develops as a result of the destruction of the erythrocytes by the malaria parasites. An important symptom is an enlarged spleen, a condition used as a measure when investigating endemic malaria (the Spleen Index, i.e. percentage of an enlarged spleen projecting downwards over the costal arch in a child). A high proportion of very large spleens (roughly reaching the navel) in the population (50%) indicates severe endemic disease, but differentiation from kala-azar, schistosomiasis and ancylostomiasis must be made.
Other symptoms, such as violent diarrhoea, are often not recognized, even though they are almost characteristic of malignant tertian malaria. There is a great danger to the patient (especially in Europe) if this disease picture is misinterpreted as malignant tertian malaria can lead to coma (respiratory insufficiency and acute renal failure) followed by death (see also EHRICH et al., 1983). Microthrombi in the capillaries blocking the bloodstream are responsible for this. According to the latest findings these develop as a result of a lower oxygen tension in the organ capillaries than in the peripheral blood and because of the formation of knobs projecting from the surface of infected erythrocyte membranes.
Black-water fever is a disease that occurs secondary to malaria. An accumulation of haemoglobin in the renal tubules leads to haemoglobinuria, mostly after repeated attacks of malignant tertian malaria. Following malaria a slight fever often develops suddenly in an apparently completely healthy patient. This increases, often after a dose of quinine, and there may be shivering with temperatures of up to 40°–41 °C. The urine which is then passed is dark red to black-brown in colour (hence black-water fever) (see MOHR, 1972).

An important principle in malaria regions is that **in any case of fever malignant tertian malaria must be considered possible** and precautionary treatment must be given (see also algid malaria, below). The same principle applies to all patients suffering from feverish conditions who have lived in malaria regions during the past 1–2 years. Malaria has in no way been eradicated; in some places it is even increasing locally, partly because of a lack of financial resources to control the mosquitoes, partly because of the effects of human activity (building of reservoirs, irrigation canals, e.g. in Turkey, India).

Algid malaria, which EICHENLAUB et al. (1984) have recently discussed must also be mentioned; an acute attack of malignant tertian malaria can take a fever-free course (algor = cold). However, typical symptoms of malignant tertian malaria occur such as headaches, loss of appetite, watery diarrhoea, weakness, dizziness, fall in blood pressure, collapse, and cerebral symptoms. The **clinical history is of great importance,** and should show that there has been a recent visit to malaria regions. Such cases are presumably mostly not recognized.

GARNHAM (1971) has pointed out the very remarkable fact, with regard to malaria in West Africa, that the local indigeneous population with a high prevalence of the Duffy factor seems to be immune to *Plasmodium vivax*. As far as is known today, *P. vivax* cannot penetrate into the erythrocytes if the erythrocyte membrane is negative for the Duffy blood-group antigen (MILLER et al., 1975); this principle does not apply to the other species of *Plasmodium*. The regions which were colonized from West Africa, e.g. Surinam in South America and Haiti in the Caribbean, also show this resistance to *P. vivax* (in contrast to the original inhabitants of Ecuador). *P. vivax* is absent there, as in the USA where the inhabitants of West African origin are immune to this parasite. Half-castes on the other hand are much less resistant to this form of malaria. West Africans show the normal susceptibility to *Plasmodium ovale*.

It is worth noting that sickle-cell anaemia (a hereditary disease, Hb-S for short) protects against *P. falciparum*. However, only the heterozygous Hb-S carriers are protected (about 60 million worldwide), because, although homozygous Hb-S carriers are immune to malaria, they succumb early to the anaemia, mostly in childhood. Thalassaemia (Mediterranean anaemia) is another genetically induced abnormality of the erythrocytes, in this case caused by a decrease in the rate synthesis of haemoglobin. In heterozygous individuals it leads only to a mild anaemia (Thalassaemia minor), but in homozygous individuals it leads to a very severe disease (Thalassaemia major) from which children die in their early years. The frail, misshapened erythrocytes with a low haemoglobin content are either not infected by *P. falciparum* or the parasites perish in them. It is assumed that the malaria parasites "make life difficult for themselves" by inducing an oxidative action (e.g. through peroxides) in Thalassaemia erythrocytes (BETKE, 1985).
A deficiency of glucose-6-phosphate dehydrogenase similarly protects against *P. falciparum* infection (about 100 million cases worldwide).

Transmission. The malaria parasites are transmitted only by female mosquitoes of the genus *Anopheles,* in which there are numerous species with very different habits. This must be taken into account if one wants to carry out successful malaria control by eradication of the vector (species sanitation). The larvae live in water, and, depending on the species may live in pools, streams, coastal swamps, tree hollows, etc. *Anopheles* mosquitoes fly at dusk and at night. It is recommended that people should avoid going out into the open air in malaria regions as far as possible and should use mosquito nets well away from their bodies in sleeping quarters. Intrauterine infection of the foetus can take place through small placental haemorrhages, but not through active migration of the parasites. On the other hand, according to observations made in West Africa, *Plasmodium* infections of the placenta are relatively frequent (about 20%), especially in the first pregnancy; they may sometimes lead to reduced birth weight, to abortion or to stillbirth (MCGREGOR et al., 1983).
There is also a danger of developing malaria after blood transfusions. Blood donors must therefore be screened for latent infection either by their history or if necessary by serological methods. The possibility of latent infection must be considered for up to 15 months with malignant tertian malaria, up to 3 years with benign tertian malaria and up to 25 years with quartan malaria, but in individual cases far longer periods have been reported. Plasmodia remain infectious for at least 5 days in stored blood kept at 4°–6°C, but in odd cases an infection has still occurred even after 14 days. Immunological checks on blood donors for malaria antibodies are necessary in suspected cases.

Control. There are three effective ways of controlling malaria:

1. The systematic elimination of the mosquitoes (because without mosquitoes, there can be no malaria), from houses and from their breeding sites
2. The destruction of the parasites in the human body with drugs (see below)
3. Individual protection

Very effective insecticides are available today for the first control measure, and through their systematic use whole areas can be cleared of mosquitoes. Individual protection consists of the use of suitable clothing and mosquito nets (Lindsay and Gibson 1988), the impregnation of textiles (canvas strips etc.) with insecticides and also of the use of repellents.
In the fight against *Anopheles* mosquitoes the primary aim should always be to eliminate the breeding places. These measures are ultimately not only more economical than the use of chemical preparations but are mostly also effective immediately and in the long-term. This recommendation basically applies to all kinds of insects that are vectors of parasites. Where such measures cannot be effective, insecticides are initially used to control the arthropods (grub-, contact-, systemic poisons). These must be adapted to suit the life cycle of the arthropods in question and combined with biological procedures; they should not be used in accordance with a rigid schedule (Krieg, 1984).

In addition there is the phenomenon of resistance to insecticides, which will be briefly discussed here for the other kinds of parasites transmitted by arthropods as well. This phenomenon makes it much more difficult to control the vectors of plasmodia (see p. 101), and also of leishmania (see p. 34), Filaria (see p. 270) and other parasites which have a similar epidemiology (arthropod-borne diseases). These include other biting insects, such as lice and bugs. At the present time there are more than 250 kinds of arthropod with local strains resistant to chemical insecticides.
More than 51 species of *Anopheles* already show signs of resistance. Initially this was only directed against DDT, but other chemical compounds have also proved to be ineffective now (e.g. in Turkey where there is multiple resistance in *Anopheles sacharovi* and *A. hycranus*). Malathion was still considered to be effective until recently (Ramsdale et al., 1980), but resistant strains of *Anopheles stephensi* have now been seen in Pakistan (Rathor and Toqir, 1980). The use of fish which eat mosquito larvae is, therefore, increasingly being recommended again and practised.

The signs of resistance probably always occur as a consequence of gene selection, and the circumstances which lead to the selection probably differ considerably. The reasons for the occurrence of resistance are partly chemical, when whole groups of chemical substances being used simultaneously, and partly anatomical in nature. Changes in behaviour are particularly worth noting; thus, for example, many insects have altered their preferences for sites in housing areas and animal quarters, which may be due to the amount of light, temperature, and moisture content. In addition there have been changes in vitality and the capacity for detoxification, i.e. a change in the metabolic state (Konigk and Putfarken, 1984).

The increasing concern about poisoning the environment through the use of excessive amounts of chemical control agents has greatly promoted biological pest control. This can be very successful, especially for use over large areas, e.g. through the release of an excess of sterile breeding partners. Insect populations can die out as a result of sterilizing the males (e.g. by irradiation; see also p. 13).

Another biological control measure consists of the use of an extract from *Bacillus thuringiensis* H14. This strain was discovered in the near East by GOLDBERG and MARGALIT (1977) (quoted by FRANZ and KRIEG, 1982) as the new variety *Bacillus thuringiensis israelensis* (serotype H14) which kills exclusively the larvae of dipteras, the dose required depending on the target species. The possibility of biological control of malaria vectors opens up whole new dimensions in the use of insecticides, because this preparation does not harm other insects, vertebrate animals or human beings (see also p. 283).

The human malaria parasites do not develop in domestic or useful animals, so apart from the mosquitoes, no other parasite reservoir exists which might be of practical importance for the control of malaria. Some species of monkeys have been recognized as parasite carriers, however. To what extent this source of infection is epidemiologically of practical importance remains to be clarified. It is worth noting that the douroucouli monkey *Aotus trivirgatus* is receptive to normal and drug-resistant strains of *P. falciparum* and has therefore become of interest for experimental chemotherapeutic studies (according to the drug resistance problem).

Diagnosis by Microscopy. In cases of acute disease it is essential to confirm the diagnosis through the microscopic demonstration of the parasite using GIEMSA-stained blood smears. At least one thin blood smear and one "thick film" are required for this (see p. 306). The "thick blood film" makes it easier to detect the parasite. However, the appearance of the parasites is so altered using this method that the typical structure of the plasmodia in the erythrocytes, as shown in Plate XI, is only partly retained (see p. 94). The changes are so characteristic, however, that the experienced investigator will have no difficulty in recognizing the parasites. Blood films must always be done before the start of any specific chemotherapy. The general view that the best chance of finding malaria parasites is during or towards the end of an attack of fever must be refused however; parasites can be expected in the blood preparation at virtually any time between two attacks of fever. The erroneous view that the demonstration of *P. falciparum* is only confirmed if crescent-shaped gamontes are present, and that the appearance of this stage in the blood must be awaited, can be most dangerous for the patient. It is sufficient to demonstrate the small *P. falciparum* ring-stage (diameter about one-fifth of the erythrocytes) and also multiply infected erythrocytes (see Plate XI). Segmenting forms, as seen with *P. vivax* for example, are not to be expected here. *Babesia* species infections represent a possible source of error in the diagnosis of malaria (see p. 108).

Diagnosis by Immunobiological Methods. The microscopic demonstration of parasites in the thin film and "thick film" are still used to confirm the diagnosis of acute malaria, as antibodies first occur 1–3 weeks after the start of the feverish disease. The significance of negative or weakly positive titres during the feverish stage may be underestimated and the patient may die from malignant tertian malaria before significant titres are attained. The indirect immunofluorescence test (IIFT) using cultured forms of *P. falciparum* as antigen and also a direct haemagglutination test are valuable aids for epidemiological studies and for the detection of previous disease (e.g. in blood donors), because antibodies do not disappear immediately after drug treatment but mostly persist with low titres for years.

The IIFT is the most widely used and most tried and tested method to date. Titres suggestive of malaria lie between 1:64 and about 1:4000; titres from 1:64 to 1:256 suggest a *Plasmodium* infection but mostly not an acute case (rising titres should be noted); at 1:256 and above the patient is, or was, suffering from malaria, roughly within the previous 3 months. The ELISA can also be used according to VOLLER et al. (1980).

Serological investigation should always be carried out by special laboratories. This also applies to the new methods for detecting malaria antigen using monoclonal antibodies. Greater specificity is expected with these methods.

Chemotherapy. Drugs for malaria should prevent or eliminate disease and suppress relapses, and should also reduce the morbidity and mortality in malaria regions. In addition, many active substances can suppress the infection of the mosquitoes and prevent the spread of epidemics. However, since the various developmental stages of the plasmodia differ metabolically, different substances are required in order to eliminate all stages (sporozoites, praeerythrocytic and erythrocytic schizonts and gamontes). In addition to chemotherapy, chemoprophylaxis has a large role to play in malaria control, because several substances remain in the body for a fairly long time. Thus, for example, chloroquine and pyrimethamine are suitable for both treatment and prophylaxis. Mechanical protection against mosquitoes is also increasingly being recommended as a prophylactic measure (see above).
The following aspects should be taken into account in the chemotherapy of the different development stages of the malaria parasites:

1. The malaria parasites in the pre-erythrocytic cycle (Plate XI, Ⓑ) are eliminated with pyrimethamine and proguanil, but this is only successful with *Plasmodium falciparum*. With *P. vivax* and *P. malariae* these products only bring about delayed development at non-toxic doses.
To date no drug has been found effective against the sporozoites; perhaps vaccination may prove possible in the future (see ENDERS 1984; GODSON, 1985).

2. The asexual parasite stages (trophozoites and schizonts) in the peripheral blood (cycle Ⓒ) are eliminated with chloroquine (for use in pregnant women also) and the related 4-aminoquinolines and quinine hydrochloride (attention must be paid to side effects); secondline drugs are proguanil and pyrimethamine, but these act more slowly. Combined preparations of sulphonamides and pyrimethamine are used for chloroquine-resistant *P. falciparum* infections (see p. 105–107).

3. The sexual stages of *Plasmodium falciparum* circulating in the peripheral blood (gamontes) can only be eliminated with the 8-aminoquinolines (e.g. primaquine); the gamontes are affected by pyrimethamine and proguanil in such a way

that no sporozoites can develop from them in the mosquito (Plate XI, Ⓐ). These drugs are also suitable for chemoprophylaxis.

4. The persisting hypnozoites in *Plasmodium vivax* and *P. ovale* infections (see above) are apparently eliminated with primaquine (15 mg daily for 14 days or 30 mg for 7 days, depending on tolerance; toxic effects possible, including haemolysis). Persisting hepatic stages from the pre-erythrocytic cycle have become increasingly unlikely.

Chloroquine is still considered to be the drug of choice for acute malaria (initial dose 600 mg base, then 300 mg after 6 hs, then 300 mg daily for next 1–3 days). In suspected cases of malignant tertian malaria the treatment must also commence with the parenteral administration of chloroquine i.m. (possibly slow intravenous infusion). Attention must be paid here to the signs of resistance described below (see also map p. 106).
The prophylactic dose of chloroquine is 2 × 250 mg (= two tablets) per week. Children should receive 5 mg/kg body weight per week, with 25 mg pyrimethamine per week. Treatment should commence 1 week before starting a journey to endemic regions.
If parasites remain in the peripheral blood even after the administration of an effective preparation monitored for several days, then a drug-resistant strain must be presumed. Chloroquine-resistant strains of *P. falciparum* initially occurred in limited local areas but they have already spread over fairly wide areas in some cases, e.g. in Brazil, Columbia, Ecuador, Guyana, Panama, Venezuela, India, Southeast Asia, East Africa (Northern Malawi, Tanzania, Kenya) and West Africa (see map p. 106).
A distinction is drawn between three degrees of resistance to chloroquine in *P. falciparum:*

RI: The parasites recur in the peripheral blood between the 7th and 28th day after their incomplete disappearance.
RII: The parasites do not disappear from the peripheral blood, but with the onset of treatment there is a reduction of less than 25% in the previously existing parasitaemia during the first 6 days.
RIII: Only a slight reduction in the parasitaemia occurs under treatment, or there may even be an increase in the density of the parasites.

In *P. falciparum* cases resistant to chloroquine, a combination of pyrimethamine (25 mg) with sulphadoxine (500 mg) is effective. In cases with marked chloroquine resistance the treatment is commenced with quinine (slow intravenous infusion) and continued with pyrimethamine-sulphadoxine. In cases of multiple resistance quinine therapy must be continued with tetracycline. However, cases of resistance to these combined preparations have also been reported (Höfler, 1980a). The new

---

[1] Because considerable side effects or deaths were observed in a number of cases the German Federal Health Authority decreed in March 1985 that the combined preparation Fansidar should be used only for the treatment of patients with chloroquine-resistant falciparum malaria and for prophylaxis only in regions with a high risk of chloroquine-resistant malaria.

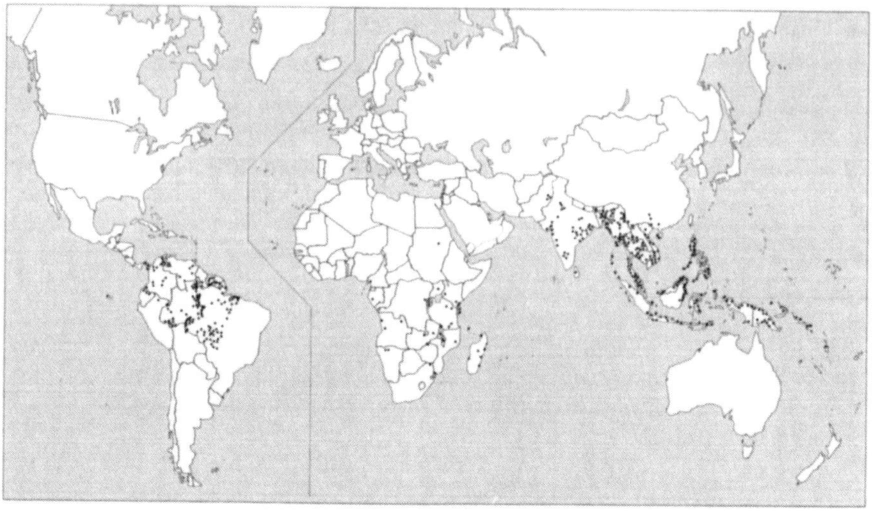

Distribution of chloroquine-resistant strains of *P. falciparum* (WHO, 1986)

preparation mefloquine (see below) has therefore become of great importance (adults: at the start 750 mg = 3 tablets, after 6–8 hrs 500 mg, after another 8 hrs 250 mg, max. 1.5 g; children: 1 × 25 mg/kg body weight) (LÓPEZ ATUÑANO and W. H. WERNSDORFER, 1979; HARINASUTA et al., 1983) (for short-term prophylaxis, 1–3 weeks of treatment with 250 mg/45 kg body weight once weekly, children between 1/4 and 3/4 of a tablet) (see p. 107) (Mechanism of chloroquine resistance in malaria see WARHUST, 1988).

Two treatments which are effective against multiresistant strains of *P. falciparum* are being tested in clinical trials: a combined preparation of mefloquine, sulphadoxine and pyrimethamine and also halofantrine (AMBROISE-THOMAS et al., 1986; WATKINS et al., 1986; HORTON 1988).

An extract from the plant *Artemisia annua* L. (Qinghaosu) is being used in China for the treatment of chloroquine-resistant *P. falciparum* infections. Synthesized derivatives (artesunates) are said to show a good effect in the treatment of cerebral malaria and other complications of malignant tertian malaria (WHO, 1984c).

The longterm objective of malaria control is to protect the population at risk by vaccination. Unfortunately, the different developmental stages (sporozoites, tissue stages, schizonts and gamontes) exhibit stage-specific antigenic components. The hope of achieving a vaccine against *P. falciparum* has increased considerably in the last 3 years, however: A surface antigen obtained from sporozoites may induce the production of antibodies able to neutralize the sporozoites and suppress the infection; it is hoped to synthesize the antigen for vaccination (MILLER, 1977; NUSSENZWEIG, 1982; ENDERS and HERMENTIN, 1983; ENDERS, 1984).

Other research groups are endeavouring to obtain a vaccine against the asexual blood stages (e.g. HEIDRICH, 1985) and also against the gamontes; this would suppress the further development of the plasmodia in the mosquito (transmission blocking antibodies, e.g. MEUVISSEN, 1986).

Recommendations for the chemotherapy and prophylaxis of malaria

International name	Trade name	Formulation (base content)	Treatment dose (adults)	Prophylactic dose	Comments
Chloroquine	Resochin	Tablets 50 mg 150 mg Syrup 3.5 ml 50 mg	600 mg initial dose [a] after 6 h 300 mg after 12 h 300 mg after 24 h 300 mg	300 mg per week	To date resistance only in *P. falciparum*; total dose should not exceed 1500–1800 mg [a]
Pyrimethamine + sulphadoxine	Fansidar	25 mg pyrimethamine + 500 mg sulphadoxine = 1 tablet	3 tablets once children ½–3 tablets	1 tablet (25 mg pyrimethamine + 500 mg sulphadoxine) per week	Resistance possible in *P. falciparum*; inadequate against *P. vivax*
Quinine (quinine sulphate)		Tablets 100 mg 200 mg	20–25 mg/kg		Resistance possible against *P. falciparum*
(quinine chloride)		Tablets 250 mg 500 mg Ampoules 250 mg/ml	 over 24 h as 2–3 slow infusions in NaCl or glucose solution		
Mefloquine	Lariam	Tablets 250 mg	750 mg initial dose after 6– 8 h 500 mg after 8–16 h 250 mg	Only for short-term prophylaxis (1–3 weeks), see p. 106 [b]	Multiple resistance only in *P. falciparum*

Following the recommendations for the prophylaxis and therapy of malaria of the German Association for Tropical Medicine (1983), Recommendations for the prevention of malaria, Deutsches Ärzteblatt 80: 35–40, August 1986.

[a] Recommendations of the WHO Technical Report Series No. 711 (1984c), Advances in malaria chemotherapy, Geneva.

[b] In Oceania, Indochina, Amazonia, East Africa.

Babesia species

Isolated cases of babesiosis in humans have been observed in Europe since the year 1957. The first reported cases involved patients who had undergone splenectomy who developed babesiosis and died. Since 1969 there have been reports from the USA that a rodent species, *Babesia microti*, can lead to latent infections or at most mild forms of disease. These blood parasites were not always correctly identified, however, and were then taken to be the malaria parasite *Plasmodium falciparum*. This resulted in a possible source of error for the diagnosis of malaria.

The representatives of the genus *Babesia* belong to the Piroplasmia, a subclass of the Sporozoa, amongst which highly pathogenic parasites of many domestic and other animals are to be found. In particular are included parasites of cattle (e.g. *Babesia bovis, B. divergens*), and also of rodents (e.g. *B. microti*), which often suffer from a latent localized infection.
Babesia sp. are transmitted by ticks, which ingest the intra-erythrocytic stages when they feed on blood. Sexual development takes place in the tick, followed by asexual multiplication and the development of sporozoites in the salivary glands. When the next meal is taken in, the parasites are then transmitted to the new host (see illustration, p. 109).
Babesia sp. can also develop in man, however, and can lead to severe disease. Infections are increasingly being seen and can no longer be considered a rare event (Yugoslavia, France, USSR, Ireland, USA, Mexico). They have become of particular interest because the parasites can easily be confused with the malaria parasite, *Plasmodium falciparum;* both live in erythrocytes and are morphologically very similar (especially in the thick film preparation).
Babesias differ from plasmodia within erythrocytes in a few very important ways. Babesia do not have any pigment; they are roughly pear-shaped, but can also be ring-shaped (about 1 μm) and after division are often present in the erythrocytes in pairs or tetrads ("Maltese cross"), located marginally (*B. divergens*) or centrally (*B. microti*). They do not form any schizonts, as plasmodia do. There are also no gamontes (see above). The erythrocytes do not show any stippling, such as occurs in *Plasmodium falciparum* infections (MAURER's dots, see p. 98).

Clinical Symptoms. Two forms of babesia infection have been observed in man, one which generally takes a fatal course and is caused by *Babesia bovis* and *B. divergens*, and a mostly latent form due to *B. microti*. The first form of the disease, mostly observed in Europe,

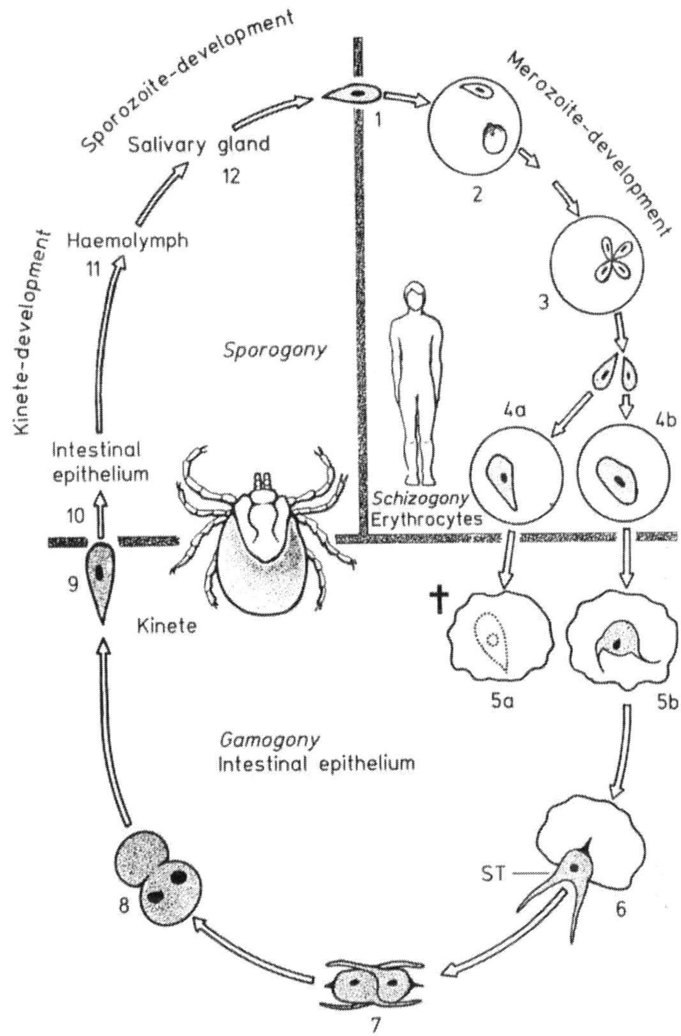

Developmental cycle of *Babesia microti* (in part from MEHLHORN and SCHEIN, 1984).

1	Sporozoite from tick saliva (*Ixodes* species)
2	Multiplication in erythrocyte by binary schizogony resulting in the formation of merozoites (also in lymphocytes?)
3	Erythrocyte containing characteristic Maltese cross stage
4 a	Merozoite; disintegrating merozoite in tick intestine (*5 a*)
5 b–8	Gamogony with the formation of " radiating" bodies (*6*) in the intestinal epithelium of the tick
9	A kinete develops from the zygote
10–12	Asexual multiplication of the kinetes in the tick; numerous sporozoites develop in the salivary gland (see MEHLHORN et al., 1986)

occurred in splenectomized patients, and the second in the USA in patients with an intact spleen. In babesiosis, symptoms include headache, loss of appetite, retching, shivering, fever, sweating attacks, a raised pulse, indeterminate pain in the back, around the kidneys and in the muscles, and facial pallor. These symptoms might also indicate a recent *Plasmodium falciparum* infection. The patient history may in itself help to differentiate between the two diseases (e.g. visit to malaria regions on the one hand, walking in woods and meadows, possibly observation of tick bite, time of year, etc., in malaria-free regions on the other) (WERNSDORFER, G., 1984).
It was assumed for some time that only splenectomized patients could develop babesiosis. In actual fact several such babesia infections with a fatal outcome have been recorded. Meanwhile, however, numerous cases of predominantly latent *Babesia microti* infections have also been observed in patients with a normal spleen. This makes it likely that the immune state of the infected person also determines the effects of the babesia infection. Latent babesia infections can lead to a severe, sometimes fatal disease in the recipient after blood transfusions, as is shown by reports from the USA (MARCUS, 1983).

Transmission. The transmission of *Babesia* is by hard ticks (Ixodidae e.g. *Ixodes ricinus, Dermacentor reticulatus*); malaria is transmitted by *Anopheles* mosquitoes. However, whereas malaria is today confined to tropical and subtropical regions, babesiosis can occur worldwide, corresponding to the distribution of the hard ticks by which it is transmitted.
The parasite reservoir is to be found amongst small mammals (e.g. *Microtus* species), as KRAMPITZ and BÄUMLER (1978) have shown. In connection with the frequent latent *Babesia microti* infections in the American island of Nantucket, KRAMPITZ et al. (1986) investigated the question of whether latent babesia infections arising from small mammals were also to be expected in the Federal Republic of Germany. Systematic serological investigations (IIFT and ELISA) of about 800 forestry workers in Bavaria, who had often been bitten by ticks, showed that asymptomatic *B. microti* infections were likely in at least two cases. Although attempts to isolate the parasites have been unsuccessful to date, this observation is worth noting and makes further research necessary.

Diagnosis. A differential diagnosis can be made only through the **microscopic demonstration** of the parasites in a GIEMSA-stained thin or thick blood film since infections with *Babesia* may be masked by therapy-resistant cases of malaria and thus overlooked.
WERNSDORFER, G., (1984) has pointed out that differentiation between malaria and babesia trophozoites carries an increased risk of error if no gamontes are found. The babesia parasites can sometimes be demonstrated by intraperitoneal inoculation of golden hamsters with the patient's blood. Then serological demonstration is possible (IIFT, ELISA).

Chemotherapy. Diamidine compounds (e.g. diminazene = Berenil), pentamidine isethionate 2.5–3 mg/kg body weight for several days (see p. 92)) are effective drugs.

Plate XII ⇨

Ciliates

Balantidium coli

Plate XII

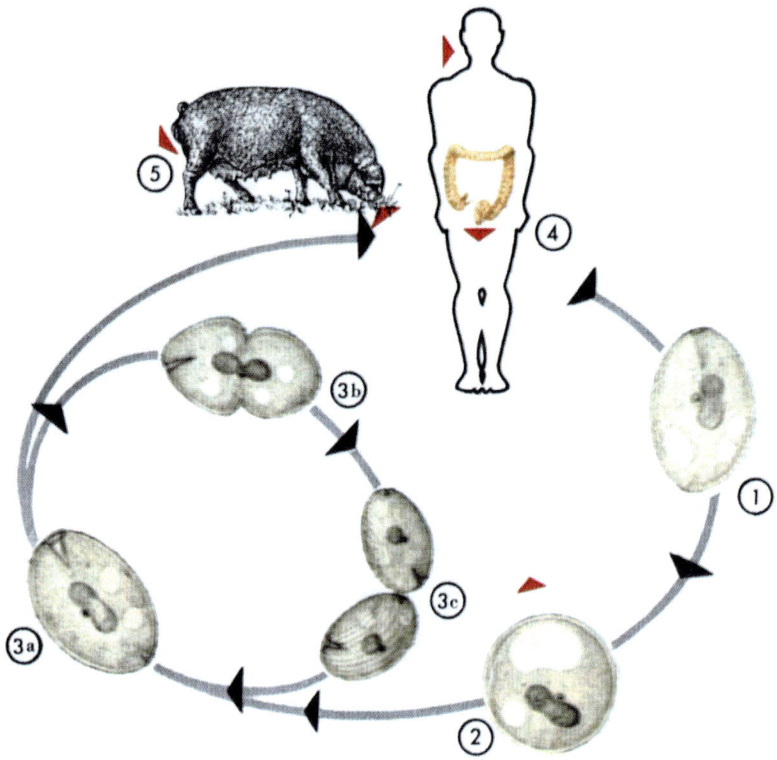

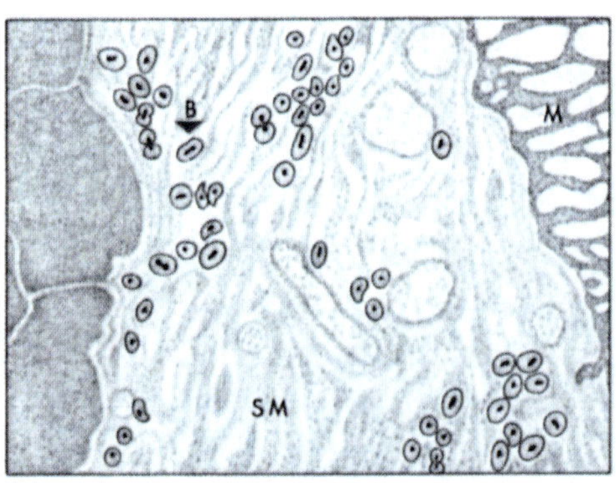

I

Balantidium coli (Malmsten, 1857) Stein, 1862

Pathogen of balantidial dysentery

1	Trophozoite (also *3 a*)
2	Cysts (see also Plate XXXII, *d*)
3 a–c	Multiplication by transverse division
4	Man: *Balantidium coli* in the large intestine
5	Pig as parasite reservoir

Multiplication takes place only in the large intestine; there are no free-living trophozoites.

I	Section through ulcer of large intestine in balantidial dysentery; numerous balantidia (*B*) are in the submucosa (*SM*). On the *left* of the picture, muscularis; on the *right*, mucosa (*M*) (Staining: Heidenhain's iron-haematoxylin)

The pathogen of balantidial dysentery, *Balantidium coli,* is a ciliated protozoon. These intestinal parasites are found relatively rarely in man, whereas the same species is to be found regularly in pigs, mostly as a harmless inhabitant of the intestines. The species is of world-wide distribution, but in man is found more frequently in many regions of Eastern Europe, Asia and America (e.g. USSR, East Asia, Indochina, Philippines, Texas, North and South Carolina). The causal relationship in balantidial dysentery of man is probably similar to that in many cases of amoebiasis. A change in the intestinal wall is induced after the parasites' penetration via production of proteolytic enzymes. This process is followed by action of pathogenic bacteria. Both together lead to the actual damage.

Morphology and Biology. *Balantidium coli* (Plate XII, *1*), which at approx. 50–150 × 50–70 μm is the largest protozoon pathogenic to man, multiplies asexually by transverse division (Plate XII, *3b, c*). The trophozoites move by cilia characteristic of these protozoa which cover the entire surface of the organism. The bacteria and similar materials which act as food are swept in via a mouth funnel. *Balantidium coli* possesses two different nuclei, a kidney-shaped macronucleus and a round micronucleus, which are also present in the cysts. The sexual process although uncommon, consists of conjugation, during which reciprocal exchange of nuclear material takes place after two nuclei divisions. The spherical resting stages (cysts, diameter about 50 μm; Plate XII, *2*) are excreted along with vegetative stages in the faeces; cysts occur more rarely in man than in pigs however. Infection occurs through oral ingestion of the cysts. Asymptomatic carriers are far more common than cases of disease.

Clinical Symptoms. The occasionally acute, but generally chronic symptoms of disease, which are due to infection of the large intestine, are very similar to those of amoebic dysentery. The stools may contain blood and numerous leucocytes. The parasites are able to penetrate into the vessels of the submucosa, occasionally reaching up into the mesenteric lymph nodes, and in rare, very severe cases parasites have even been found in the lungs. They sometimes cause deep ulcers in the mucosa and muscularis. Bacteria are presumably also involved in this. In a case described by NICHOLSON (1978) the 50-year-old female patient complained of diarrhoea with blood-stained stools. In addition symptoms of pneumonia occurred – coughing, chest pain and fever. After treatment with metronidazole for 4 days the stools became normal; all of the symptoms disappeared (see below). It is worth noting that a deficiency of vitamin C can be detected in the blood plasma and the urine, which regresses again after specific treatment. The additional administration of vitamin C promotes the action of the drugs (KHAMTSOV, 1973).
Abdominal colic, tenesmus, nausea and vomiting are common symptoms. The uncharacteristic symptoms may persist for years, because the cause is often not recognized. In such cases the disease can have a fatal outcome.

Transmission to man occurs relatively rarely. Patients are mostly to be found amongst those who deal with pigs in their jobs (swineherds, butchers etc.). The resistant cysts are ingested with contaminated food (cf. *Ascaris lumbricoides,* Plate XXVIII). Besides pigs, monkeys, chimpanzees and orang-utans may be parasite-reservoirs and may sometimes become severely ill. Rabbits, cats and rats can be infected experimentally.

As a **prophylactic measure** it is advisable to avoid contact with domestic and wild pigs (especially in slaughter houses) because of the possibility of infection with *Balantidium coli*. Almost all animals are infected, but they probably never develop disease as a result of it.

Diagnosis by Microscopy. This is carried out by the examination of a fresh sample of faeces, possibly diluted with physiological saline. The vegetative forms can easily be recognized even at low magnification because of their size, activity and typical mode of movement (rotatory propulsion). The use of concentration techniques for the demonstration of cysts is recommended. *Balantidium coli* can also be cultured in the media used to grow for amoebae. Serological methods are of no practical importance.

Chemotherapy. It is advisable to treat as for intestinal amoebic dysentery (800 mg metronidazole on each of 5 days; caution is needed in pregnant women, see also p. 54). In addition, sulphadiazine and the antibiotics tetracycline (contraindicated in pregnancy) and paromomycin (about 25 mg/kg body weight) are effective.

Helminths

Helminths

As a rule the nature of the pathological symptoms in helminthiasis is related to the number of worms present and to the reaction of the individual to the worm infection. Subclinical worm infection is far more common than acute disease. The severity of a helminthiasis is generally directly related to the number of larval stages that have been ingested or which have otherwise entered the body (sometimes still within the egg casing, sometimes as free-living larvae). As a rule the parasitic helminths do not multiply in man. Exceptions are *Hymenolepis nana* (see p. 181) and *Strongyloides stercoralis* (see p. 233). An apparent multiplication takes place in *filariae p. 270,* and also in *Enterobius vermicularis* (see p. 218). With *E. vermicularis* the eggs become infectious within hours of being laid and can then lead to auto-infection. Sexually mature filariae produce microfilariae constantly and over a long period of time, and these appear either in the peripheral blood or in the subcutaneous tissue (see pp. 271, 281).
The transmission of many helminths to man takes place via contact with the soil (soil-transmitted helminths), through food contaminated with faeces, or through infected foodstuffs eaten raw (*Trichinella, Clonorchis, Opisthorchis, Paragonimus, Anisakis* and others). Some are transmitted by insects (filariae), and others have the capacity as larvae to actively penetrate human skin (*Schistosoma, Ancylostoma*).
For the microscopic demonstration of the parasites attention should be given to the **prepatent period,** i.e. the interval between infection with the worm and the first appearance of parasites in sites where they are accessible for microscopic examination (eggs, larvae or microfilariae in blood, tissues or stools). This period of time varies depending on the type of parasite and ranges from 2 weeks (e.g. *Strongyloides*) to many months (e.g. filariae). This means that microscopic demonstration will not be successful before the prepatent period has elapsed (Table p. 313).

Trematodes (Flukes)

Trematodes (flukes) do not have any body segmentation (cf. cestodes, p. 176). They possess suckers (always two for distomes, echinostomes and schistosomes) with which they can attach in the host. They are sometimes hermaphrodite, sometimes with separate sexes. The trematodes of man always undergo their larval development in one or in two intermediate hosts, whereas in the final (definitive) host the parasite becomes the sexually mature helminth (alternation of generations, Digenea). The first larval stages, miracidia, hatch out of the eggs, and, attracted by so-called miraxone, bore percutaneously into snails, where they develop further and multiply as sporocysts and rediae. Cercariae are finally formed depending on the type of parasite. These may develop into metacercariae in a second intermediate host (fish, snail, ant or plants) or – in the case of schistosomes – bore percutaneously into the definitive host, to then become sexually mature parasites. The metacercariae reach the alimentary tract of the definitive host via the oral route.

The distomata which affect man live in the definitive host in close contact with epithelial cells (intestines, pulmonary alveoli, bile ducts). Schistosomes on the other hand live in close contact with the endothelium in the vascular system of the host.

The chemotherapy for trematode infections has undergone a decisive improvement with the availability of new drugs that can be given orally. For example, praziquantel has an unusually broad spectrum of activity and eliminates virtually all species of trematodes – sometimes with a single dose – including schistosomes (HARNETT 1988). There are special drugs which can also be taken orally for two species of this genus, oxamniquine for *Schistosoma mansoni*, and metrifonate for *S. haematobium*. Only praziquantel is effective against *S. japonicum*.

The trematodes can be classified into four groups, according to the organs which the sexually mature helminths seek out in man:

1. Intestinal trematodes (intestinal flukes) (pp. 121–133)
2. Liver trematodes (liver flukes) (pp. 135–150)
3. Lung trematodes (lung flukes) (pp. 151–157)
4. Blood trematodes, schistosomes or Bilharzia (blood flukes) (pp. 159–171)

Plate XIII ⇨

Intestinal Trematodes (Intestinal Flukes)

Fasciolopsis buski

Plate XIII

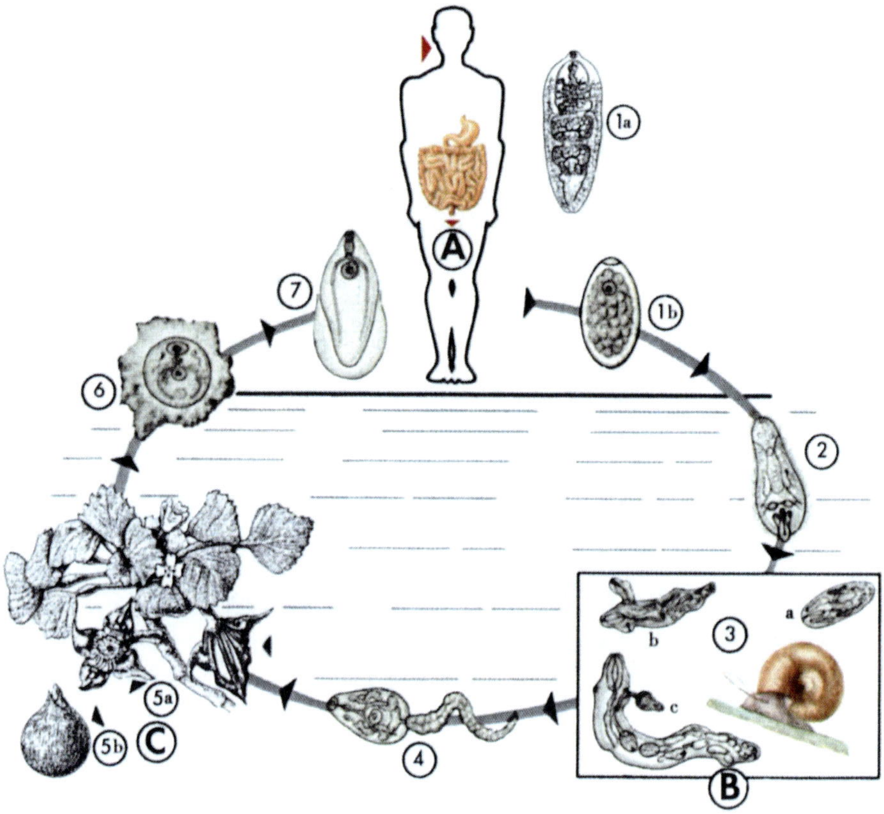

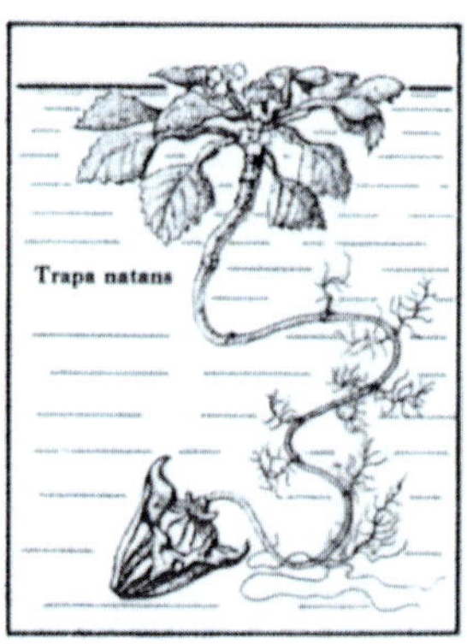

I

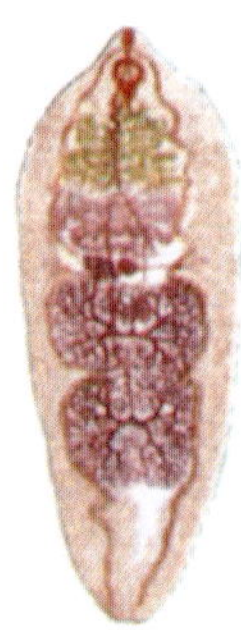

II

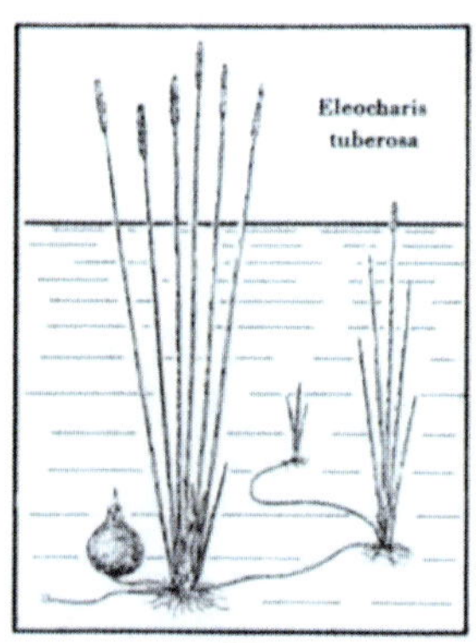

III

Fasciolopsis buski (LANKESTER, 1857) ODHNER, 1902

Giant intestinal fluke

Ⓐ Definitive host: man, intestinal fluke in the small intestine

1 a Sexually mature intestinal fluke (see also *II*)
b Egg
2 Free larval stage (miracidium) in the water

Ⓑ First intermediate host: water snail

3 a Sporocyst
b Mother redia with hatching daughter redia
c Daughter redia with hatching cercaria
4 Free cercaria

Ⓒ Second intermediate host: water plants

5 a Fruits of the water caltrop *Trapa natans* (see also *I*)
b Tuber of *Eleocharis tuberosa* (see also *III*)
6 Metacercaria from a water chest nut
7 Young hatching fluke (in human intestine)

I *Trapa natans,* water caltrop "host plant" for metacercaria (see under Ⓒ *5 a*)

II *F. buski,* sexually mature fluke (approximately natural size)

III *Eleocharis tuberosa* water chest nut (see *5 b*)

(See also Plates XXXII, *g* and XXXIII, *w*)

The giant intestinal fluke *Fasciolopsis buski* (Plate XIII, *1 a*), a parasite typical of East Asia, is one of the largest species of trematode which affects man. It is mainly found in India, Bangladesh, Thailand, China and Taiwan. Its occurrence is linked to certain species of snails Ⓑ, which act as intermediate hosts, and to waterplants Ⓒ, to which the cercariae become attached and mature into infective metacercariae.

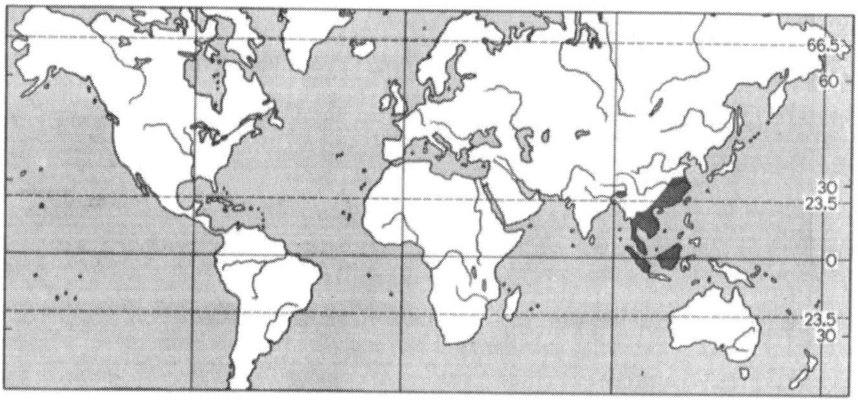

Distribution of *Fasciolopsis buski*

Morphology and Development. The elongate oval intestinal fluke (Plate XIII, *1 a II;* about 3–7 cm in length) inhabits the small intestine. The size of individual flukes is inversely proportional to the severity of the infection. The relatively large eggs (130–140 × 80 μm) are always numerous and are therefore generally easy to find in the stools (Plate XIII, *1 b*). The ciliate larvae hatch in the water (miracidium; Plate XIII, *2*), and actively bore through the body surface into snails Ⓑ of the genera *Segmentina* and *Hippeutis,* where they develop into sporocysts (Plate XIII, *3 a*). Rediae, daughter rediae (Plate XIII, *3 b, c*) and cercariae are produced in succession. The cercariae are released (Plate XIII, *4*) and attach to freshwater plants of the genera *Trapa* and *Eleocharis* Ⓒ (Plate XIII, *5 a, b*) and develop into metacercariae (Plate XIII, *6*). These plants are often cultivated in ponds and pools which are fertilized with human excreta. When the fruits of these water plants are peeled with the teeth, the metacercariae are introduced into the human gastro-intestinal tract Ⓐ (Plate XIII, *7*).

Clinical Symptoms. These are closely related to the severity of the helminth infection. Sexually mature parasites inhabit the small intestine, mostly the duodenum and jejunum. As a rule a few parasites do not cause symptoms; however, numerous intestinal flukes – more than 1000 have been found in some cases – lead to various unspecific symptoms. Symptoms develop 1–2 months after the initial helminth infection and mostly consist of violent diarrhoea, abdominal pain, loss of weight and the progressive development of generalized weakness. Depending on the severity of the infection and the reaction of the host, oedema and ascites may occur. Jaundice, blood stools, anaemia, fever and other non-specific symptoms may follow and lead to death in extreme cases. Disorders of growth and development are observed in children. The symptoms of the disease are attributed to a general toxic effect of metabolic products of the helminths.

Transmission. The mode of transmission is closely linked with the eating habits of the population in East Asia, who for example, like to eat the fruits of the water nut (Plate XIII, *1*), sometimes in candied form, which they bite open and peel with the teeth. The tubers of *E. tuberosa* are also sources of metacercariae.
Pigs represent an important animal reservoir. Dogs and cats may also be infected, but they probably do not have an important epidemiological role to play. The most important first intermediate hosts are snails of the species *Segmentina hemisphaerula* in China, North Vietnam, and Taiwan, *S. trochoideus* in India, and, amongst other species, representatives of the genus *Hippeutis*.

Diagnosis by Microscopy. This can be performed by direct examination of the stools for eggs. Concentration techniques make it easier to find the eggs (see p. 308).

Diagnosis by Immunobiological Methods. There are virtually no diagnostically useful antibodies demonstrable in intestinal fluke infection.

Chemotherapy. Niclosamide (adults and children over 6 years, 1×2 g; children aged 2–6 years, 1 g; children under 2 years, 0.5 g; 1 tablet = 0.5 g) and praziquantel (1×40 mg/kg body weight) have proved of value (see p. 193).

Plate XIV ⇨

Echinostoma ilocanum

E. echinatum

(=*E. lindoënse*)

Plate XIV

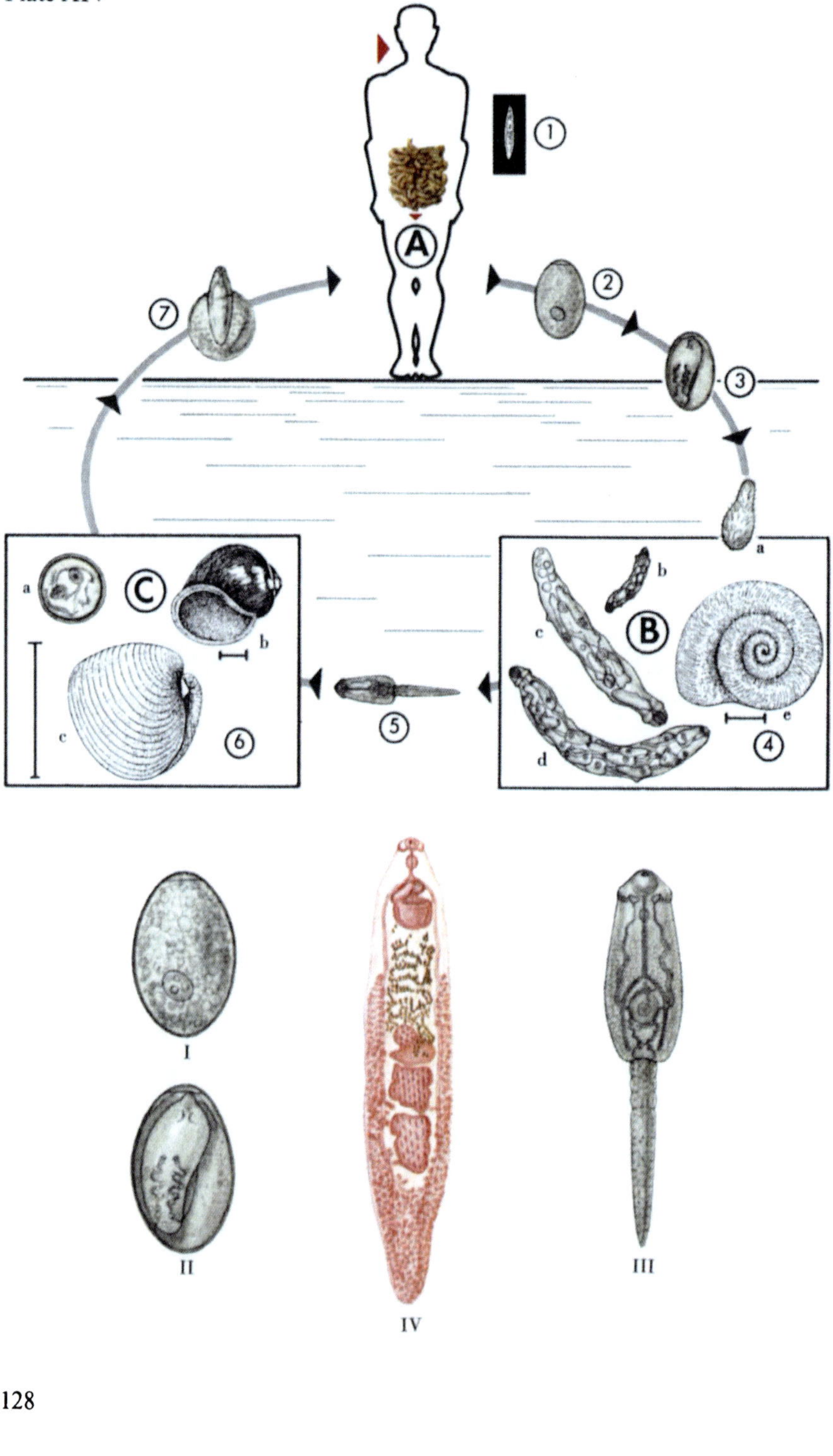

Echinostoma ilocanum (GARRISON, 1908) ODHNER, 1911
E. echinatum (ZEDER 1803) KANEV 1985

Intestinal flukes

Ⓐ Definitive host: man

1 *Echinostoma ilocanum,* sexually mature intestinal fluke (approximately natural size; see *IV*)
2 Freshly laid egg
3 Egg with miracidium

Ⓑ First intermediate host: snails of the genus *Gyraulus* (*4 e*)

4 a Free miracidium (actively bores into snail)
b Young mother redia
c Fully grown mother redia
d Daughter redia
5 Cercaria which has been released

Ⓒ Second intermediate host: snails of the genus *Pila* (*6 b*), and also *Gyraulus* species (see above); mussels of the genus *Corbicula* (*6 c*)

6 a Metacercaria
b *Pila*
c *Corbicula lindoënsis*
7 Hatching metacercaria

I Egg of *Echinostoma ilocanum* (about ×250)

II Egg with miracidium

III Cercaria

IV Sexually mature intestinal fluke (about ×10)

The two species of small intestinal fluke, *Echinostoma (Euparyphium) ilocanum* and *E. echinatum* (at least five further species of the genus exist along with these) are of particular interest because the two intermediate hosts Ⓑ and Ⓒ are mostly snails. The parasites are found in East Asia, especially in Canton, India, the Philippines and Java. *E. echinatum* is found particularly frequently on Celebes (Sulawesi), but also in Brazil (DÖNGES 1988). Members of the family Echinostomatidae are characterized by a distinct set of spines around the oral sucker.

Morphology and Development. The *Echinostoma* species which live in the small intestine have a characteristic reddish-grey appearance (*E. ilocanum* 3–6.5 mm, *E.* echinatum about 9.8 mm; Plate XIV, *I, IV*) and are covered with numerous small spines. These occur in transverse rows, starting with a collar behind the head (they are particularly large here) and extending down to the middle of the body. The oral sucker is located in the middle of the collar. The larger ventral sucker is in the mid-line just behind the intestinal bifurcation. The operculate eggs (about 95 × 65 µm; Plate XIV, *2, I*) are immature when passed in the faeces.
Within 6–15 days, but possibly after 6 weeks (depending on the temperature), a miracidium develops in the egg (Plate XIV, *3*), which hatches whilst still in the water (Plate XIV, *4a*) and bores into the first intermediate host Ⓑ. The first intermediate hosts are snails of the genus *Gyraulus* (often *G. chinensis*, syn. *convexiusculus*, see Ⓑ; in Brazil *Biomphalaria glabrata* for *E. echinatum*). In the region of the digestive glands, the miracidium develops into a special kind of sporocyst, the miracidial sporocyst, in which parent rediae develop (Plate XIV, *4b, c*). Within these, second-generation rediae are produced (daughter rediae, Plate XIV, *4d*), in which the cercariae develop (Plate XIV, *5, III*). Once these are released, they seek out snails, mostly the species *Pila chinensis*, and also mussels of the genus *Corbicula*, which in many areas (e.g. Celebes) form part of the daily diet of the indigenous people (Plate XIV, Ⓒ). In the molluscs the cercariae develop into metacercariae (Plate XIII, *6a*) (about 120–130 µm in diameter). Apparently metacercariae can develop in the first intermediate host. In the intestine of the definitive host Ⓐ (besides man various animals can also be included here, e.g. pigs, dogs, cats and monkeys) the metacercariae excyst and develop directly into sexually mature intestinal flukes. They are mostly found in the region of the jejunum.

Clinical Symptoms. The clinical symptoms are closely related to the number of parasites in the intestine. These are mostly non-specific (see also *Fasciolopsis buski*, p. 121) and primarily consist of diarrhoea associated with severe dehydration, slight abdominal pain and signs of general toxicity, such as headache and anaemia. However, symptoms can be very severe in hyper-sensitive patients and with massive infections. The helminths become firmly attached between the intestinal villi, erode the intestinal epithelium with their spines and lead to inflammation of the mucous membranes. An eosinophilia often occurs temporarily (up to 38% has been recorded). The helminths produce numerous eggs over several months, and these may sometimes be carried into other organs (e.g. CNS).

Epidemiology. Infection with *Echinostoma* species probably occurs solely through the consumption of uncooked snails and mussels containing metacercariae. These are often eaten by the indigenous population prepared in the form of salads. Dogs, cats and rats form the animal reservoir in nature.

Prophylactic Measures. These consist of the avoidance of meals of raw mussels and snails and the careful disposal of infected faeces from people and animals, so as to prevent infection of the snails, mussels and water, e.g. rice fields.

Diagnosis by Microscopy. The eggs of *E. ilocanum* and *E. echinatum* (= *E. lindoense*) are comparatively easy to find on careful examination of the stools, because the number of eggs laid is generally very high. Concentration techniques are recommended (see p. 308).

Chemotherapy. The helminths can be reliably eliminated with praziquantel (25 mg/kg body weight daily for 1–3 days or 3 × 25 mg/kg body weight in a 24 hrs period).

Other Species of Intestinal Trematodes

In addition to the representatives of the genera *Echinostoma* and *Fasciolopsis,* a few other species of intestinal trematode also occur in man. These are relatively small (roughly between 1 and 10 mm) compared with *Fasciolopsis buski* (up to 75 mm; see Plate XIII) and as individual specimens do not cause any symptoms. If there is massive invasion and hypersensitivity in the patient, clinical symptoms may also become apparent; as described above, these are ever non-specific. Infection mostly results in a congestive enteritis. The consequences are tiredness, loss of appetite, abdominal pain, watery stools and associated fluid depletion. The most common species are *Heterophyes heterophyes, H. nocens, Metagonimus yokogawai* (Plate XXXIIII, *r, s*), *Watsonius watsoni* and *Gastrodiscoides hominis* (see following tables). *Heterophyes* species infections are not uncommon in China, Japan, Korea, and also in the Balkan countries and in Egypt. *Metagonimus yokogawai* is found in Northeast Asia including the Amur basin and in the Bal-

Summary of the development of the most important intestinal trematodes

	Fasciolopsis	*Metagonimus*	*Echinostoma*	*Heterophyes*
First intermediate hosts: snails (genera)	*Segmentina* *Hippeutis*	*Melania* *Semisulcospira*	*Gyraulus* *Pila*	*Pirenella* *Cerithidia*
Miracidium	×	×	×	orally with eggs
Sporocyst	×	×	Miracidial sporocyst	×
Mother redia	×	×	×	×
Daughter redia	×	×	×	× (?)
Cercaria	×	×	×	×
Second intermediate hosts	Water plants	Carp-like fish – *Plectoglossus, Salmo* etc.	Pila species *Corbicula* mussels	Fish – *Mugil* species
Metacercaria	×	×	×	×
Definitive hosts: small intestine	Man pigs, dogs and others	Man fish-eating mammals, pelican	Man monkeys, dogs, cats, rats,	Man cats, dogs, rats and others

Intestinal trematodes: details of sizes

	Adult worms		Eggs
	Length (mm)	Width (mm)	(μm)
Echinostoma ilocanum	5.5– 9.2	0.6– 1.5	~ 96 × 63
Fasciolopsis buski	max. 75	30	~135 × 80
Gastrodiscoides hominis	5 –10	4 – 5	~160 × 70
Heterophyes heterophyes	1 – 1.7	0.3– 0.6	~ 25 × 15
Metagonimus yokogawai	1 – 2.5	0.4– 0.7	~ 28 × 16
Watsonius watsoni	8 –10	4 – 5	~125 × 78

kans. These species basically develop first in snails (first intermediate host) as for *Echinostoma ilocanum*. Freshwater fish act as the second intermediate host for metacercariae of the genera *Heterophyes* and *Metagonimus,* and probably the fruits of water plants (as with the giant intestinal fluke, see Plate XIII) for *Watsonius watsoni* and *Gastrodiscoides hominis*. The consumption of raw fish can thus lead to infection with *Heterophyes heterophyes* and *Metagonimus yokogawai*. For infection with *Gastrodiscoides hominis* and *Watsonius watsoni* the fruits of water plants have to be eaten raw (see table).

Plate XV ⇨

Liver Trematodes (Liver Flukes)

Clonorchis sinensis

Opisthorchis felineus

Plate XV

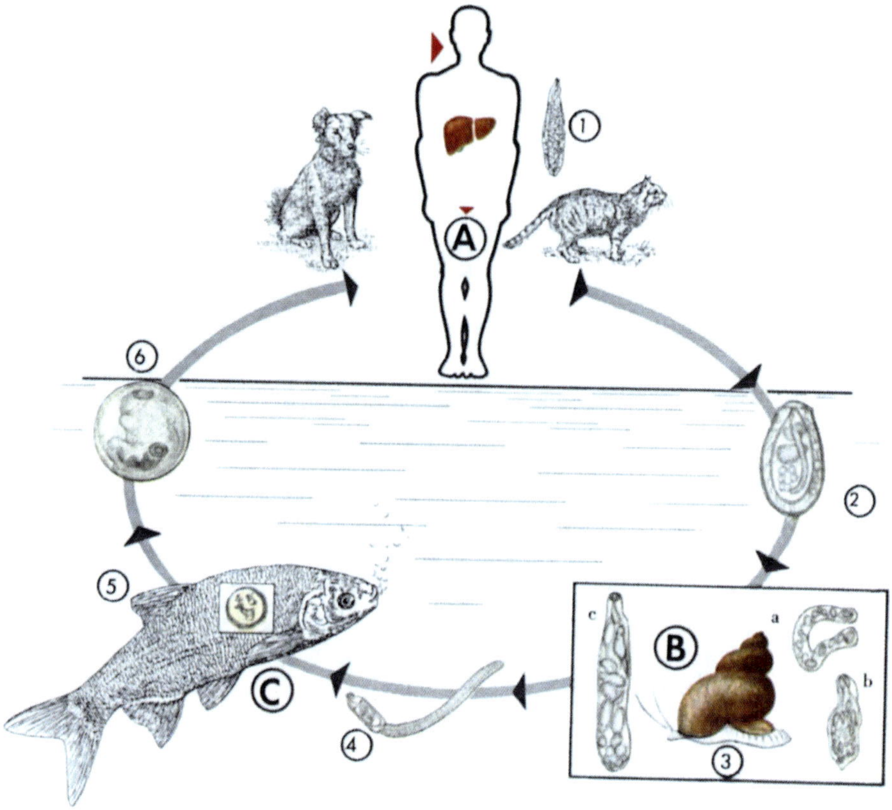

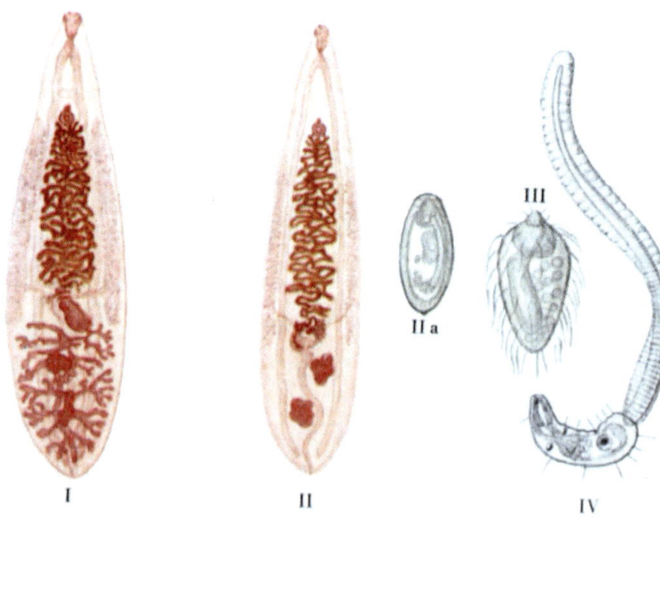

Clonorchis sinensis (Cobbold, 1875) Looss, 1907

Chinese liver fluke

Opisthorchis felineus (Rivolta, 1884) Blanchard, 1895

Cat liver fluke

Ⓐ Definitive host: man, along with cats and dogs (also other domestic and useful animals)

1 Sexually mature liver fluke (natural size)
2 Egg (with miracidium) of *Clonorchis sinensis*

Ⓑ First intermediate host: snails of the genus *Bithynia* and others

3 a Young sporocyst
b Parent redia
c Daughter redia with rudimentary cercaria
4 Free cercaria

Ⓒ Second intermediate host: mostly cyprinids

5 Fish with metacercaria
6 Metacercaria (about ×100)

I *Clonorchis sinensis* (about ×5)

II *Opisthorchis felineus* (about ×7)

II a Egg, with miracidium, of *O. felineus* (about 28 μm)

III Free miracidium from a snail

IV Cercaria in typical swimming position

(See also Plates XXXII, *i, k* and XXXIII, *t, u*)

The Chinese liver fluke, *Clonorchis sinensis* is a widespread parasite in East Asia. It is extraordinarily common in many regions of China and is also found in Japan, Korea, Taiwan and Indochina. The number of people infected with this liver fluke is estimated at almost 20 million. The area of distribution of the fluke is related to certain river systems because the cercariae which are released from snails (first intermediate host Ⓑ) penetrate into freshwater fish (second intermediate host Ⓒ), in which they develop into metacercariae. Invasion of the definitive host takes place when the host eats raw fish Ⓐ.

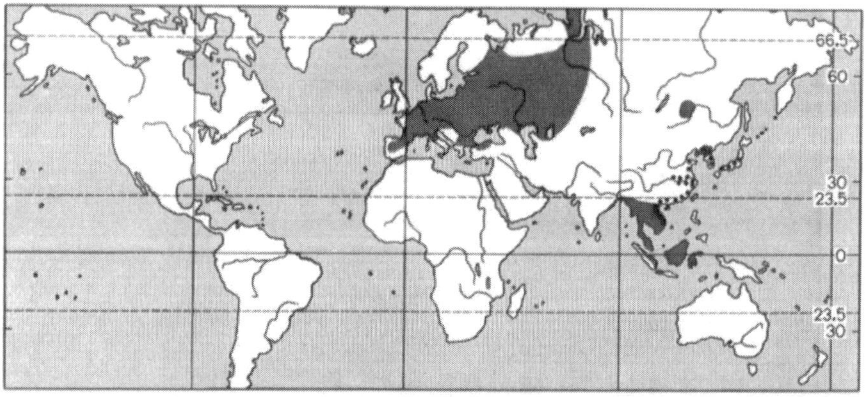

Distribution of ▇ *Opisthorchis felineus*, ⛆ *Clonorchis sinensis* (from Tischler, 1969)

Morphology and Development. *Clonorchis sinensis* (about 10–20 mm, Plate XV, *1, I*), has a roughly lanceolate, leaf-like structure and in the living state appears almost translucent. The surface is smooth. The position of the ovary and of the branching, paired testes in the posterior quarter of the body is characteristic. The convoluted uterus extends throughout the entire anterior part of the body. The yellowish-brown, very small eggs (about 30 µm; Plate XV, *2*), have a collar-like swelling which gives them a characteristic jug-shaped appearance.

The egg has a "lid" at the upper pole. The development of the larvae commences in the uterus. The deposited eggs already contain a developed miracidium (Plate XV, *2*). The eggs pass into the small intestine with the bile and out into the open in the faeces. Further development takes place in water, where the eggs are eaten mostly by snails of the genera *Bithynia* and *Semisulcospira* Ⓑ. The miracidium hatches from the egg in the snail and develops into a sporocyst (Plate XV, *3a*), then rediae (Plate XV, *3b, c*), followed by the cercariae which are released, emerge from the snail and enter the water (Plate XV, *4, IV*). Cercariae bore through the skin of their second intermediate host Ⓒ, a freshwater fish – almost always a cyprinid. In the muscles the cercariae develop into metacercariae (Plate XV, *5, 6*). They are also found under the scales and in the subcutaneous connective tissue. The final host, e.g. man Ⓐ, becomes infected by eating raw fish. The metacercariae or

young flukes which have been released in the small intestine actively migrate to the bile ducts, especially the distal regions, where they then become sexually mature. The first eggs may appear in the faeces about 14 days after invasion (prepatent period), but generally they appear later.
The life-span of these helminths, which is said to sometimes reach 25 years in man, is particularly remarkable.

Clinical Symptoms. If there is only a mild fluke infection of the liver, no obvious symptoms occur. In more severe cases dilatation of the bile ducts and thickening of their walls occurs as a result of marked proliferation of the mucosa. Consequently there may be severe damage to the liver, which becomes enlarged in cases of acute disease, and damage may lead to cirrhosis of the liver, ascites and oedema. Cancer of the liver is also occasionally seen. In cases of chronic disease non-specific gastrointestinal symptoms may occur.

Transmission takes place through the consumption of raw or inadequately cooked freshwater fish which have acted as intermediate hosts. For example, about 33%–35% of the population in Korea where new fish is routinely ingested, suffer from liver fluke infection. The rural population is affected to a greater extent than the city dwellers, and men more than women (locally for example 54% versus 9%) because of their more frequent consumption of raw fish. The custom of eating dried and salted fish uncooked also leads to infection because the metacercariae survive this treatment. In many parts of Asia even babies are fed on raw fish, so that *C. sinensis* can already be found in 3-month-old infants.
Apart from man, other possible definitive hosts which must be considered are pigs, dogs and cats, the latter being particularly susceptible. They often harbour numerous helminths without showing any overt signs of disease, yet constantly excrete fluke eggs (animal reservoirs). Liver fluke infection is avoided only if adequately heated or cooked fish (also fish remains) are eaten or used for feeding.

Diagnosis by Immunobiological Methods. The ELISA has proved useful as a screening test (FELDHEIM and KNOBLOCH, 1982). Cross-reactions with other trematodes (*Fasciola, Paragonimus, Schistosoma*) occur however (RIM, 1986).

Diagnosis by Microscopy. Repeated microscopic examination of the faeces and duodenal juice for the helminth eggs can be used for diagnosis. The eggs may be very numerous, but are frequently overlooked because of their small size (about 30 × 16 μm). Concentration techniques are therefore recommended (see p. 308).

Chemotherapy. Praziquantel has proved of value for both *Clonorchis* and *Opisthorchis* infection (3 × 25 mg/kg body weight on each of 2 successive days or 20 mg/kg body weight for 3 days; RIM et al., 1981; LÖSCHER et al., 1981; WEGNER, 1984).

Opisthorchis felineus (RIVOLTA, 1884) BLANCHARD 1895

The cat liver fluke, *Opisthorchis felineus* (Plate XV) is closely related to the Chinese liver fluke. For its development it is also dependent on certain water snails as first intermediate hosts Ⓑ and on freshwater fish as second intermediate hosts Ⓒ. Snails *Bithynia leachi* and numerous species of the carp family (Cyprinidae) act as intermediate hosts. The distribution of these parasites is therefore limited, in a similar way to that of the Chinese liver fluke, to certain river and lake regions; they are predominantly distributed in the temperate zone. For example, well-known foci of *Opisthorchis* occur in the coastal regions of the Baltic Sea, along the Weichsel-river, in the Baltic provinces, in the Danube region, in Central Russia, especially in North Siberia, in India and Indochina as well as in Japan. As indicated by the name of the fluke, cats are frequently affected. However, the parasite can also develop in man, in dogs and some fish-eaters (e.g. seals).
The species *Opisthorchis viverrini* STILES and HASSALL, 1896 is a frequent parasite in Southeast Asia, especially in the northeastern part of Thailand (at least 7 million people affected), and in Benegal. According to SADUN (1955) about 25%–45% of the population in these regions are affected. This species is differentiated from *O. felineus* by certain anatomical characteristics (e.g. greater proximity of the ovaries to the testes, the special formation of the ootype, rather smaller egg size: 27 × 15 μm).

Morphology and Development. The morphology of the cat liver fluke (size 8–12 mm; Plate XV, *II*) largely resembles that of the Chinese liver fluke. It is differentiated from it by the lobed testis, which lies in the posterior quarter of the body. The yellowish-brown eggs of *Opisthorchis* species are somewhat narrower (30 × 12 μm, Plate XV, *IIa*) than those of *Clonorchis sinensis*, and possess a somewhat less clearly outlined cap (Plate XV, *2*).
The *developmental cycle* which was described by H. VOGEL in 1934, is identical to that of the Chinese liver fluke (see p. 137). The sexually mature worm likewise lives in the biliary tract, and is rarely, and only with severe infections, to be seen in the pancreas. The prepatent period is between 3 and 4 weeks. The lifespan of the parasite may be 15–20 years.

Clinical Symptoms. These are closely related to the severity of infection. A few flukes (50–60) are frequently not noticed at all, but can cause slight enlargement of the liver, plus biliary tract and gall bladder inflammation. Proliferation of the epithelium of the biliary tract, desquamation of tissue and jaundice may occur. Eosinophilic leucocytosis (up to 40%) also

occurs. Additional symptoms include diarrhoea and flatulence, pains in the stomach region, and gall bladder problems of various kinds. Patient history can be very helpful here (nutritional customs). Biliary tract and pancreatic carcinoma occasionally occur in chronic infections (see also p. 139).
An infection with *O. viverrini* also causes no definitive symptoms when mild to moderate, but due to the long lifespan of the parasite, chronic disease may develop and this can lead to recurrent cholecystitis and malignant change. Biochemical parameters (e.g. rise of serum proteinase inhibitors, α_1-antitrypsin, α_1-antichymotrypsin, α_2-macroglobulin) offer criteria for the assessment of the disease. They indicate the degree of impairment of liver function (Changbumrung et al., 1982). *Opisthorchis* infection is to be found in many refugees from Indochina.

Transmission. The occurrence of cat liver flukes in man is always related to certain nutritional customs, e.g. the consumption of fish salads made from uncooked freshwater fish. As fishermen often eat such meals they are frequently carriers of cat liver flukes. Therefore care should be taken to consume only boiled or fried fish; the metacercariae die on heating to 55 °C. Since cats and dogs acquire the flukes by eating fresh fish and fish waste, these should only be fed to animals after cooking. Domestic animals, therefore, deserve attention as possible parasite carriers (see for this also p. 125).

Diagnosis by Microscopy and Chemotherapy. See *Clonorchis sinensis* (p. 139).

Plate XVI ⇨

Dicrocoelium dendriticum

Plate XVI

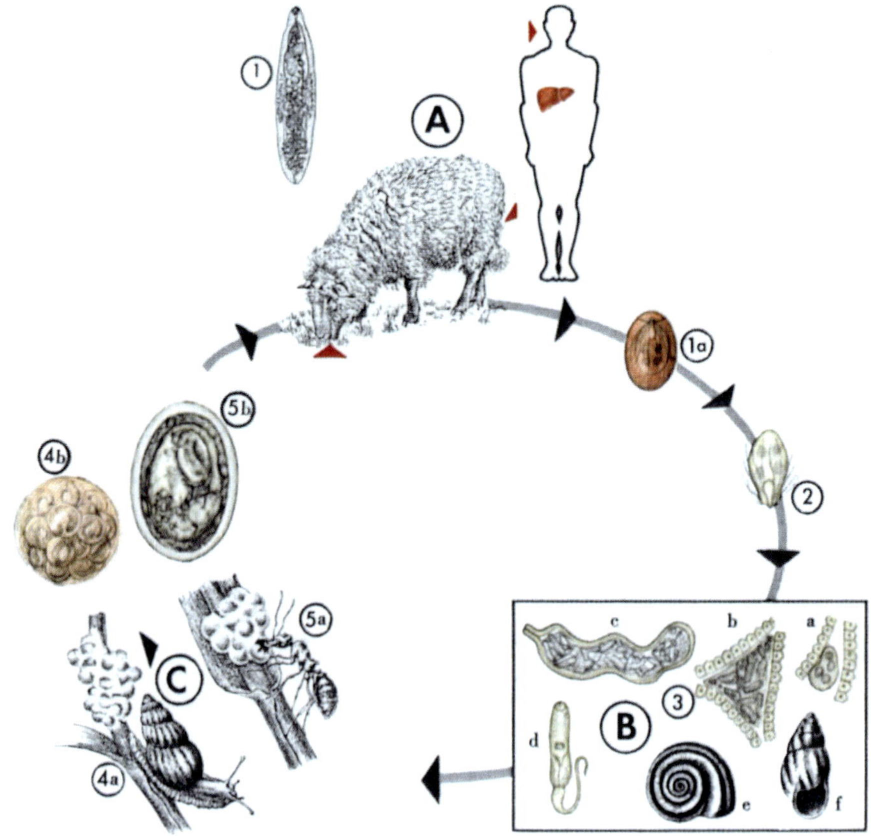

I

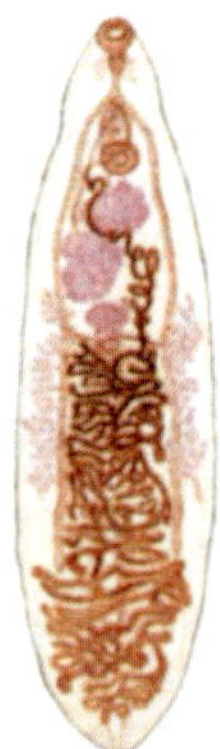
III

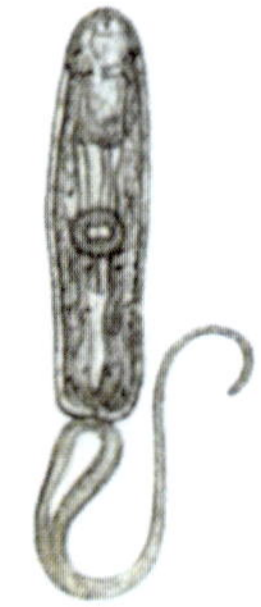
II

Dicrocoelium dendriticum (RUDOLPH, 1818) LOOSS, 1899

Lancet liver fluke

Ⓐ Definitive host: chiefly sheep and cattle; man as (accidental) secondary host

1 Sexually mature liver fluke (see *III*)
a Egg with developed miracidium

Ⓑ First intermediate host: land snails (*Helicella, Zebrina*)

2 Hatched miracidium (from snail)
3 a Young sporocysts
b Older sporocyst stages (mother sporocyst)
c Daughter sporocyst
d Individual cercaria (cercaria vitrina)
e Shell of *Helicella ericetorum*
f Shell of *Zebrina detrita*

Ⓒ Second intermediate host: ants

4 a Slime-balls deposited on a grass stalk by the snail
b Individually deposited slime-balls from a snail
5 a Slime-balls consumed by ants (*Formica fusca*)
b Mature metacercariae from ants

I Hatching miracidium

II "Cercaria vitrina"

III Sexually mature liver fluke (approx. × 5)

(See Plates XXXII, *h* and XXXIII, *q*)

The small liver fluke *Dicrocoelium dendriticum* is primarily a parasite of ruminants, but is occasionally also found in man Ⓐ. As with many fluke species snails are the first intermediate hosts Ⓑ. The second intermediate host in this case is an ant Ⓒ. The occurrence of the small liver fluke is linked to calcium-rich soils because the snail intermediate hosts live in such areas. This fluke frequently occurs in North Africa (Egypt, Algeria), in Siberia, Turkestan and South America. It is less common in North America. In West Africa the species *D. hospes* Looss (1907) occurs instead of *D. dendriticum* (Lucius and Frank, 1978).

Morphology and Development. *Dicrocoelium dendriticum* (size about 5–12 mm; Plate XVI, *I, III*) has a narrow curved blade-like shape. In life it appears pale reddish with a darker interior structure – the uterus – which at times is filled with eggs. The parasite is found in the biliary tract of the liver and in the gall bladder. The relatively small dark-brown eggs (38–45 μm, Plate XVI, *1a*) already contain a fully formed larva (miracidium) when laid. However, the miracidia are not released in water. Land snails Ⓑ, of the genera *Zebrina* and *Helicella* in Europe, and *Cionella* in North America, must ingest these eggs. Within the snail, the ciliate larvae (miracidia, Plate XVI, *2, I*) hatch out, and transform first into mother sporocysts (Plate XVI, *3a, b*) which form daughter sporocysts (Plate XVI, *3c*). Inside these the cercariae develop (Plate XVI, *3d, II*). The cercariae accumulate in the snail where they are coated with mucus and eventually are deposited as slime-balls (Plate XVI, *4a, b*) and must then be eaten by ants of the species *Formica fusca* Ⓒ. The cercariae induce a behavioural change such that the ants do not return to their ant hills but continue to browse on the food plants and in this way are eaten by the definitive hosts along with the plants. (With *D. hospes* the snails of the genus *Limicolaria* are the first intermediate hosts; the second intermediate host is presumably the ant of the genus *Camponotus* sp.)
In the abdominal cavity of the ant the majority of the metacercariae develop within cysts, which have an oval shape and are about 365 × 250 μm (Plate XVI, *5b*). The mature, opaque metacercariae finally lie involuted in the cyst case. More than 300, but generally only 50–60, such cysts can find room in one ant. Development in the ant takes about 38–56 days at a temperature of 26 °C (see also p. 184). The metacercariae hatch in the gastrointestinal tract of the definitive host Ⓐ through a narrow opening at one pole of the cyst. They then migrate through the choledochous duct into the liver, which they reach within 2 hrs. They remain in the bile ducts of the liver where they reach sexual maturity. The first eggs appear in the faeces of rabbits or sheep 50–56 days after invasion of the liver (prepatent period).

Clinical Symptoms. Only with massive infection does the fluke cause symptoms and illness in sheep and in man. Single parasites generally remain unnoticed. Infection becomes apparent in animals by their having a poor appetite, and failure to gain weight. In man liver enlargement, anaemia and upper abdominal pain occur amongst other non-specific signs and symptoms.

Transmission. The parasites are transmitted to the definitive host following the uptake of infected ants with food (see Plate XVI, *5 a, b*). This is achieved, according to HOHORST and GRAEFE (1961), in the following manner. After the uptake of cercariae by the ant (Plate XVI, *5 a*), one cercaria (rarely two) always penetrates as a "brain worm" into the subpharyngeal ganglion of the ant. There it develops into a cyst of varying appearance with an extra-ordinarily thin wall. This causes a change in the behaviour of the ant. Grazing animals must eat ants infected with cercariae in their food. Such ants, immobilized by the falling outside temperature, remain for hours, singly or in clusters, at the tips of plants eaten by sheep. This behaviour contrasts with that of uninfected ants. Mature metacercariae (Plate XVI, *5 b*) may be in the ganglion but can also be situated at any other site in the ant's body (KLOFT, 1978). The striking behaviour makes it possible to detect and then collect the ants in the field. Experimental feeding of such ants to sheep, rabbits and other hosts regularly leads to liver fluke infections. Man probably acquires the infection by inadvertent ingestion of ants. The characteristics of the epidemiological relationships explained above make it clear that this liver fluke occurs very rarely in man, and then mainly in children.

Diagnosis by Microscopy. The very small eggs may be detected as a consequence of their large number, mainly by direct microscopic stool examination. When clinical suspicion is high the use of a concentration technique is recommended (see p. 308–310). During investigation of biliary secretions, eggs are to be expected first in the B bile (gall bladder bile). With weakly positive results it is possible that the eggs represent a spurious infection and are "passers-by" following the eating of a meal of liver. Therefore, in such cases the investigation should be repeated several days later, checking the eating habits of the patient.

Chemotherapy. As with other intestinal and liver flukes, praziquantel appears to be active against *Dicrocoelium dendriticum* (see p. 139, *Clonorchis sinensis*).

Fasciola hepatica Linné, 1758

Sheep liver fluke

The large liver fluke *Fasciola hepatica,* a close relative of the large intestinal fluke of man, is primarily a frequent and widespread parasite of cows, sheep, goats and many other herbivores. Correspondingly it occurs in man in regions where it is usual to eat raw aquatic vegetation, and where suitable snails occur as intermediate hosts. Such regions include France and Corsica, Portugal and Madeira, North and South Africa, several countries of South America (Brazil, Peru, Chile) as well as parts of Asia.

Morphology and Development. Almost the entire surface of the fluke, which is up to 40 mm long and 13 mm broad, is covered with conspicuous integumentary spines, by which the species can readily be identified especially in histological preparations. A distinct head cone and shoulders are characteristic of *F. hepatica.* The large oval eggs have a small operculum, are light yellow to brown in colour and measure 130–150 × 53–90 µm. The immature eggs pass via the biliary tract into the intestine to be voided with the faeces. The "ripening time" in water at the optimal temperature of 22°–25 °C is 9–15 days. The miracidium then hatches out of the egg, and for further development must penetrate a water snail within 8 hrs. Snails of the genus *Lymnaea* serve predominantly as intermediate hosts. The miracidium bores through the epidermis and penetrates into the snail where it is transformed into the young sporocyst.

Summary of the developmental pathway of *Fasciola hepatica*

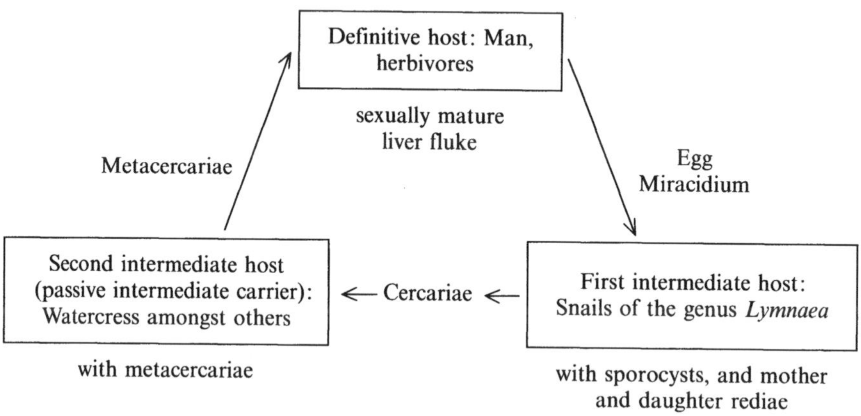

Within 3 weeks the sporocysts produce the mother rediae (potentially in unlimited numbers), and in another week daughter rediae and cercariae are formed. The mature cercariae leave the snail, and after a few hours' sojourn in the water detach their tails, attach to various aquatic plants (grasses and water cress) and encyst to form metacercariae (see p. 149). Infection results from the consumption of these plants. The encysted metacercariae remain viable for a long time in the damp atmosphere, but die rapidly when they become dry.

It is significant for animal husbandry that the metacercariae can survive for about 28 days on stored hay and on other food plants at –10 °C. They can survive a temperature range of between about –5 °C and +10 °C with high air humidity and remain capable of infection for up to 70 days.

The metacercariae or the young liver flukes begin their migration in the definitive host by penetrating the wall of the small intestine into the peritoneal cavity to the liver. After having penetrated the liver capsule they spread throughout the liver parenchyma in the biliary pathways. Here the young flukes mature within 1–2 months. The eggs pass into the intestinal canal with the bile and can then be found in the stools.

Clinical Symptoms. Disease manifestations begin about 1 month after infection with fever, general malaise, fatigue, loss of appetite and loss of weight. The patient complains of pains in the right upper abdomen, predominantly in the liver region, irregular fever and digestive disorders. In severe cases there may be nodule formation in the liver, hepatomegaly, and even unilateral paralysis. The scale of injury ranges from simple inflammatory reactions to malignant tissue change. Flukes in ectopic sites (e.g. in the region of the CNS) can cause manifestations very uncharacteristic of *F. hepatica*. The blood eosinophil count is frequently very high (over 50%).

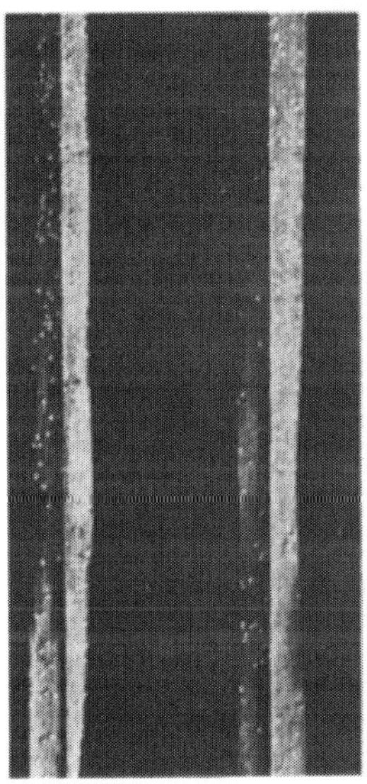
Metacercariae on blades of grass

The viability of the fluke is quoted as being from 1 to 20 years.

Epidemiology. The route of infection for man, in the majority of cases, is through the consumption of watercress. The proportion of infected people always varies according to nutritional customs. This explains the fact that in France, for example, *F. hepatica* infections are relatively frequent, whereas in Germany they are extremely rare. From this it follows, that prophylactically, the consumption of raw watercress should always be avoided.
COUMBARAS (1966) reported that the indigenous population in Algeria and Morocco eat only cooked watercress, whereas the French newcomers eat it raw in salads in the usual manner. The latter suffered from fascioliasis, the indigenous population did not.

The ascorbic acid requirement of the parasite is remarkable, leading to a corresponding deficit in the host. As ascorbic acid is necessarily involved in many metabolic processes, the deficiency causes reduced liver protein and glycogen synthesis (GAMEEL, 1982a, b).

It is noteworthy that some snails (e.g. *Bulinus truncatus*), which are castrated by the parasite attack, become larger and live longer than non-infected snails. The primarily semi-aquatic intermediate hosts like *Lymnaea truncatula* prefer wet media which facilitate the water-borne dissemination of cercariae after infection with *F. hepatica*. This and similar behavioural changes are frequently observed in definitive and intermediate hosts due to parasitic attacks (HAAS, 1984) (see for example p. 147).

The snails of the genus *Lymnaea,* which serve as intermediate hosts, are represented by different species according to geographical region, e.g. *Lymnaea truncatula* in Europe, *L. tomentosa* in Australia. This can result in the habitat of the individual snail species being very different (e.g. aquatic, semi-aquatic).

Diagnosis by Microscopy. Liver fluke eggs may be found by examining the stools and duodenal fluid (B bile). In the initial stage of the disease the stools frequently contain no eggs. They first appear after about 2–3 months. Liver fluke eggs can be found in the stools after eating infected bovine liver ("false fascioliasis"). When this is suspected the patient must consume liver-free meals for several days before stool samples are taken again. Only then, if eggs continue to be found in the stools, is liver fluke infection confirmed.

Diagnosis by Immunobiological Methods. Serological methods (e.g. complement fixation test, indirect immunofluorescence test, ELISA and skin test; antigen from adult worms) are recommended in laboratory practice. Positive results can be obtained before the occurrence of eggs in the stool or bile, i.e. at the time of the migration, and even several months after clinical cure (MINNING, 1969).

Chemotherapy. Praziquantel, which has activity against other liver fluke species, is also frequently helpful, according to recent observations, in *Fasciola hepatica* infections in man (always 5 × 15 mg/kg body weight for at least 3 days). On the other hand, it is probably not helpful in infections in herbivores (WAHN and MEHLHORN, 1984). Dehydroemetine used to be the drug of choice (1.5 mg/kg body weight i.m. daily for 10 days) but because of possible cardiac side effects it should always be given under clinical supervision.

Plate XVII ⇨

Lung Trematodes

Paragonimus westermani

P. kellicotti

P. africanus

Plate XVII

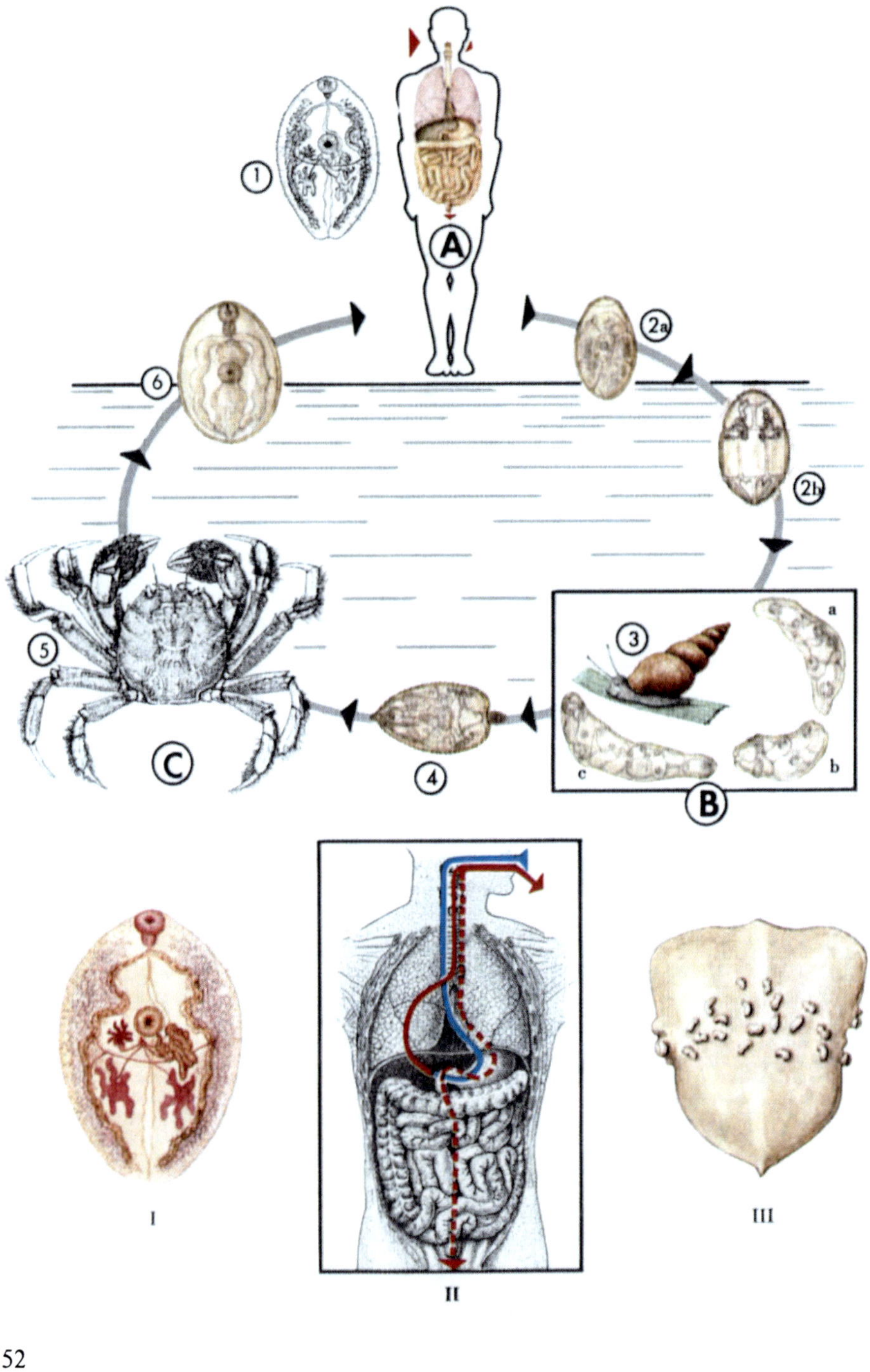

Paragonimus westermani (KERBERT, 1878) BRAUN, 1899
P. kellicotti WARD, 1908; **P. africanus** VOELKER and VOGEL, 1965

Lung fluke

Ⓐ Definitive host: man (also various domestic and fur-bearing animals)

1 Sexually mature lung flukes (see also *I*)
2a Freshly laid eggs (egg cells with yolk cells)
2b Hatched miracidium

Ⓑ 1 First intermediate host: water snails of the genera *Brotia*, *Semisulcospira* and others

3a Sporocyst
3b Mother redia
3c Daughter redia
4 Free cercariae

Ⓒ 2 Second intermediate host: crayfish and crabs

5 Chinese crab (*Eriocheir sinensis*)
6 Metacercariae from crab muscle

I Sexually mature lung fluke (approx. ×4)
II Migration pathway of young lung fluke in the definitive host:

blue Metacercariae arrive in the gastrointestinal canal;
red Fluke, having bored through the small intestinal wall and migrated through the diaphragm into the lung; mature eggs are coughed up with sputum or swallowed and then appear in the faeces.

III Crab heart affected by metacercariae

(See also Plates XXXII, *f* and XXXIII, *v*)

There are several species of human lung flukes. The best known are *Paragonimus westermani (=P. ringeri), P. africanus* VOELKER and VOGEL, 1965, and *P. kellicotti* WARD, 1908. In total there are more than 20 species known at the present time. Development occurs in certain species of snails Ⓑ acting as the first intermediate host, and a second intermediate host, crabs and crayfish Ⓒ. Human infection therefore follows the consumption of raw crab or crayfish meat. The classical distribution of *P. westermani* includes Japan, Korea, Taiwan, China (especially in Manchuria) and the Philippines, but it is also to be found in India. In China there exist at least 14 different species of *Paragonimus*. Refugees coming from Southeast Asia must always be considered as victims of this parasite. *P. kellicotti* is found in Canada as well as in North and Central America, and parts of South America. In many species two or more strains with different morphological and biological characteristics occur (HE et al., 1981).
Since 1943 reports on lung fluke infections from many parts of tropical West Africa, from the Congo region, from Nigeria, but especially from the Southern Cameroons, Gabon, Liberia and Libya, have accumulated. *P. africanus* and *P. uterobilateralis* were described from these regions by VOELKER and VOGEL (1965; see also SACHS and VOELKER, 1969).
The difficulties of species differentiation may sometimes be overcome by the use of electrophoretic methods (zymodeme, see p. 11; YOSHIMURA et al., 1969; YOKOGAWA, 1969).

Morphology and Development. The lung fluke (Plate XVII, *1;* about 7–12 × 4–7 mm) is plump, nearly egg-shaped, with an oval shape in cross-sections. It appears reddish-brown in life, like a coffee bean. The oral sucker lies subterminally, and the somewhat larger ventral sucker is almost in the middle of the slightly flattened ventral side (Plate XVII; *1*). The surface is covered with numerous prominent spines.

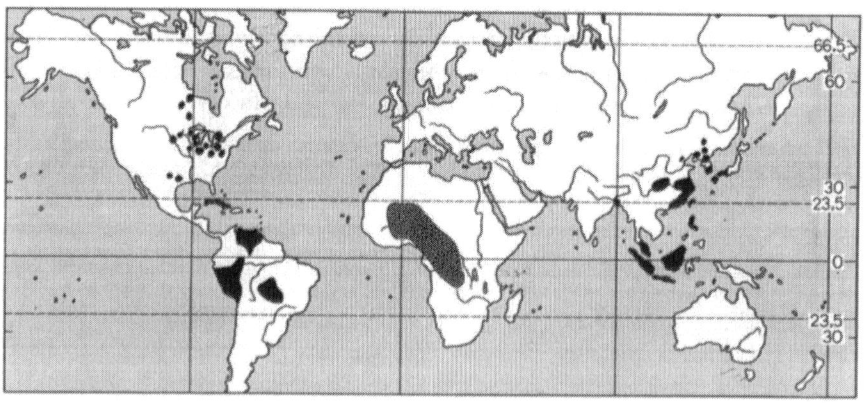

Distribution of *Paragonimus: dark areas, P. westermani* in East Asia, *P. kellicotti* in America; *grey areas, P. africanus* in Africa

The golden-brown eggs (about $90 \times 60\ \mu m$), which generally reach the exterior with the sputum, have an operculum. They initially contain only the ovum with 5–10 yolk cells (Plate XVII, *2a*). Once the egg is laid a miracidium develops within it. This development takes 3 weeks at temperatures of 25°–30 °C. The hatched miracidium (Plate XVII, *2b*) penetrates actively into snails, e.g. the genera *Semisulcospira, Hua, Thiara, Brotia* and *Melania* Ⓑ. For *P. kellicotti* the snail species is *Pomatiopsis lapidaria*. Within the snail the sack-like sporocyst develops containing the first generation rediae (Plate XVII, *3a, b*). The mother rediae form daughter rediae (*3e*) which, after further development, always liberate 20–30 cercariae.

The short-lived cercariae (Plate XVII, *4*) are ellipsoid in shape and have a characteristic stumpy tail. Their body surfaces are covered with fine spines and their movements in water resemble those of a free-living leech. Using their tails the cercariae swim or are carried by currents to an encounter with crayfish or crabs, second intermediate hosts (Plate XVII, *5*). In the muscles of the latter they establish themselves, boring in with the aid of acone-shaped stylet Ⓒ (with *P. westermani* the second intermediate hosts *Eriocheir japonicum* and several other species; with *P. kellicotti* they are predominantly *Cambarus* species). In the crab the cercariae encyst and develop into metacercariae. They are found both in the leg and tail muscles, but with the North American crabs they are located in the heart region, when they are confined to a ligamentous region (Plate XVII; *III*). These crustaceans can also acquire cercariae by eating infected snails.

The metacercariae consumed with a crab-meat meal (Plate XVII, *6*) reach the small intestine in the definitive host, hatch out of their shells, and bore their way as young flukes through the intestinal wall in the duodenum. From here they migrate through the diaphragm into the thoracic cavity, and then mainly penetrate the lung. In the lung they become enclosed in 1–2 cm connective tissue cysts (often single in man, but there may be two or three in animals) and become sexually mature. The first eggs are found in the sputum 2.5–3 months later. Aberrant young flukes may colonize other (heterotopic) sites in the body, for example the liver, spleen, kidney or brain. The lifespan of the worms is often very long and can be more than 20 years.

Clinical Symptoms. These are closely related to the development and number of lung flukes which have penetrated the host. The symptoms vary with the site of the worms. The lung is the site of predilection, and in this case the symptoms are mainly severe chronic cough and piercing pain in the chest, associated with dyspnoea. Many of the symptoms are reminiscent of tuberculosis. In the sputum, which is often reddish-brown or bloody (haemoptysis), the characteristic operculated eggs can be detected as well as numerous erythrocytes. The number of lung flukes in man is seldom greater than 10.

Along the route of migration, which in the normal course of events leads to the lung, extrapulmonary damage can occur when individual worms lose their way and become sexually mature. The eggs then liberated lead to an inflammatory reaction, e.g. in the regions of the peritoneal or pleural cavities. Dead worms similarly lead to ectopic (i.e. far from the lung, for example in the brain or spinal cord) tissue reactions, the origin of which is usually not understood. From this there results an extraordinarily diverse clinical picture, which,

under some circumstances, may no longer resemble that of a lung fluke infection (see YOKOGAWA et al., 1960). With the help of computed tomography and bronchoscopy (the parasites lie encapsulated in cystic structures), the diagnosis of a parasitic infection of the lung can be made.
According to SHAO (1981) and CHENG et al. (1981) cases of *Paragonimus* infection commonly occur in China in which no eggs are found in the sputum, but where marked enlargement of the liver (1–7 cm below the right costal margin), high leucocytosis and eosinophilia (42–97%), with abnormal results of liver-function tests occur. The patients suffer from rapidly developing fatigue, loss of appetite, subfebrile temperatures, dry cough, distended abdomen, etc. The detection of the parasite must then take place by serological methods only (ELISA, IHAT, metacercarial membrane reaction, see below).
In experimental infections of Rhesus monkeys with *P. uterobilateralis* RACZ and VOELKER (1984) established that the cellular reactions with eosinophils, plasma cells, and mast cells are not limited locally but extend to whole lobes of the lung. This results in extensive perivascular and peribronchial infiltration of lymphocytes and plasma cells. The lymphocyte reaction is characterized by follicular hyperplasia, as also occurs in allergic reactions.

Transmission. The transmission to man is closely related to the nutritional customs of the population. Uncooked crayfish and crab meat is frequently included in salads in the affected areas and consumed raw. In a number of places, such as Korea and Japan, crustaceans are crushed and pressed to extract the juice, which is then consumed raw. This fluid is partly used in combination with other foods, and partly is prescribed for its curative power in fever and diarrhoea. As such juices frequently contain metacercariae, lung fluke infection can also arise in this way. Men more frequently consume raw crustaceans and therefore suffer more frequently from lung fluke infections than women. On the other hand, both sexes in children are about equally affected. Apart from man, various domestic and other animals which consume crayfish and crabs can become infected, e.g. pigs, dogs, mink, martens, wild cats and other cat species (see above), badgers and raccoons. Such animals probably infect the water in the first place by egg-containing faeces. They are often at risk from their habitats, as rapidly flowing mountain streams contain numerous crustaceans besides the snails which serve as the first intermediate hosts. – With adequate information on infection sources and avoidance of eating raw crayfish and crabs, and other raw crab products, infection with lung fluke can largely be avoided. A further measure for the prevention of lung fluke infection is the consistent eradication of snails, crayfish and crabs, which act as the intermediate hosts. Mammals, which can serve as reservoirs of the parasite, are very difficult to control. Therefore, controlling lung fluke infection is a hard task. Snails can be infected by eggs or miracidia from the sputum and stools of fluke carriers. For experimental lung fluke infection rhesus monkeys are mainly used. According to the observations of VOELKER and SACHS (1984), the animals remain infected for up to 10 years. After this period the lung flukes show degenerative changes (uterus empty, ovaries and testes atrophied, eggs malformed).
A remarkable observation from Japan is that pigs can behave like a paratenic host (see p. 2). Metacercariae are able to invade muscle tissue of pigs and remain infective. Human infection is obviously acquired by eating raw meat from such animals (MIYAZAKI 1976).

Diagnosis by Microscopy. Sputum samples (often blood-stained) and stools can be examined for eggs. Apart from the golden-brown, operculated eggs (Plate XVII, *2a*), one generally finds CHARCOT-LEYDEN crystals (see pp. 311, 324).

Diagnosis by Immunobiological Methods. When clinical suspicion is high and no eggs are detected, the diagnosis can be further supported by the complement fixation test (CFT), a flocculation test, the indirect haemagglutination test (IHAT), and ELISA, as well as a metacercarial membrane reaction and the intradermal test. Antigen is prepared from adult worms. However, cross-reactions can occur with a schistosome infection. The IgE level is often markedly raised and may fall with praziquantel therapy after 6 months. The IgG level falls after 2–6 months (KNOBLOCH and LEDERER, 1983).

Chemotherapy. Praziquantel has been shown to be effective (always 3 × 25 mg/kg body weight on 2 days). On account of its high tolerance this drug should be regarded as the treatment of choice. The side effects are always mild (e.g. headache, vomiting, nausea) and are related to the intestinal tract and to the CNS. Temperature elevation may occur, but it always settles following cessation of therapy without additional treatment.

Plate XVIII ⇨

Blood Trematodes

Schistosoma haematobium

S. mansoni

S. intercalatum

S. japonicum

S. mekongi

Plate XVIII

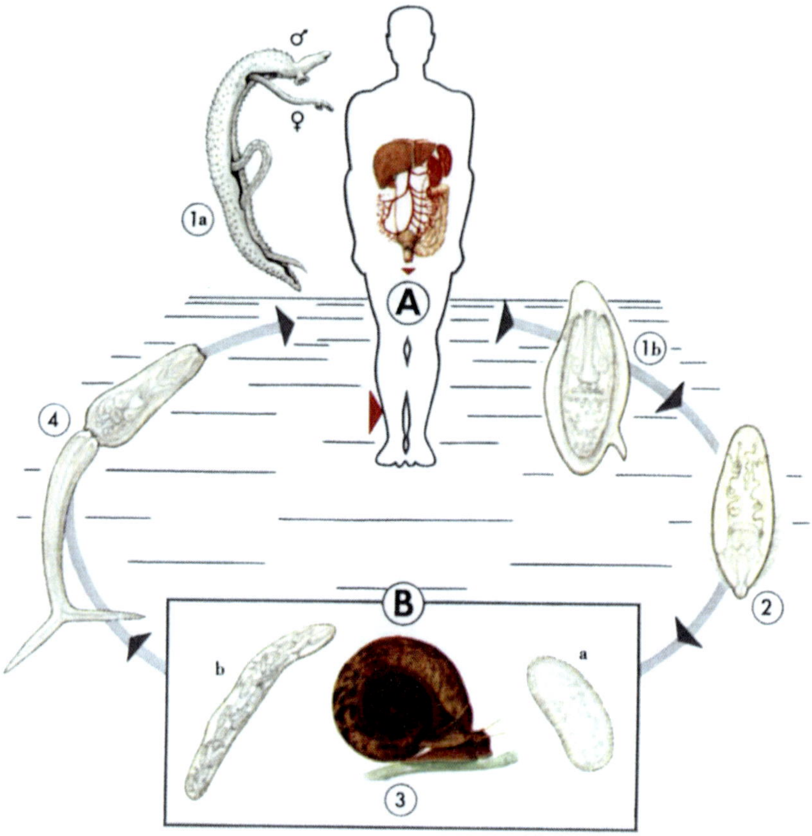

I S. haematobium Weinland 1858

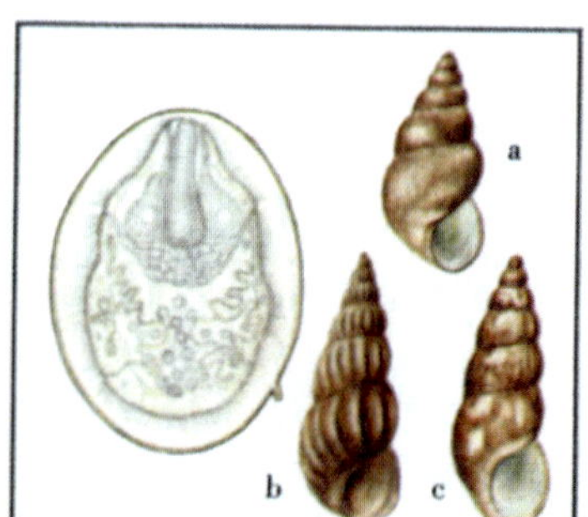

II S. japonicum Katsurada 1904

Schistosoma (Bilharzia) Weinland, 1858

Blood flukes

Ⓐ Definitive host: man; the mature worms live in the mesenteric blood vessels

- *1 a* Pair of mature flukes, *S. mansoni* Sambon, 1907
- *1 b* Mature egg of *S. mansoni* (lateral spine)
- *2* Miracidium

Ⓑ Intermediate host: water snails (e.g. *Biomphalaria glabrata*)

- *3 a* Primary sporocyst (mother sporocyst)
- *3 b* Secondary sporocyst (daughter sporocyst)
- *4* Free cercariae ("forked-tail cercariae")

I *S. haematobium* Weinland, 1858; egg with miracidium (terminal spine)
Shells of intermediate host species:

- *a* *Bulinus truncatus* (North Africa)
- *b* *Bulinus globosus* (West Africa)

II *S. japonicum* Katsurada, 1904; egg with miracidium (inconspicuous lateral spine or knob)
Shells of intermediate hosts belonging to the genus *Oncomelania*

(See also Plates XXXII, *e* and XXXIII, *p*)

Schistosomes (Blood Flukes)

Schistosomiasis or bilharziasis is a disease of warm countries. It is especially frequent where intensive agriculture first becomes possible through artificial irrigation. In the irrigation canals live the snails in which the parasite undergoes part of its development and in which the stages infectious for man are formed. Consequently, the disease has recently become known in the vernacular as "snail fever" (e.g. in the Philippines).

According to the estimates of the WHO (1984) there are at present about 200–300 million people infected in 74 countries, especially children and young people (only 114 millions according to STOLL, 1947) but more than 600 million live in high risk regions. From these data it is obvious that schistosomiasis is one of the most threatening parasitic diseases of warm countries. Through tourism in the widest sense the number of European and North American cases of schistosomiasis increases constantly, and it is estimated at present that there are 400,000 (imported) cases in the USA (see DOUMENGE et al. 1987).

As early as 1910, Sir A. RUFFER found calcified eggs in the kidneys of two Egyptian mummies (about 1,250 and 1,000 years B.C. respectively). In 1851, in a hospital in Cairo, Theodor BILHARZ discovered the first schistosomes in the portal vein of a patient, at that time without differentiating the species.

Schistosomiasis is caused essentially by at least four different dioecious species of the genus *Schistosoma:*

Schistosoma haematobium (BILHARZ, 1852) WEINLAND, 1858; cause of urinary schistosomiasis
Schistosoma mansoni SAMBON, 1907; cause of intestinal schistosomiasis
Schistosoma intercalatum FISHER, 1934; cause of intestinal schistosomiasis
Schistosoma japonicum KATSURADA, 1904
and *Schistosoma mekongi* VOGE, BRUCKNER and BRUCE, 1978; causes of Asiatic intestinal schistosomiasis

The chief distribution of *S. haematobium* covers Northern and Eastern Africa and extends from Morocco over Algeria and Tunisia up to Egypt, from the Nile to South Africa, especially Natal. The parasite is found in North Ethiopia, Uganda, Zimbabwe and in the Congo region. In West Africa, Liberia, Sierra-Leone and Ghana are affected. Smaller foci occur in Arabia, parts of Israel, Iran, Iraq, Cyprus and Madagascar.

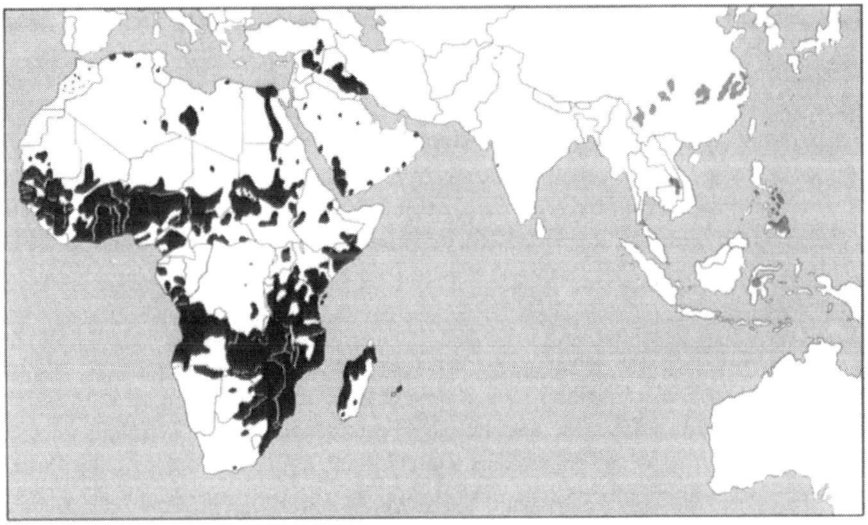

Distributions of *S. haematobium* (*dark*), *S. japonicum* (*hatching*), and *S. mekongi* (*grey*) (only in southern parts of Laos and Cambodia) (WHO, 1985)

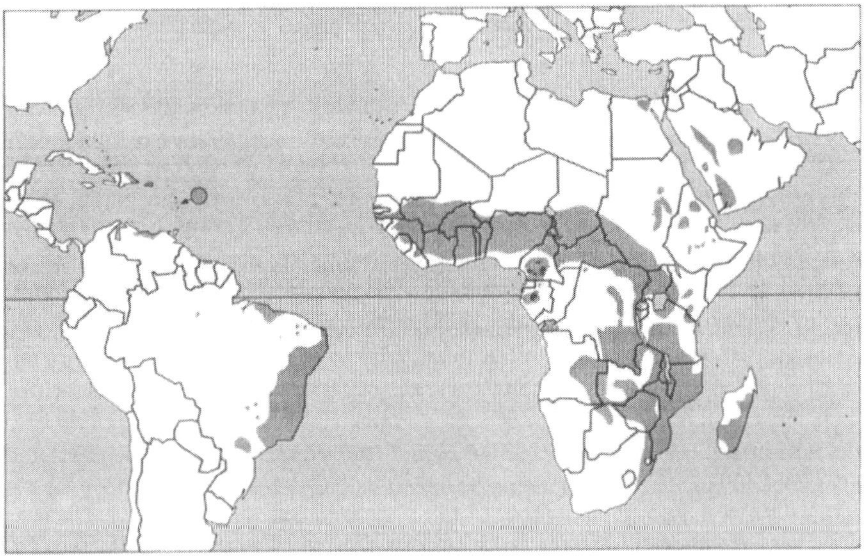

Distribution of *S. mansoni* and *S. intercalatum* (only Africa) (WHO, 1985) (see p. 164)

S. japonicum is limited to East Asia. This parasite is found predominantly in the region of the Yangtze Valley, in the Chinese provinces of Hunan, Hupeh, Anhwei, Kiangsu and Kiangsi, in smaller regions of Japan and in the South Philippines (Leyte, Mindanao). There are no recognizable morphological differences between some strains of *S. japonicum* but there are physiological ones. There are at least four geographically different races of this species, which occur in Taiwan, China, Japan and the Philippines. *S. japonicum* is closely related to *S. mekongi,* which occurs in the southern part of Laos and Cambodia in the region of the Mekong and its tributaries, dispersed in small foci, in which the population at times shows a high frequency of infection (up to 90%).

S. mansoni is not as frequent in Africa as *S. haematobium.* It is endemic in the Nile Valley, from which the affected areas have extended over Central, East and West Africa. *S. mansoni* was also imported to South and Middle America and discovered there by PIRAJA DA SILVA. Venezuela and East Brazil are predominantly affected.

S. intercalatum, also a cause of intestinal schistosomiasis, occurs predominantly in Central Africa, in the Congo Basin, and in West Africa. The eggs, about 170 × 60 μm, are distinguished by a terminal spine. They are similar to those of *S. haematobium* but are on average more slender (see figure on p. 168).

Two additional species of *Schistosoma* which can occur in man must be mentioned: *S. bovis* (SONSINO, 1876) Blanchard 1895, and *S. mattheei* VEGLIA and LE ROUX, 1929. Both species are primary parasites of vertebrates: *S. bovis* is the cause of intestinal and bladder schistosomiasis in cattle and sheep, e.g. in Italy and Africa; *S. mattheei* is the cause of bladder schistosomiasis in ruminants in South Africa. The eggs of both species have terminal spines (see p. 168).

Morphology and Development. The males with their long, lamelliform shape are unusual in that their margins fold inwards to form a canal. The female is held enclosed within this canal (hence the description "paired fluke"). Because of this the body of the male appears split longitudinally. These worms are unlike other trematodes of man in that they have separate sexes. The sexually mature worms (depending on the species, 10–20 mm long) live mainly in the mesenteric vessels of the intestine as well as in the hepatic veins (*S. mansoni, S. mekongi,* and *S. japonicum*), or in the vessels of the bladder (*S. haematobium*).

The eggs, which in contrast to other trematodes (see for example pp. 123) are not operculated, pass out into the capillaries, whence they pass through the intestinal or bladder wall into the lumina of these organs, and finally are transported into the open with the excreta. The eggs of *S. mansoni* are excreted with the stools and are about 150 μm in length, bearing a characteristic lateral spine (see Plate XVIII, *1b*). Those of *S. intercalatum* (about 140 × 36 μm) have a terminal spine. The eggs of *S. japonicum* (about 85 μm) are more compact, almost spherical, and possess a small lateral hook (Plate XVIII, *II*). As a consequence of their rather rough surface, stool particles often adhere to the eggs, making them more difficult to recognize than the eggs of *S. mansoni* (HE et al., 1980; for further information see table on p. 171).

S. mekongi can be differentiated from *S. japonicum* by several morphological and biological characteristics. The adult worms are smaller (male about 16 mm, female about 12 mm, typical papilla arrangement at the posterior body opening, etc.). The embryonated eggs are almost spherical and at 40–45 µm are smaller than the eggs of *S. japonicum*. Both show a rudimentary knobbed lateral spine.

Development of the different *Schistosoma* species is for the most part similar and is always dependent on the presence of certain water snails (intermediate host Ⓑ). The eggs passed with the excreta already contain a larva (miracidium; Plate XVIII, *2*), which hatches out into the water, and only survives for about 48 hrs. During this time it must find a suitable snail (hence they use glutathione as bait; see HAAS, 1984), if it is to develop further into the tube-shaped mother sporocyst (Plate XVIII, *3a*), within which the daughter sporocysts arise (Plate XVIII, *3b*). The daughter sporocysts give rise to the characteristic infectious larval stages, the cercariae (forked-tail cercariae; see Plate XVIII, *4*). After about 3–15 weeks (depending on the species and temperature; "larval prepatency") the cercariae get into the water via the respiratory cavity of the snail. According to PFLÜGER et al. (1984) the optimum temperature range for *S. haematobium* is 18 °C and 32 °C. For further development the cercaria must penetrate through the skin of a definitive host Ⓐ within a few hours (maximum 48 hrs) or it perishes. In the course of penetration it throws off its tail. In the host it forms a surface coat (see p. 12) as protection against the defence forces of the host (for details of the penetration mechanism see GRANZER, 1984; HAAS et al., 1984).

By way of blood vessels the young fluke (schistosomulum) is passively carried to the right ventricle of the heart. From there it passes through the lung arterioles, lung veins and the left ventricle to the large systemic arteries. The next objective is to reach the portal veins, where the flukes mature. Only after pairing do the females become sexually mature and the adults migrate out of the portal system in the mesenteric vessels. The eggs which are laid reach the capillaries adjacent to the intestine or bladder. Others reach the liver, where they cause the formation of granulomata. Due to the inflammatory processes in the intestinal or bladder walls the eggs are able to break through the vessels (action of their excreted proteolytic enzyme) and pass into the intestinal or bladder lumen. They can then be detected in stools or urine. Eggs may be found in ectopic sites in almost all organs including the CNS and sex organs.

Clinical Symptoms. Cercariae penetrating the skin cause itching. In sensitized people, allergic reactions also occur at the entrance point, in the form of red spots and papules (cercarial dermatitis). These manifestations regress within a week. With the migration of the young schistosomes through the lung vessels irritation of the bronchioles can sometimes occur. After about 4–7 weeks clinical manifestations occur which are produced by the adult worms and the deposited eggs.

The symptomatology of intestinal schistosomiasis manifests itself in many organs. It can be accompanied by diarrhoeic stools with bloody mucus so that confusion with, for example, amoebic dysentery is possible. In severe cases obstruction in the region of the portal vein can lead to enlargement of the liver and spleen, combined with liver fibrosis. The symptoms are more marked at the beginning of egg laying. The majority of eggs gets into the capillaries of the ileocolon and colon. They are aligned within the capillaries and lead to widening of the vessels. The consequences of this are blood stasis and blocking of vessels. Eggs may be found in all sections of the intestinal wall, but predominantly in the submucosa.

The eggs induce local acute granulomatous reactions (see Plate XVIII, *11, 12*), which lead to focal scarring, and often to a more diffuse chronic inflammation. "Pseudotubercles" occur as small pale elevations. The mucosa is initially congested, oedematous and infiltrated with fine haemorrhagic spots. With progressive disease different grades of ulceration, swelling and polyp formation occur. Surface biopsies of the rectum are a valuable diagnostic aid, and in asymptomatic cases eggs can often be found in the mucosa and submucosa of the rectal tissues.

Finally, it is only in light or latent parasitic infections that chronic disease generally occurs. Continuous impairment of hygiene leads to considerable loss of general health, which can in turn lead to psychological and physical retardation. Long-standing worm infection in the tissues (5–25 years) regularly leads to tissue damage and blood loss. The dramatic picture of acute disease is lacking (Bell et al., 1973). According to De Paola and Winslow (1967), however, this disease picture may manifest itself only in combination with other, simultaneously occurring liver-damaging factors, such as viral hepatitis or nutritional deficits (see Warren, 1972). Patients can remain asymptomatic for a long time, but a recognized infection should always be treated on account of possible sequelae. Infection can persist for more than 30 years with decreasing excretion of viable eggs (Harris et al., 1984).

In bladder schistosomiasis (infection with *S. haematobium*), urinary symptoms are prominent. Such symptoms include haematuria, a burning sensation in the urethra, and urgent desire to urinate. The urinary sediment contains erythrocytes, leucocytes, and usually the eggs with their characteristic terminal spines (see Plate XVIII, *1*). Even the ovaries can be affected. In severe cases malignant bladder tumours can develop.

Japanese schistosomiasis (infection with *S. japonicum;* Katayama disease) has a similar symptomatology to *S. mansoni* infection. In severe infections, besides periodic diarrhoea, there may be enlargement of the liver and spleen, which may lead to liver fibrosis and portal vein congestion, and result finally in ascites. With juvenile patients, in severe infections growth as well as psychological and sexual development are considerably disturbed. In parasite infections of long duration eggs are no longer regularly found in the stools.

The clinical manifestations of *S. mekongi* infection are very similar to those of *S. japonicum* infection. The adult worms stay predominantly in the mesenteric venous plexus. Essential symptoms are those of hepatosplenomegaly, frequently associated with portal hypertension.

In patients with unclear symptomatology who live in schistosome-infected regions (travel history should also be noted) stool and urine samples should be regularly investigated for *Schistosoma* eggs.

Cercarial dermatitis should also be mentioned. This can occur in fish breeders and bathers. Cercariae of trematode species from certain birds (e.g. *Trichobilharzia szidati* Neuhaus, 1952) are also able to penetrate the skin of man, but do not become sexually mature, and die off. In sensitive persons a painful dermatitis can be induced (see also p. 165).

Transmission. The transmission of the schistosome parasites never occurs directly from man to man. Water snails always serve as intermediate hosts (in *S. haematobium*, e.g. *Bulinus truncatus, B. globosus, B. africanus;* in *S. mansoni*, e.g. *Biomphalaria (=Australorbis) alexandrina* and *B. pfeifferi* in Africa and *B. glabrata* in South and Central America, in *S. intercalatum* in Cameroon and Gabon, e.g. *Bulinus forskalii* and *B. crystallinus* (Jelnes and Highton, 1984); in *S. japonicum,* local species and varieties of the genus *Oncomelania* (China, Southern Japan, Taiwan, Southern Philippines). Intermediate hosts for *S. mekongi* are relatively small water snails of the species *Tricula aperta* (2–4 mm). The natural animal reservoirs for *S. mekongi* are dogs, but the parasite is able to infect different species of laboratory animals. The snails produce different substances that stimulate the miracidia, and are generally known as miraxones (these include amongst other things vitamin C according to DISKO and MIELCAREK, 1984). Miraxones assist the miracidia in finding the snails. This host-finding is not, however, species specific. The definitive host in *S. haematobium* and *S. mansoni* is almost always man. Rats are also regarded as definitive hosts for *S. mansoni*. *S. japonicum* can also develop in different domestic animals (e.g. dog, cat, pig, cattle, water buffalo). The animal reservoirs for *S. intercalatum* include goat and sheep (FRANDSEN et al., 1978). The white mouse and the rat *Mastomys natalensis* (which according to LÄMMLER et al., 1968, can easily be infected experimentally with *S. mansoni*) are considered to be very suitable laboratory animals. For *S. haematobium* the golden hamster is the most suitable research animal. For *S. japonicum* the hamster and the mouse are suitable laboratory animals (up to 70% or 50% infection rates, respectively; prepatency 54 or 48 days; LICHTENBERG et al., 1977; SCHMIDT and RAUB, 1984).

Prevention. In the prevention of schistosome infection the following measures are necessary:

- Systematic control of snails which are involved as intermediate hosts
- Avoidance of infested water, or the use of suitable protective clothing if contact with cercaria-infested water is unavoidable
- Prevention of contamination of water with faeces and urine from infected persons and animals (*S. japonicum* infection)
- Avoidance of spreading infected snails, e.g. through importation of water plants (WALKER, 1978)

In order to implement these provisions in endemic regions, extensive health education for the indigenous population is necessary, so as to understand the epidemiological relationships. Health education is therefore at the spearhead of the campaign measures promoted by the WHO, increasingly carried out under the slogan "People cause Schistosomiasis" ("water-borne disease").

The danger to the population in endemic areas is growing because the increasing water and energy requirements lead to the development of new breeding places for the snail, e.g. the Volta dam lake in Ghana (almost all of the inhabitants are infected), and the Aswan dam lake in Egypt. Through artificial irrigation the agriculturally useful areas are enlarged, but simultaneously the danger of schistosomiasis is increased. In Egypt a change in the relative occurrence of *S. mansoni* and *S. haematobium* has resulted through the presence of the dam. Whereas before the building of the Aswan dam, bladder schistosomiasis predominated over intestinal schistosomiasis (74.3% to 3.2%), now the intestinal form has considerably increased (1979, 73% to 2.2%; ABDEL WAHAB et al., 1979).

Destruction of the snails is possible by the systematic use of molluscicides, for example niclosamide, pentachlorphenol and copper salts. Resistance to these preparations has not yet been observed in spite of years of use. Biological control measures such as molluscicidal plants like *Phytolacca dodecandra,* flies which preferentially seek out snails as hosts for their brood, and snail-eating molluscs have been investigated, but so far without useful results. However, Tirgari (1986) was able to kill up to 94% of the snails of the species *Bulinus truncatus,* the intermediate host of *S. haematobium,* in a research aquarium, with cultured and transplanted larvae of the fly *Sepedon sphegea* (Sciomyzidae).
The investigations of PFLÜGER et al. (1984) have clearly shown that control measures, taking account of snail biology and schistosome infection, can be more rationally designed now than previously. Thus, he recommends for instance that in Egypt molluscicides against *S. haematobium*-infected snails should be introduced for the first time after the winter period, beginning in February, and for the second time in July.

Diagnosis by Microscopy. The detection of schistosome eggs in faeces and urine is relatively simple. In bladder schistosomiasis the eggs can be concentrated by centrifuging the urine. In the urinary sediment (24-hrs urine collection) the characteristic eggs (with terminal spines; see figure below) are easy to find. Between 12.00 and 14.00 h is an especially favourable time for the collection of eggs. A simple and semi-quantitative detection method consists of the use of a filter (paper or nylon). It permits the number of eggs in 5 or 10 ml urine to be estimated and from this to draw some conclusions about the density of the infection. Moreover, the filter can be stained for the easier recognition of the eggs. This method is also suitable for field research (FELDMEIER et al., 1979), and user-ready kits are available.

In intestinal schistosomiasis, the eggs (with lateral spines in *S. mansoni,* Plate XVIII, *1b;* with terminal spines in *S. intercalatum*) are present in large numbers in the bloody mucus which adheres to the faeces (the eggs of *S. japonicum*

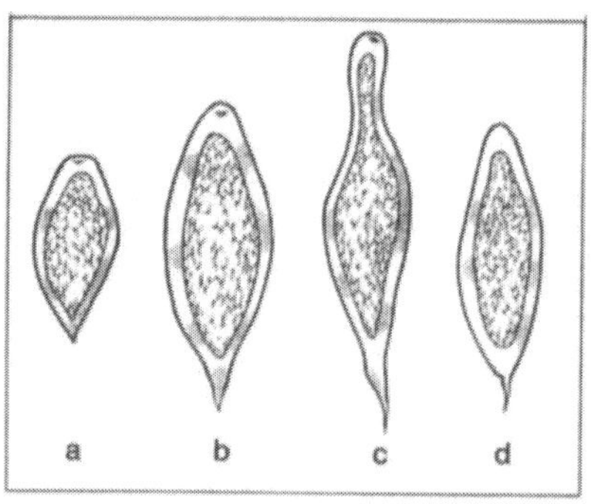

Highly schematic diagram of eggs of various African *Schistosoma* species with terminal spines. *a S. haematobium, b S. mattheei, c S. bovis, d S. intercalatum* (approx. ×100) (after SCHWETZ, 1951)

have inconspicuous lateral spines). If the disease is suspected on clinical grounds and microscopy of urine and stool has proved negative, a crush preparation of mucous membrane from the rectum or bladder should be examined.

The association of the terminal-spined eggs of *S. haematobium* with urine and the lateral-spined eggs of *S. mansoni* with faeces is not to be regarded as absolute. In a small fraction of cases, under certain circumstances, and in particular areas, *S. haematobium* eggs may be found in the faeces and *S. mansoni* eggs in the urine. Microscopic identification is especially difficult in the rare cases in which *S. intercalatum* eggs as well as *S. haematobium* eggs are present in the urine instead of in the faeces. Accurate knowledge of the morphology of schistosome eggs is thus of the greatest value.

Differentiation between the terminal-spined eggs of *S. intercalatum* and *S. mattheei* (the latter occurs very rarely in man) and *S. haematobium* can be facilitated by ZIEHL-NEELSEN acid-fast staining. This is positive only with *S. intercalatum* (acid fast).
VOGEL (1965) reported from Liberia that in 6- to 15-year-olds the excretion of *S. haematobium* eggs is at its highest, but then declines with increasing age. The density of egg excretion in *S. mansoni* infections, however, remains approximately the same. This fact could be useful in diagnosis. In all cases it is recommended that repeated samples should be examined if they are initially negative.

The Miracidium hatching method may be useful in light infections. A faecal sample (about 5 g) is mixed by stirring with 250 ml physiological saline, removal of coarse particles with a sieve and the suspension poured into a tall glass cylinder or a sedimentation glass. After sedimentation this purification process is repeated until the fluid above the sediment is clear. The vessel is placed in the refrigerator overnight. Next morning warm tap water is added, until a temperature of 30° – 40 °C is reached. With the vessel left in sunlight or under a powerful electric lamp the miracidia hatch out within a few minutes to hours. They can easily be recognized by their rapid, purposeful movements if examined against a dark background, for example, by dark-field conditions.

Diagnosis by Immunobiological Methods. For the detection of schistosomes a number of indirect procedures has been developed, some of which, however, require living parasites. Methods include the complement fixation test (CFT), the indirect haemagglutination test (IHAT), the indirect immunofluorescence test (IIFT), a latex flocculation test (LFT), and a skin test. Living or frozen (–196 °C) eggs are used in the circumoval precipitin test, which is highly specific (ISMAIL et al., 1983). Living parasites are also necessary for the miracidia immobilization test and the Cercarien-Hüllenreaction (CHR). For diagnosis of individual cases CHR, IIFT and IHAT are preferred today. For the IIFT formol-fixed or lyophilized cercariae, as well as frozen sections of adult worms, are used. For epidemiological studies JANITSCHKE et al. (1981) recommend ELISA which can be automated and saves time and cost.

It is possible to differentiate to a limited extent between patients with acute or chronic schistosomiasis by the use of different antigens. With an acute infection strongly positive reactions occur with cercarial antigens as well as adult antigens. The latter turn out clearly positive in long-lasting infections (ELISA method; LUNDE et al., 1979). Cross-reactions with schistosome antigens occur in *Fasciola* infections (in IIFT up to 1:80). The titre usually rises transiently after successful therapy (e.g. with praziquantel) as a consequence of increased antigen release. The starting point for immunization is the activation of macrophages so that they immediately kill the cercariae (the schistosomula), which have penetrated into the host (KUBELKA et al., 1983; MOHR and RACZ, 1983). This apparently occurs when cryo-preserved attenuated schistosomula are used as vaccine, as the experimental results of BICKLE and JAMES (1978) have shown. Cercariae attenuated by radiation treatment (50,000 rad γ-radiation from a cobalt source over 25 min) also produced immunity in mice which was evidently immunologically species specific (SMITHERS and TERRY, 1969, 1976; WARREN, 1972).

KÖSTER and SEITZ (1985) collected results with nine monoclonal antibodies against the cercariae of *S. mansoni*. The antibodies are distributed homogeneously over the surface of the cercariae without a recognizable pattern. Each of the nine monoclonal antibodies (4 IgA, 4 IgG, 1 IgM) can induce a CHR (see p. 311 and SETHI et al., 1984).

A naturally acquired schistosome infection also induces a certain level of immunity.

STURROCK et al. (1983), for example, systematically examined 306 school children, between 6 and 16 years of age, for *S. mansoni*. Most were treated with hycanthone (1.5 mg/kg body weight), and observed over 2 years, with particular reference to reinfection. Those children who had cytotoxic antibodies as well as an eosinophilia clearly had increased protection against reinfection (VOGEL, 1962; BUTTERWORTH et al., 1982, 1984).

Chemotherapy. Drug treatment has become very straightforward because of the discovery of the activity of praziquantel (GÖNNERT and ANDREWS, 1977; WEGNER and THOMAS, 1980; ANDREWS, 1981). It constitutes an orally administrable, only slightly toxic drug, which usually needs to be given on only one day, and is suitable for mass therapy as well as individual therapy (on 1 day, 1 × 40 mg/kg body weight or 2 × 20 mg/kg body weight; with *S. japonicum* and *S. mekongi* 1 × 60 mg/kg body weight or 2 × 30 mg/kg body weight). Praziquantel is also active in cases in which resistance against other drugs (oxamniquine, only against *S. mansoni;* metrifonate, only against *S. haematobium*) has occurred (see FELDMEIER et al., 1982; MEHLHORN et al., 1982).

Prophylactic Measures. Experimental studies of the use of vaccination as a prophylactic measure have already provided some promising results, using cryo-preserved attenuated schistosomula as antigen. The immunity produced inhibits the penetration of additional cercariae, but is not active against adult worms (BICKLE and JAMES, 1978).

The most important schistosome species of man

	S. mansoni	*S. intercalatum*	*S. japonicum*	*S. haematobium*
Disease picture	Intestinal schistosomiasis	Intestinal schistosomiasis	Intestinal schistosomiasis	Urinary schistosomiasis
Location in the body	Mesenteric veins, intestine, liver	Mesenteric veins, intestine	Mesenteric veins, intestine, liver	Pelvic veins (urogenital system)
Adult worm				
Surface	Coarse tubercles, many hooks	Tubercles, hook free	No tubercles, smooth	Fine tubercles, many hooks
Size of male (mm)	6–10	11–15	12–20	10–15
Number of testes	6–12 (small, clusters)	3–5	6–8 (ovoid, in a row)	4 (large, clustered)
Size of female (mm)	7–14	13–24	~20	13–24
Number of eggs in uterus	Mainly 1–4, up to 10	10–20–50	50 or more	20–30 or more
Shape of eggs	Elliptical, lateral spine (pp. 160, 168)	Elliptical, terminal spine (slimmer than *S. haematobium*)	Oval, small lateral knob (p. 160)	Elliptical, terminal spine (pp. 160, 168)
Size of eggs (μm)	~150 × 60	~140 × 50	~90 × 55	~150 × 55
Intermediate host (genus)	*Biomphalaria*	*Bulinus*	*Oncomelania*	*Bulinus* and *Physopsis*
Parasite reservoir	Monkeys, nile rats, water rodents	Rodents, ungulates	Domestic animals, rodents	Monkeys (?), rats, pigs

Plate XIX ⇨

Cestodes (Tapeworms)

Diphyllobothrium latum
Diphyllobothrium pacificum
Dipylidium caninum

Plate XIX

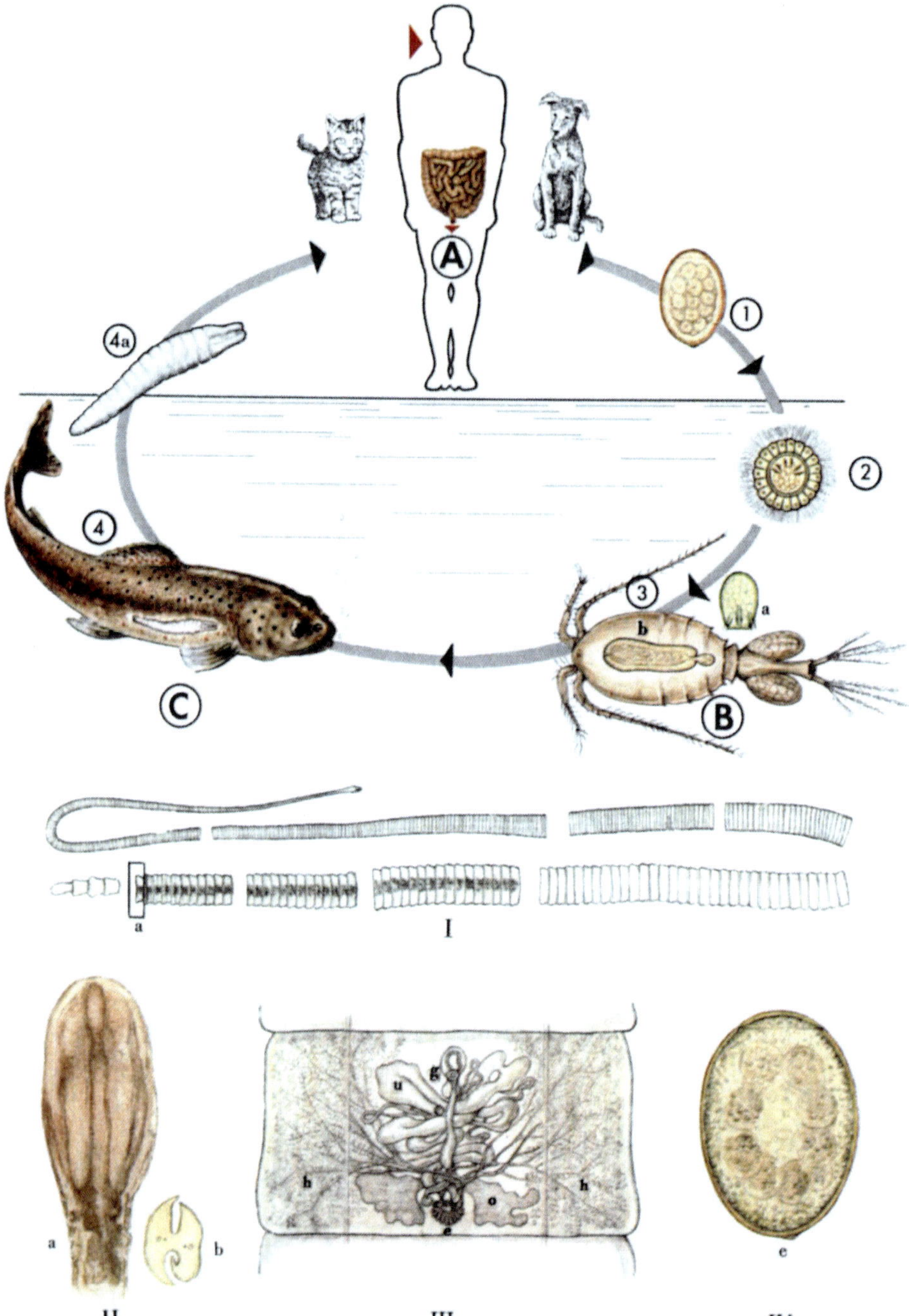

Diphyllobothrium latum (LINNÉ, 1758) LÜHE, 1910
D. pacificum NYBELIN, 1931

Fish tapeworm

Ⓐ Definitive host: man, dog, cat (and also fish-eating domestic and wild animals)

Site of the tapeworm: small intestine

1 Deposited egg
2 Ciliate larva (the coracidium) with six-hooked embryo (oncosphere)

Ⓑ First intermediate host: small crustaceans (copepods)

3 a Six-hooked larva, youngest stage from a *Cyclops*
3 b Procercoid in a *Cyclops*

© Second intermediate host: predatory fish or carp-like fish (paratenic host)

4 Trout with plerocercoid (sparganum)
4 a Isolated plerocercoid

I Sexually mature fish tapeworm; segments of tapeworm with proglottids of varying degrees of maturity

a Mature segment with rosette-shaped uterus (see *III*)

II *a* Spatulate scolex
b Scolex in cross-section; the two lateral sucking grooves (bothria) are clearly visible

III Mature tapeworm segment (proglottid)

e Ootype
h Testis
o Ovary
g Genital pore
u Uterus

IV Individual egg from the stool with operculum

(See also Plates XXXII, *q* and XXXIII, *l*)

Cestodes

Human cestodes (eucestodes) differ from trematodes in having body distinct divisions head (scolex), neck (proliferation zone) and main body (strobila). A tapeworm lacks a digestive tract, but instead lives as an adult worm in the gut of its host. Nutrition is obtained by active transport through the body surface (tegument), the surface area being substantially increased by structures resembling microvilli, the microtriches. In addition, the worm has a special surface coat consisting of mucopolysaccharides, which evidently works in combination with an ATPase (adenosine triphosphatase) to protect the parasite from the digestive enzymes of the host (see also pp. 12, 165).
The scolex is often relatively very small (0.5–1 mm) and is equipped with anchoring organs (suckers or hooks). The strobila body extends from the very short proliferation zone (behind the scolex), the proglottid progressing from an immature to a mature state along the length of the worm. Annular and longitudinal muscles allow the proglottids to move actively. Each proglottid has a complete set of female and male sex organs, although self-fertilization is not usual. The eggs are produced in the mature proglottids, and contain the six-hooked larva (oncosphere). The eggs escape either via a special sexual orifice to the outside or are released by degeneration of the segment in a second (intermediate) host where the eggs develop. The larval stage in the intermediate host has either single (e.g. cysticercus, cysticercoid) or multiple (e.g. hydatid, coenurus) protoscolices, depending on the species of the cestode. The larva (e.g. cysticercus) develops into a sexually mature worm in the definitive host.

Man is the definitive host for several tapeworms (species of the genera *Diphyllobothrium, Taenia, Hymenolepis* and *Dipylidium*); man can also be the intermediate host for *Echinococcus* species and sometimes also for *Taenia* species.

Transmission to the final host is mostly by the consumption of infected raw meat from slaughtered animals (*Taenia* species) or fish (*Diphyllobothrium* species), or through accidentally swallowed insects (*Dipylidium caninum*)), which act as intermediate hosts. Only the dwarf tapeworm (*Hymenolepis nana*) can be transferred directly from man to man via its eggs.

Diphyllobothrium latum (LINNÉ, 1758) LÜHE, 1910
D. pacificum NYBELIN, 1931

There are at least two species of broad or fish tapeworm that infect humans, namely, *Diphyllobothrium latum* and *D. pacificum* and these develop as follows before they become sexually mature in man Ⓐ: Initial development is in certain small crustaceans Ⓑ, and then in freshwater (*D. latum*) or saltwater fish (*D. pacificum*) Ⓒ. Consequently the distribution of *D. latum* is related to certain river systems and lakes (similar to that of the Chinese liver fluke). Occurrence in man is then related to certain dietary customs. Man is infected, as are certain domesticated and wild animals, by the consumption of uncooked fish.
D. latum occurs in bays in the Baltic Sea and in the Volga basin, as well as in Finland, in the area of Lake Constance, and in lake areas in Switzerland (with substantially decreasing frequency), in Italy, in the Danube delta, in the Near East, in Siberia, and in Manchuria, as well as in Japan and North America. *D. pacificum* is quite common in South America, e.g. North Peru (LUMBRERAS et al., 1982).

Morphology and Development. The somewhat spatula-shaped scolex is 2–3 mm across and has two bothria (Plate XIX, *IIa, b*). The main body (strobila) can be up to 20 m long; the broad or fish tapeworm is thus one of the longest tapeworms. The mature segments (proglottids) are about 10–15 mm across and 3–5 mm long (Plate XIX, *III*). The eggs are about 70 × 50 µm and are released singly through a special uterine orifice and thus appear in the stools. The mode of release contrasts with that in bovine or porcine tapeworms (Plate XXI, *2*). The eggs have a smooth surface and a lid-like operculum at one end and a small knob at the other end. They have a substantial number of yolk cells in addition to the egg cell (Plate XIX, *I, IV*).
The mature eggs produce ciliate larvae (coracidia, about 50 µm in diameter), which already contain embryos (oncospheres), each of which has six hooks. These develop in the first intermediate host, namely small copepods of the genera *Cyclops* and *Diaptomus* Ⓑ (Plate XIX, *3a, b*), which take in the tapeworm larvae with their food. The larvae bore through the gut wall in the copepod and enter the coelom, where they develop into procercoids (Plate XIX, *3b*). They remain there until the copepod is eaten by a fish (such as one of the carp species or other carnivorous fish) which acts as the second intermediate host Ⓒ. Then the larvae migrate in the muscles of the fish, the procercoid developing into a plerocercoid (sparganum) (Plate XIX, *4, 4a*). If the second intermediate host is eaten by a carnivorous fish, the plerocercoids migrate in the new intermediate (paratenic) host, which can accumulate numerous larvae. The plerocercoids develop into sexually mature tapeworms following further transmission to a suitable definitive host Ⓐ (in

addition to man, this may be the dog, cat, or fox). The eggs of *D. pacificum* are smaller than those of *D. latum* (about 55 × 40 μm). The second intermediate hosts for *D. pacificum* are salt-water fish in the region of the Peru coast. The uterus in *D. pacificum* has four to seven loops on each side. The main definitive host for *D. pacificum* is the seal.

Clinical Symptoms. The sexually mature tapeworm injures the final host less as a result of its metabolic products than as a result of taking up vitamin B_{12} (cobalamine), the anti-anaemia vitamin. Deficiency of this vitamin leads to megaloblastic anaemia of a pernicious type. However, this only occurs if the tapeworm is located near the exit from the stomach. Eskimos are quite frequently parasitised by *Diphyllobothrium* pacificum, but this type of anaemia has not been recorded in them. It appears that they get microcytic hypochromic anaemia, which occurs more frequently in women and children than in men and which is thought to be due to iron deficiency. Elimination of the tapeworm leads rapidly to a return to normal of the hemoglobin level. In *D. pacificum* infection the commonest symptoms are meteorism, flatulence and diarrhoea (LUMBRERAS et al., 1982).

Plerocercoids (spargana) from related species of tapeworm that do not become sexually mature in man, can sometimes be found in man and cause disease. In this case, they migrate as in the second intermediate host, for example, into the abdominal cavity and musculature, where they can produce local swelling and inflammation (sparganosis). This occurs mainly in the USSR, Japan, China, and Indo-China.

Transmission. Man acquires the broad or fish tapeworm by the consumption of raw fish containing plerocercoids. If fish-eating domestic or wild animals carry the tapeworm, they can contaminate water with their faeces (containing eggs) and lead to the infection of copepods and fish. Infection in man is prevented by eating well-cooked fish foods.

If appropriate education is given to population groups at risk (e.g. fishermen), the number of affected people can be reduced substantially, as has been demonstrated in the Archangelsk region. There the incidence fell from 28% to 3% within a year following treatment. Similar observations have been reported from Finland. This reduction in the incidence of the tapeworm is important in the epidemiology of the disease, because man is the principal host.

Diagnosis by Microscopy. Microscopic examination of a stool specimen reveals the characteristic individual eggs. As there are usually large numbers of these broad fish tapeworm, infection is readily identified (see p. 327).

Chemotherapy. The sexually mature tapeworm can be easily eliminated with well-known drugs (see p. 193). Praziquantel has been found to have high activity: a single dose of 10–25 mg/kg is used (BYLUND et al., 1977). *D. pacificum* can regularly be eliminated with a single dose of praziquantel (10 mg/kg) (LUMBRERAS et al., 1982). Niclosamide is also often used because of its relatively low cost.

In the case of *D. latum*, pernicious anaemia can also occur, which requires special treatment. Sparganosis is apparently unaffected by chemotherapy. It is notable that niclosamide treatment (1 × 2 g) inhibits the development of the released eggs.

Dipylidium caninum (LINNÉ, 1758) RAILLIET, 1892

Dog tapeworm

Dipylidium caninum is relatively rare in man and occurs mostly in children. (It is also called the cucumber tapeworm because of the shape of its proglottids found in faeces.) The life cycle includes insects (fleas and dog hair licc) as intcrmcdiatc hosts, in which the cysticercus stage arises. The parasite occurs world-wide, particularly in canine and feline species.

Morphology and Development. The adult worm is between 15 and 50 cm long and 2–4 mm wide. The longish segments measure about 20 mm, are yellowish to reddish in colour and resemble cucumber seeds. The scolex (width about 0.5 mm) bears four suckers with three to seven small rows of hooks arranged on a retractable rostellum. The eggs are usually excreted with the faeces in packets of 30–40, each (25–30 µm) containing a larva with six hooks. For further development, the eggs must be taken up by flea larvae or dog lice. The oncosphere hatches out of the egg and enters the gut of the insect, passes through the gut wall and gains access to the fatty tissue. The larva develops into a cysticercoid during this migration within the insect. The cysticercoid usually survives pupation of the flea and remains within the adult insect. If an affected flea is ingested by the definitive host, the scolex evaginates from the cysticercoid in the gut and attaches to the gut wall, developing over 15–20 days into an adult worm. Characteristic features of the adult are the two genital orifices (hence the name *Dipylidium*) opening at the centre of both lateral margins of the proglottids.

Clinical Symptoms. These occur in dog tapeworm infection only if there is particular individual sensitivity or gross infection. The effects are those of general toxicity which manifest themselves, amongst other things, in gut disorders. Usually, one gets the same non-specific symptoms as in invasion by other gut worms (see pp. 191, 220). There may be weight loss and occasionally seizures or cramps indicating a toxic effect on the CNS.

Epidemiology. There is an infection hazard for man when in close contact with dogs or cats, particularly if these are infested with fleas. As infection is acquired orally by swallowing fleas, it is commoner in children than in adults. Dogs ingest infected fleas or hair lice on licking their coats.

Precautions amount to keeping dogs free from fleas and deworming them (praziquantel).

Diagnosis by Microscopy. *Dipylidium caninum* is identifiable by examining stools for proglottids and eggs. Mostly egg packets are found. In rare cases, the stool also contains individual eggs, which are often enclosed in thin embryophoric integuments (individual eggs about 25–30 μm).
Characteristically, the dried tape worm segments adhere to the anal region or coat of the affected animal, where they have the appearance of rice grains and can be confused with these. These grains swell rapidly in water, and even more so in dilute potassium hydroxide solution, allowing one to identify the characteristic shape of the proglottids.

Diagnosis of Immunobiological Methods. Such methods are of no practical significance.

Chemotherapy. See *Taenia saginata,* p. 193.

Plate XX ⇨

Hymenolepis nana
H. diminuta

Plate XX

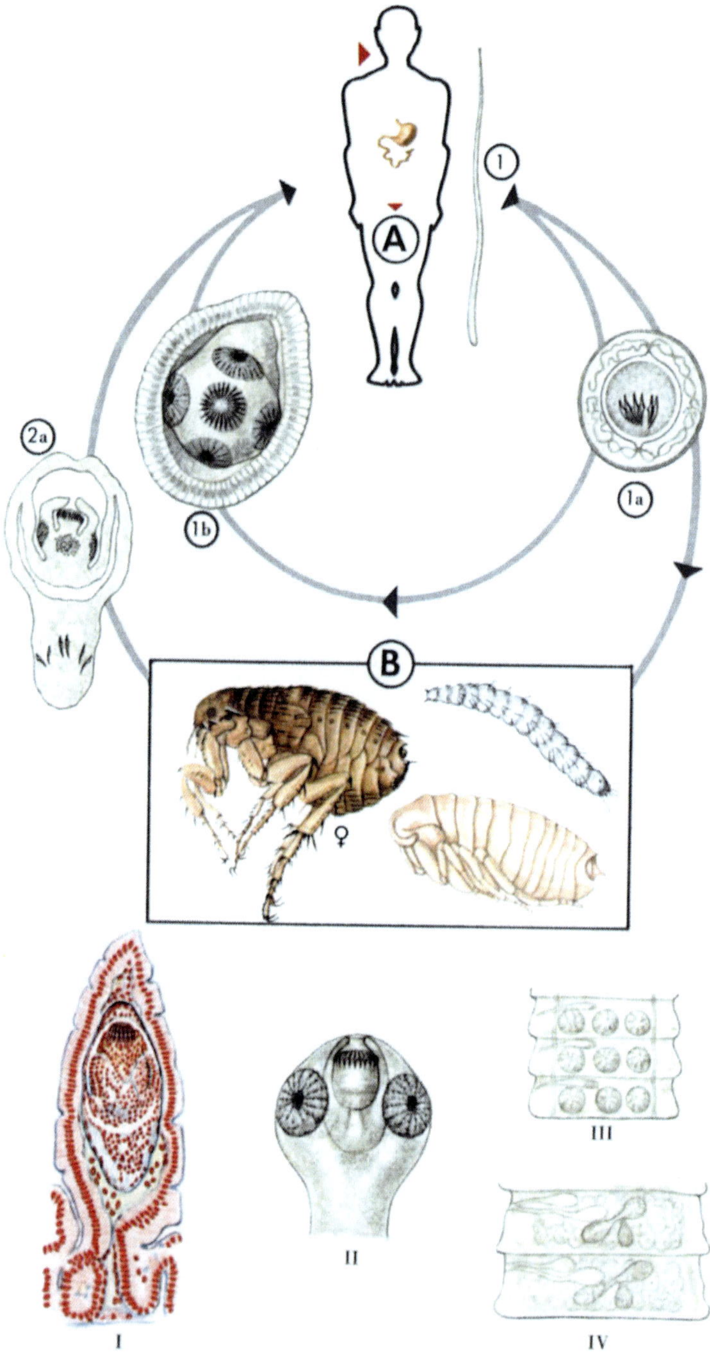

Hymenolepis nana (von Siebold, 1852) Blanchard, 1891
H. diminuta (Rudolphi, 1819) Blanchard, 1891

Dwarf tapeworm (*H. nana*)
Rat tapeworm (*H. diminuta*)

Ⓐ Definitive host: man (also dog and rodents)

1 *Hymenolepis nana,* sexually mature worm
1 a, b Development without an intermediate host
a Egg containing six-hooked larva (oncosphere) of *H. nana;* the polar filaments (*a*) are lacking in *H. diminuta*
b Cysticercoid from gut mucosa (see *I*)

Ⓑ Intermediate host: e.g. the flea (with larva and pupa). Development with intermediate host; the flea larva ingests the tapeworm egg (*1 a*)

2 a Cysticercoid from coelom in flea; tail stump with remaining (larval) hooks; circlet of hooks lacking in *H. diminuta*

I Villus in mouse small intestine with cysticercoid (cross-section)

II Scolex of *Hymenolepis nana*

III Immature proglottids

IV Gravid proglottids (about ×40)

(See also Plates XXXII, *s* and XXXIII, *n*)

The dwarf tapeworm *Hymenolepis nana* is the smallest tapeworm inhabiting the small intestine in man. Children are always more frequently affected than adults. Although it occurs world-wide, it is more frequent in hot countries. The principal areas of occurrence are Mediterranean countries, particularly Portugal, Spain, Sicily, and Egypt, as well as the Sudan, India, Thailand, Japan, South America, and Cuba, in addition to the USA.
The species *Hymenolepis diminuta,* the rat tapeworm, also occurs in man, but is rare. This species is acquired from an intermediate host (insects), in contrast to *H. nana* (see below).

Morphology and Development. *Hymenolepis nana* is only 10–45 mm long and 0.5–1 mm wide (Plate XX, *I*). Isolated tapeworms can become quite long, but the more worms are present, the smaller the size of the individual. A worm may have 100–200 segments. These are always wider than they are long and have a genital pore at the margin (Plate XX, *II, IV*). The scolex (about 0.15–0.5 mm) has a relatively small rostellum with a simple circlet of 20–30 hooklets (lacking in *H. diminuta*) and four suckers (Plate XX, *II*). A mature segment contains three large testes. The mature uterus takes up the major part of the proglottid. The eggs (30–50 μm) are oval to spherical and almost colourless, and they occur individually in the faeces. The spherical oncosphere with six hooks is enclosed in the embryophore (with polar filaments) and the egg case. Between the two cases, there is a broad zone filled with fluid that gives the egg its characteristic appearance (see Plate XX, *1a*).
The eggs of *Hymenolepis diminuta* lack the polar filaments. The spheroid eggs are about 65 × 75 μm in size, while the sexually mature tapeworm is up to 70 cm long. A distinctive feature of the development is that the larval stage (cysticercoid) of *H. nana* can develop directly in the small intestine of the definitive host as well as in an intermediate host (insect). *H. diminuta* must have an intermediate host; the cysticercoid develops in insects (flour beetles, dust beetles, etc.). If larva-bearing eggs of *H. nana* enter the human gastrointestinal tract, the oncospheres are released into the duodenum and attach to the villi, where each develops into a cysticercoid (Plate XX, *1b, I*). After about 4 days, they move back into the gut lumen and attach to the mucosa. Within 2–3 weeks, the cysticercoid develops into a tapeworm.
After about 4 weeks, the first eggs may be found in the stools (prepatent period). Partial digestion of the mature proglottids releases the eggs, which can readily be identified from their characteristic appearance (Plate XX, *1a*). In carriers of the dwarf tapeworm, one sometimes finds that the lumen of the ileum contains several developmental stages simultaneously (eggs shortly before larval release, free embryos, and young tapeworms), whereas the cysticercoids of *H. nana* occur within the gut villi. According to observations of Kilkinov (1967), some cysticercoids may also be found outside the villi in the lymphatic follicles and in the mesentery. Such invasion probably explains the uncharacteristic pains sometimes found in dwarf tapeworm infection.

Clinical Symptoms. These are generally slight or entirely lacking in infection with the dwarf tapeworm. Only in case of heavy infection there is likely to be pain in the region of the gastrointestinal tract or abdominal pain, diarrhoea, and other non-specific symptoms. These occur more prominently in hyper-sensitive people. It is notable that mice relatively quickly develop pronounced immunity, which does not permit reinfection.

Epidemiology. Infection with *H. nana* arises from the ingestion of food contaminated with eggs, but often also through auto-infection. In the latter one often finds a very heavy infection, particularly in children. Contact infection from man to man is frequent, because the mature eggs can develop without an intermediate host. As the eggs are relatively resistant, direct contact is the most likely mode of spread. This is indicated by the higher incidence in urban populations – about twice that in rural ones (see FORESI, 1967). Infections with *H. diminuta* are relatively rare, occurring in children rather than adults, because the mode of infection is via insects, which have to be ingested.
Immunosuppression favours development, and it can allow massive cysticercoid production. Any *Hymenolepis* infection should thus be eliminated before starting of immunosuppressive treatment (see also *Strongyloides,* p. 233).

Diagnosis by Microscopy. For dwarf tapeworm infection, this is based on examining the stool, where the characteristic eggs (Plates XX; *1 a,* and XXXII, *s*) can be detected quite readily. The salt concentration method (see p. 309) is also to be recommended in the search for *H. nana* eggs.

Chemotherapy. Suitable chemotherapeutic agents are niclosamide or praziquantel (see p. 193). However, it must be remembered that the wall of the small intestine may contain cysticercoids as well as sexually mature tapeworms. Consequently, the first therapeutic dose should be followed 1 week later by half the dose, and then the treatment should be repeated again after 3 weeks.

Electron-microscope studies of the mechanism of action of praziquantel have shown that the proliferative zone behind the scolex is severely damaged. This part then shows marked vacuolization in the tegument (BECKER et al., 1980, 1981).

Plate XXI ⇨

Taenia saginata
T. solium

Plate XXI

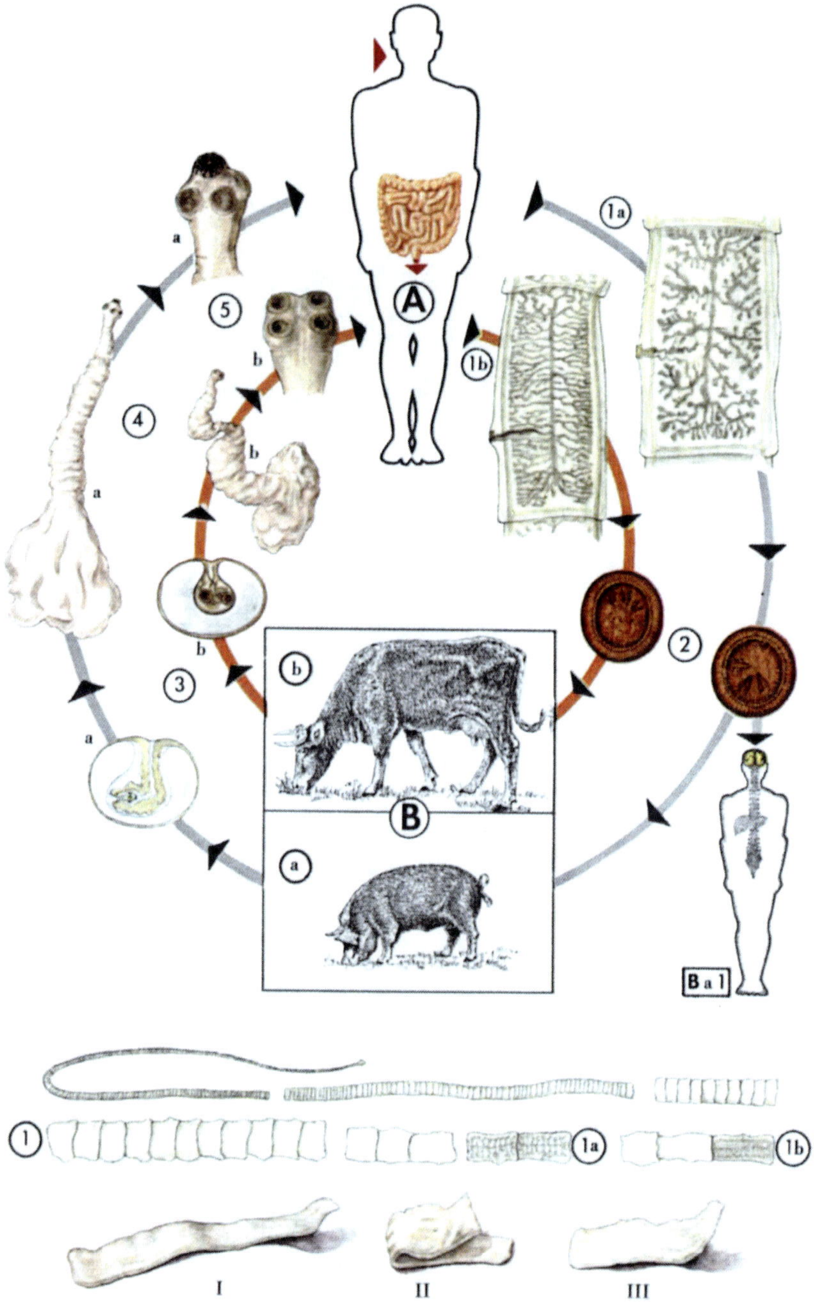

Taenia saginata GOEZE, 1782

Beef tapeworm

T. solium LINNÉ, 1758

Pork tapeworm

a Pork tapeworm (grey pathway)
b Beef tapeworm (yellow pathway)

Ⓐ Definitive host: only man, in small intestine

- *①* Tapeworm (see table below)
- *1 a* Gravid proglottids of *T. solium*
- *b* Gravid proglottids of *T. saginata*
- *2* Tapeworm egg (embryophore containing the oncospheres = six-hooked larvae) (species difficult to distinguish morphologically)

Ⓑ Intermediate hosts:

- *a* Pig, exceptionally also man (*Ba1;* cysticercosis)
- *b* Cow
- *3–4* Cysticerci in various stages of scolex inversion
- *3 a* Cysticercus cellulosae of *T. solium* (with circle of hooks and four suckers); onset of inversion
- *b* Cysticercus bovis of *T. saginata* (with four suckers; no hooks)
- *4* Everted cysticercus stages of *T. solium* (*a*) and *T. saginata* (*b*)
- *5* Scolices
  - *a* Pork tapeworm (with circle of hooks)
  - *b* Beef tapeworm (without circle of hooks)

I–III Mobile stages of freshly separated proglottids

(See also Plates XXXII, *r*, and XXXIII, *k, m*)

The beef tapeworm *Taenia saginata* (= *Taeniarhynchus saginatus*) and the pork tapeworm *T. solium* are the best known tapeworms of man; they invade the small intestine Ⓐ. The larval stage of the first species develops in cattle Ⓑ to cysticercus bovis (also called metacestode, Plate XXI, *3b*), while the second species develops in the pig Ⓑ to cysticercus cellulosae (Plate XXI, *3a*). These tapeworms mature sexually only in man (definitive host). However, man can also act as the intermediate host, because oral ingestion of the eggs leads them to become cysticerci, mostly in the case of *T. solium* larvae, very rarely in *T. saginata* larvae. In such a case, one gets the decidedly dangerous disease picture of cysticercosis. This can lead to severe functional disorders and even to death if the larvae, for example, enter the central nervous system (neurocysticercosis, Plate XXI, *BaI*). The two species occur world-wide, but the occurrence and frequency are very much dependent on the living and eating customs of the population (see pp. 191; world-wide there are about 45 million carriers of *T. saginata* and about 3.5 million of *T. solium,* with about 50,000 deaths a year due to cysticercosis; see below). *T. solium* has become rare e.g. in Europe and North America on account of systematic meat inspection.

Morphology and Development. Larval development begins in the intermediate host following oral ingestion of eggs. The egg (38 × 32 μm) has an outer layer, beneath which is the yellow-brown radially striated embryophore that surrounds a six-hooked larva (Plate XXI, *2*). The larva (oncosphaera) is released into the gastrointestinal tract, penetrates the wall of the small intestine, and passes via the venous blood and the liver, heart and lungs into the main circulation. It thus may be carried into almost any organ, but in particular the skeletal muscles.
After 2–4 months, the cysticercus stage (size 3–10 mm) develops in the intermediate host, and remains infectious for 1–2 years depending on the strain and the host animal. Intrauterine transfer to the foetus is possible from the third month of pregnancy. The preferred organs are the tongue, larynx, diaphragm, muscles of the back and thigh, as well as the heart and peritoneum, although the parasite can also be demonstrated in the liver, lungs and brain. The cysticercus vesicle is enclosed in a connective tissue capsule and already contains the scolex, which has four suckers in the beef tapeworm and in addition has a double circle of hooks in the pork tapeworm (Plate XXI, *3a, b, 5a, b*).
If the cysticercus enters the gastrointestinal canal in man Ⓐ, the gastric juice dissolves the outer capsule, the scolex is everted in the small intestine, as it rests in a large vesicle mostly containing fluid (Plate XXI, *4*). The parasite anchors itself in the upper part of the small intestine by means of its attachment apparatus, the suckers. The proglottids develop from the region directly behind the scolex, and within 3–4 months they give rise to a chain of proglottids 3–4 m in length (*T. solium*) or 6–10 m or even more (*T. saginata*) (Plate XXI, *1*). The scolex is only 1–2 mm in size. These tapeworms can persist for many years in the gut, according to Beier (1983) up to 35 years. Pawłowski and Schultz (1972) state that the life

span is restricted only by that of the host. The eggs are usually pressed out at the break off line within the proglottid (up to about 80,000 per segment and up to 720,000 daily).
A further difference between the bovine and porcine species concerns the number of uterine branches in a gravid proglottid. In *T. solium* there are 8–12 (Plate XXI, *1 a*), but there are 20 or more for *T. saginata* (Plate XXI, *1 b*). The most reliable feature for identifying *T. saginata* is the vaginal sphincter, which is lacking in *T. solium.*

LE RICHE and SEWELL (1977) used thin-layer enzyme electrophoresis for differentiation by observing the migration speed of glucose phosphate isomerase, which was higher for *T. saginata* than for *T. solium.*

It is noteworthy that the mature proglottids, when shed, move spontaneously on the surface of faeces etc. They thus appear like independent worms, and frequently there has been confusion with "unknown worm species" (see Plate XXI, *I–III*).

Clinical Symptoms. Symptoms of a serious nature are very rare in man following infection with the beef or pork tapeworms, and when they do occur they are mostly very non-specific. The more frequent symptoms are (HORNBOSTEL, 1959) weight loss, digestive pains, epigastric pain, colic (pancreatic symptoms) and loss of appetite. Sometimes, however, there may be pronounced hunger or nervous complaints as well as anal pruritis. Less commonly, there is orthostatic circulatory disturbance and a tendency to collapse. Such symptoms vanish when the worm is eliminated. If the patient complains of these symptoms and no other cause can be found, shed tapeworm segments should be looked for. It is not likely that the tapeworm will cause any severe damage to the gut mucosa, although there may be slight inflammation in the wall of the small intestine. No changes can be detected in the red-cell blood picture. In rare cases, there is a slight eosinophilia. The average frequency of beef tapeworm infection in the population of the Federal Republic of Germany is 0.5%–2%, varies with the local dietary customs. Systematic meat inspection has substantially reduced the tapeworm incidence in man.

Cysticercosis (Cysticercus cellulosae) develops when eggs of *Taenia solium* are ingested orally by man (for example, with unsatisfactorily washed salad vegetables) or because of antiperistalsis in worm carriers. The larvae can reach almost all organs, but they have a preference mainly for the muscles and CNS. Cysticercosis in the muscles can be relatively asymptomatic. If there is a severe attack on the muscles, rheumatic pains occur. If the CNS is attacked (often also the eye), one gets all conceivable neurological and mental symptoms (neurocysticercosis). If the cysticerci enter the eye, they can sometimes be observed directly in the front chamber. Ocular cysticercosis occurs in about 20% of cases and can lead to loss of vision. If the cysticerci become calcified, they can be detected within the muscles by X-radiography as shadows about 1 cm in length. Sometimes cysticerci in the brain lead to strange dysplastic growth forms with root-like branching and grape-shaped vesicles, which VIRCHOW called Cysticercus racemosus.

Transmission. Tapeworm infection occurs from consuming raw pork or beef containing the cysticerci, e.g. as beef steak. The bovine cysticerci show a distinct tendency to attach to human skin. Consequently, persons frequently in contact with raw meat are particularly at risk. Housewives and kitchen staff have tapeworm infection more frequently than other occupational groups. The cysticerci are

rapidly killed by cooking or roasting (tolerance limit 38 °C), but they survive temperatures below 0 °C for several hours. Bovine cysticerci are more sensitive than porcine ones, as –3 °C for 24 hrs or –30 °C for 30 min is sufficient to kill them, whereas the latter require 150 hrs or more at –2 to –6 °C before they die. Raw meat should be stored before consumption for at least 24 hrs at –20 to –30 °C.
The infection rate in cattle has been found from abattoir observations to be 1%–2%, although locally it may be up to 4% (HÖRCHNER, 1983; the same applies in Japan, the USA, Canada, and Australia). In addition to cattle, other intermediate hosts are the zebra, buffalo, reindeer, and African antelope and gazelle. In East Africa, the incidence in cattle ist 10%–90% depending on the herd, and in any case often over 10%. Regions with more than 10% incidence include Ethiopia, Kenya and the Congo area.
The eggs can survive free for about 60 days in the presence of moisture and at temperatures above 0 °C, but they generally die within 12–16 days under dry conditions and at temperatures of about –30 °C.
There has been an increase in the bovine form in Western Europe on account of inadequate disposal of faeces when people camp "in the wild". This has led to increased infection in domesticated animals. The danger of infection is also increased by applying untreated sewage to pastures.
The scope for independent movement (see Plate XXI, *I–III*) allows the proglottids to migrate, for example from open dung- and cess pits, and enter the living areas of pigs and cattle. Care must thus be taken that cess- and dung pits are not located close to the stock. Cattle can lose about a third of their value from cysticercus infection (on dealing with bovine cysticercus see HÖRCHNER and ALBERT, 1979). A role is also played in the distribution of *Taenia* species eggs by coprophagic insects, because they re-excrete the ingested eggs in an undamaged form.

In handling proglottids, caution is essential, since there is a danger of cysticercosis, particularly with *T. solium* (see above on p. 191).

Diagnosis by Microscopy. Generally, an infection is detected by the spontaneously shed yellow-white individual segments or chains of them. These can be identified by the unaided eye in the stool. Microscopic examination of the faeces only sometimes leads to identification of infection because the eggs are not shed individually and occur only occasionally in the stool. *T. solium* and *T. saginata* are identified from the numbers of uterine branches in the mature proglottids (see Plate XXI, *1a, b, 5a, b*). The scolex cannot be used as a distinguishing feature (Plate XXI, *5a, b*) with current chemotherapy because the dead tapeworm is digested and no longer expelled (see below).

Diagnosis by Immunobiological Methods. Immunological methods are of considerable importance for identifying invasive disease caused by cysticerci of the pork tapeworm (Cysticercus cellulosae). Methods include complement fixation, indirect haemagglutination, and immunofluorescence. On the other hand, so far results have only been obtained with group-specific antigens. Also, hydatid antigen from *Echinococcus granulosus* gives positive results with antibodies produced to Cysticercus cellulosae (see GOTTSTEIN et al., 1986).

Chemotherapy. Niclosamide works very reliably and is well tolerated (single dose of four tablets = 2 g for adults, children ½ to 2 tablets depending on age). This also applies to praziquantel (single dose of 10 mg/kg, tablets of 150 mg). With the older drugs, the rule was that the removal of the tapeworm was confirmed by the excretion of the scolex (about 1–2 mm). The new tapeworm treatments do not require this, because the killed tapeworm is rapidly digested. Eggs can sometimes be detected in the stool or around the anus (see *Enterobius* for the technique). As niclosamide kills the worm but not the eggs, with *T. solium* care must be taken after such treatment to avoid cysticercosis (regurgitation hazard). In these cases, it is recommended that an antiemetic drug may be given before the treatment and a laxative 1–2 h after the drug has been given. Pregnancy and breast feeding are not contraindications to the treatment (BEIER, 1983).

The action of praziquantel on the tapeworm can be detected by electron microscopy within 5 min. Vesicles are produced in the proliferative zone behind the scolex. Numerous vacuoles arise in the superficial tissue, which lead to irreparable damage and the death of the parasite (BECKER et al., 1981).

Praziquantel has in addition a distinct effect on the cysticerci. Even in neurocysticercosis in man, one can observe a distinct improvement (at 50 mg/kg in two or three doses per day, continued for 15 days, with tablets of 600 mg; SPINA-FRANÇA et al., 1982). Additional treatment with steroids (e.g. 4–16 mg of dexamethasone per day for 15 days) is recommended if there are no contraindications to the immunosuppression associated with this drug (e.g. with tuberculosis, toxoplasmosis, pneumocystosis; see GROLL, 1981).

Plate XXII ⇨

Echinococcus granulosus
E. (Alveococcus) multilocularis

Plate XXII

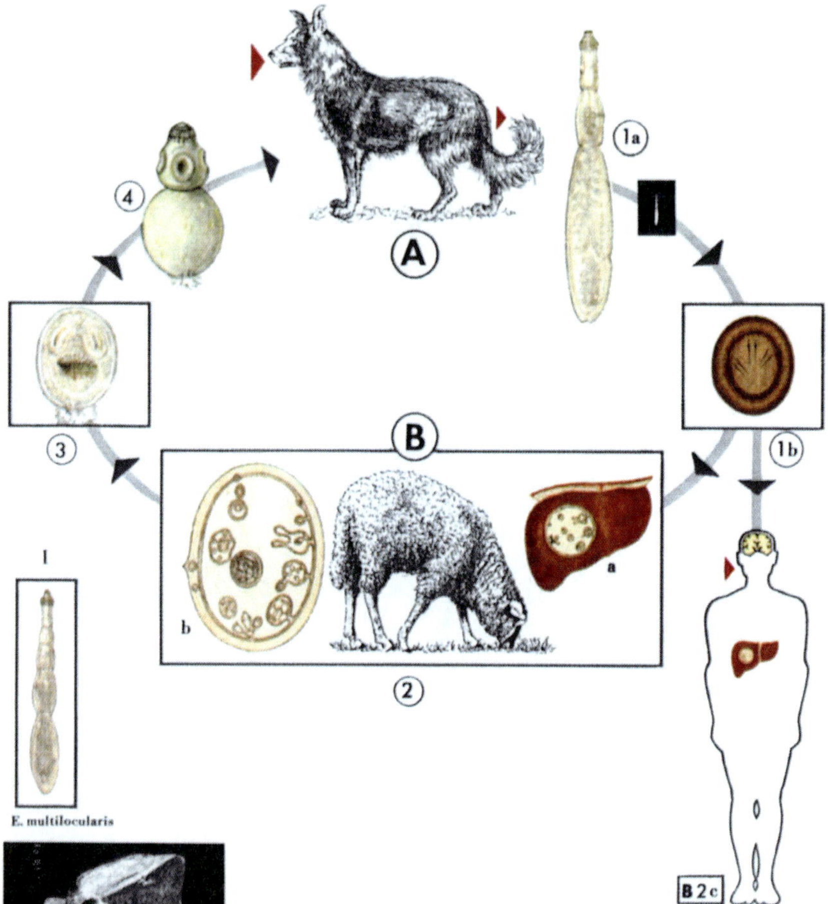

I

E. multilocularis

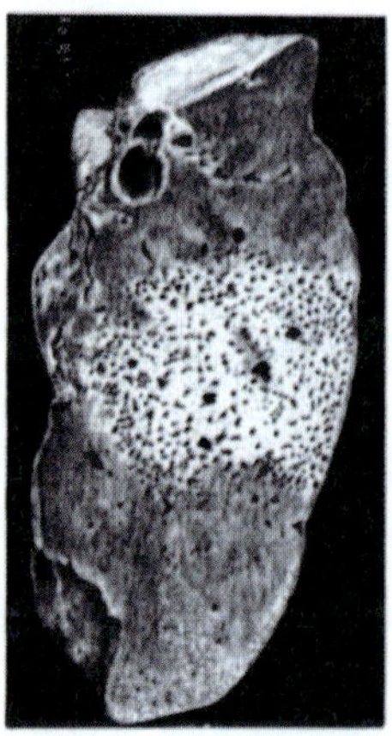

II

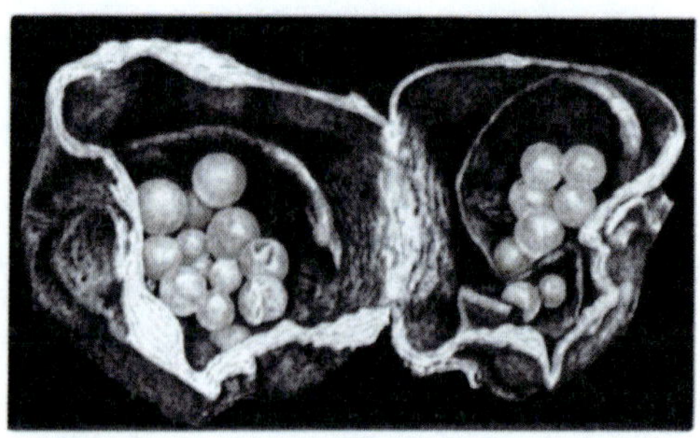

III

Echinococcus granulosus (Batsch, 1786) Rudolphi, 1805
E. multilocularis (Leuckart, 1863) Vogel, 1955

The hydatid worm
(See also the schematic survey on p. 200)

Ⓐ Definitive host: dog (and other canines)

1 a *Echinococcus granulosus* (in dark field at natural size)
b Egg containing six-hooked larva (oncosphere)

Ⓑ Intermediate host: sheep (for *E. granulosus*) or mouse (for *E. multilocularis*)

2 a Liver containing *Echinococcus* cysts (hydatid stage)
b Diagram of an *Echinococcus* cyst with daughter vesicles (i.e. brood capsules) and protoscolices (see *III*)
c Man as (secondary) intermediate host (echinococcosis); organs mainly affected are liver, brain, and lungs
3, 4 Individual protoscolices: invaginated (*3*), evaginated protoscolex (*4*); circle of hooklets and suckers visible

I *Echinococcus multilocularis,* sexually mature worm

II Human liver infected with the alveolar larvae of *E. multilocularis* (after Hamperl) (see p. 201)

III Hydatid cysts of *E. granulosus* opened, daughter vesicles visible (after Hamperl)

(See also Plate XXXIII, *o*)

The tapeworms of the genus *Echinococcus* are, strictly speaking, dog tapeworms called hydatid worm. The mature worm lives in the small intestine of the dog and of closely related animals such as the wolf, fox and cat Ⓐ. Sheep, camels, other domestic animals and certain small rodents act as intermediate hosts Ⓑ. The hydatid stage develops in these (Plate XXII, *2, III*). Man can also become the intermediate host (Plate XXII, *B 2 c,* zoo-anthroponosis) by orally ingesting dog tapeworm eggs from which the hydatid stage develops (hydatid disease) (see Plate XXII). VOGEL (1957) established that the echinococci occurring in the Federal Republic of Germany belong to two different species.
The dog tapeworm occurs world-wide. There are geographically distinct areas for the two most frequent species, *E. granulosus* (RUDOLPHI, 1805) and *E. multilocularis* (VOGEL, 1955). *E. granulosus* occurs mainly in the north of the Federal Republic of Germany and in the German Democratic Republic (Mecklenburg and Pomerania), in Yugoslavia, in Turkey and in major parts of South America (e.g. in Argentina locally there may be 100 cases per 100,000 inhabitants; VARELA-DIAZ et al., 1983), as well as in Africa and Southern Australia. *E. multilocularis* (also called the fox tapeworm, see below) occurs primarily only in the northern hemisphere and there only in certain circumscribed areas, in particular Upper Bavaria, Southern Württemberg (Swabian Alb), Baden, the Tirol, Carinthia and Styria. The parasite also occurs in Italy and Switzerland, and in the USSR in the Ukraine, in the area around Moscow, as well as in areas on the Volga, around Leningrad and Archangelsk, and in Azerbaidzhan, in addition to Canada, and the USA, especially North Dakota and Alaska. However, the boundaries have become fluid as knowledge of the distribution of the two species has extended (particularly for the fox tapeworm; see ZEYHLE, 1983).
In addition, three other species are reported to occur occasionally in man: *E. oligarthrus* (Brazil and Panama), *E. patagonicus* (southern South America), and

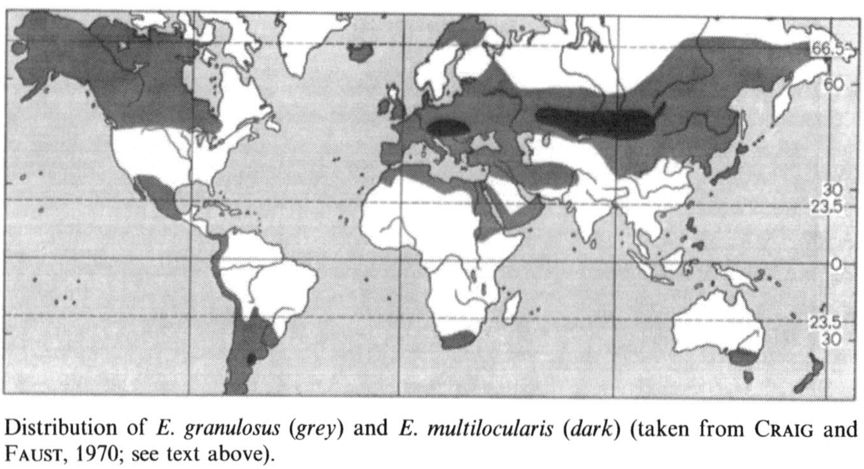

Distribution of *E. granulosus* (*grey*) and *E. multilocularis* (*dark*) (taken from CRAIG and FAUST, 1970; see text above).

E. vogeli (Ecuador, Columbia and Panama); all three species appear to be similar to *E. multilocularis* (FRANK, 1982).

Morphology and Development. *E. granulosus* (Plate XXII, *1 a*) and *E. multilocularis* (Plate XXII, *I*) can be distinguished by the morphology and biology of the mature worms (see p. 203) and the cystic (=larval) stages. The larger *E. granulosus* (length about 3.5 mm) produces a vesicle-type (unilocular) stage, the hydatid cyst, i.e. a clearly distinguishable cyst filled with fluid, which in man (mainly in the liver) can become as large as a foetal head. This cyst (Plate XXII, *2 a, b, III*) contains numerous daughter cysts each with many protoscolices, the true tapeworm structures. The cyst wall consists of an inner germinal layer and an external laminated, acellular, cuticular membrane. The inner layer may produce daughter cysts or may produce brood capsules directly. Within these, the developing scolices (the protoscolices) appear. The hydatid cyst is enclosed in a connective-tissue layer produced by the host. These hydatids in man (intermediate host) sometimes occur in the lung and CNS, although other organs can also be affected. The parasite can grow for years (hydatidosis) without causing pain, and with *E. granulosus* generally causes pain relatively late, for example associated with pressure atrophy in the liver. If operated on promptly, the cysts can readily be removed. If the cyst is damaged and hydatid fluid leaks into the peritoneum, the daughter cysts and protoscolices are disseminated and from these new cysts can arise (secondary hydatidosis, see below).

The cyst stage in the smaller *E. multilocularis* (length about 2.1 mm) has, in section, a structure consisting of numerous small vesicles, filled with a honeycomb of gelatinous, colloidal material containing the protoscolices (as a result, the genus has also been called *Alveococcus* by Russian researchers). This stage consists, however, not of a conglomerate of larger and smaller cysts but a network of partly solid, partly tubular or root-like ramifications (sometimes only 5–10 μm in diameter). These have cystic swellings, and extend throughout the affected organ, particularly when the liver is involved. Growth occurs at the ends of these tubes, in which the protoscolices later arise. The infection can also metastasize to other organs. ECKERT et al. (1983) suspected from laboratory observations that metastases can be produced by cells that split off from the main growth and are transported by the lymph or blood. This assumption has been confirmed since then by electron microscopy (MEHLHORN et al., 1983). Consequently, the parasite does not resemble *E. granulosus*, which can be removed by surgery. *E. multilocularis* is thus extremely life threatening.
Necrotic cavities in the central parts of the liver often occur. This development almost always leads to the death of the patient, and consequently, *E. multilocularis* is one of the most dangerous worm parasites of man. (The multilocular hydatid worms sometimes occurring in cattle are variants of *E. granulosus*, but they represent atypical forms.)

The mature worm is attached by its double circle of hooklets and four suckers on the scolex between the villi in the small intestine. The mature proglottids show up on the layer coating the mucosa as white spots because of their chalk-white colour. They can be identified with the naked eye. The life span of the stages is not known exactly, but in general they appear not to live for more than 100 days (VOGEL, 1957; SMYTH, 1969), or 4–6 months, or occasionally 1 year (according to FRANK, 1976). The mature segments (one segment is cast about every 14 days) each contain 200–1000 eggs (35–40 μm in diameter), and are very resistant. They survive up to 2½ years at 2 °C (in the case of *E. granulosus*). In nature, they remain infectious for at least 3–8 months, depending on climatic conditions. The final host (dog, fox, etc.) excretes the first eggs, about 35–45 days after oral ingestion of the *E. multilocularis* protoscolex; or 48–61 days with *E. granulosus* (prepatent period).

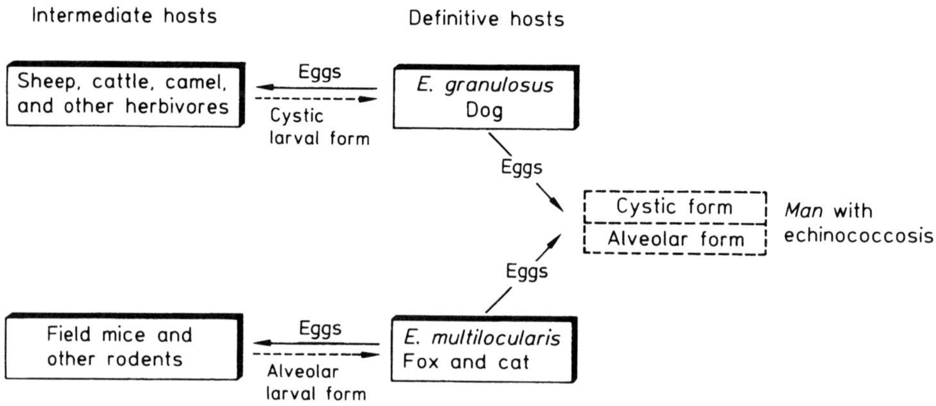

Routes of transmission for *Echinococcus* species

Clinical Symptoms. These are lacking in the dog and fox if the infection is not very heavy (foxes may have up to about 200,000 worms). However, there may be inflammation of the gut wall with non-specific pains.
In man, echinococcosis can be extremely dangerous. In case of *E. granulosus* echinococcosis often remains unrecognized for a long time, if the cysts occur, for example, in the liver or lung. This is because the cysts mostly grow very slowly and then are usually either accidentally discovered by X-radiography (mostly as round patches) or cause pain as they enlarge (pressure atrophy). If there is lung involvement, there are chest pains, with cough and expectoration. If a cyst breaks into the bronchial tree, scolices and isolated hooks may be found in the sputum.
Computered tomography and sonography have produced considerable improvements in clinical diagnosis. Almost all organs can be attacked, although the liver is involved in up to 60% of cases and the spleen or lungs in up to about 20%. The CNS and bones may also be affected. As the surrounding tissue is not directly involved, the hydatid cysts can mostly be removed by surgery, without much damage. In so far as operative removal of the cyst is possible, surgical intervention remains the treatment of choice, as previously. An important point is that protoscolices released spontaneously or by operative intervention or by percutaneous perforation of cysts can produce local secondary cysts or spread through the blood to

other organs. Therefore, care is needed during surgery. In addition, if the fluid contents of the cyst are released there is a danger of anaphylactic shock. If a cyst becomes infected, an abscess develops. Some cysts become calcified, mainly in the liver and brain. The resulting sterile cysts (acephalic cysts) arise as a result of immunological processes.
Considerably higher risk is associated with the larval stage of *E. multilocularis*, which spreads throughout the affected organ (mainly the liver) in an infiltrative fashion, like a malignancy, and causes continual tissue destruction (and abnormal results on liver function test). The result is a spongy tissue containing enumerable pinhead to lentil-sized vesicles, which contain many protoscolices, as in the case of typical hydatid cysts. Surgery is virtually valueless and consequently chemotherapy must be recommended (see below). This can at least extend life expectancy (42% have a 5-year survival rate). Serological studies can confirm the diagnosis (see below).

Epidemiology. The carriers of *E. granulosus* are mainly pet dogs. Pigs, sheep and cattle are the primary intermediate hosts, as well as camels and horses (synanthropic cycle). In the case of *E. multilocularis*, the definitive hosts are foxes (for the fox tapeworm, red fox in Europe, white fox in Alaska), as well as domestic cats, and never dogs. The intermediate hosts are rodents, field mice (*Microtus* sp.) and various voles, as well as the musk rat (cycle in wild animals – sylvatic cycle (FRANK, 1976, 1982; CAMPBELL 1988).
The survival time for hydatid cysts from slaughtered animals very much depends on the external temperature. Infectivity persists for about 80 days at 4 °C, but for only about 10 days around the freezing point, and only about 2 days at –20° to –22 °C. These data of course are only preliminary. The larval stage of *E. multilocularis* has been frozen in liquid nitrogen at –196 °C and has remained in a viable and infectious form (ECKERT et al., 1986). There are substantial differences between the various strains in experimental infection with *E. multilocularis* in the mouse.
The eggs survive for about 1 year in the open air, but they die rapidly on drying. Bilberries and similar wild fruits have been suggested as a source of infection, as they can be contaminated by foxes. The possibility of airborne spread has also been discussed (and possible risk for hunters). Eggs contaminating equipment and floors can be killed with boiling water, or better still by steam. The best preventive measures against *E. granulosus* consist in checking dogs for tapeworms and treating them with praziquantel. For *E. multilocularis*, preventive measures are aimed at reducing the fox population. This also eliminates a source of rabies, as areas endemic for *E. multilocularis* correspond to those in which rabies is endemic (MÜLLER, 1982).
Studies by ZEYHLE (1983) give more precise figures on the occurrence of *E. multilocularis*. According to him, foxes in the region of the Swabian Alb are infected at a level of 13.5% (up to 27%), while 0.5% (locally up to 15%) of field mice, acting as intermediate hosts, are affected. Dogs and cats in endemic regions should be regularly dewormed and never fed raw offal (particularly liver and lung). Offal should be given either after having been cooked or after deep-freezing (–18 °C) for at least 3 days. Up to 81.7% of dogs in certain areas of the USSR (Yamel peninsula) have tapeworm infection, as compared to between 13.9 and 70% in North Dakota (USA), 30–90% in Alaska, and 23% in Japan (see ZEYHLE, 1982).

MacPherson et al. (1984) pointed out a cycle that has largely been overlooked in Africa. The Turkana tribe in Northern Kenya suffers to an unusual extent from *E. granulosus* infection (1 in 600 inhabitants). The maintenance of the transmission cycle between man and animal is favoured by numerous circumstances such as the nomadic life style and the contamination of drinking water with the faeces of infected jackals and dogs. The custom of leaving dead people above ground allows jackals and dogs to be further infected.

Diagnosis by Immunobiological Methods. Serological methods are appropriate for identifying echinococcosis in man (hydatidosis), in addition to the previously mentioned clinical methods. Such methods include complement fixation, indirect haemagglutination (IHAT), ELISA, and skin tests (Casoni test) using sterile hydatid fluid as antigen. It must be remembered that this antigen differs in quality from host to host and depends on the size of the cyst. It must therefore be titrated (Janitschke et al., 1981; Böhle and Janitschke, 1984; Disko et al., 1984). The indirect immunofluorescence test (IIFT) can be performed on sections of *Echinococcus* cysts (containing protoscolices); this is the most sensitive method. In all cases, it must be borne in mind that these are mostly only group-specific reactions. In liver echinococcosis, about 80% of cases can be identified serologically, while in lung echinococcosis it is only 58%–60%. Negative results may be obtained, particularly in lung echinococcosis, even in cases confirmed by surgery (see Ambroise-Thomas and Desgeorges, 1979). According to Janitschke et al. (1981), IIFT should be accompanied by a second test, e.g. IHAT. The end titre in the IIFT lies at 1:320, while in the IHAT it is 1:16,000; ELISA is positive at 1:80 (for this see the recommendations in Werner, 1981). Assistance in epidemiological studies is also provided by immunoelectrophoresis, in which a characteristic precipitation line forms (Arc 5; see Varela-Diaz et al., 1983).
Cross-reactions have been observed with other cestodes (*Taenia* cysticerci) and also (in animal tests) with filariae (Hinz et al., 1981). Within certain limits, *E. granulosus* (E. cysticus) and *E. multilocularis* (E. alveolaris) can be distinguished serologically [1]. If a homologous antigen is used, higher end titres are obtained than with a heterologous antigen. Post-operatively, there is a slow fall in titre over 1–2 years, sometimes following a transient rise in antibody. If the titre persists, the patient may still have unidentified cysts. An advance in immunobiological techniques is the ability to detect circulating antigen. This can be done by ELISA using parasite-specific antibodies to detect circulating antigen (Gottstein, 1984a, b; Gottstein et al., 1984).

Diagnosis by Microscopy. Eggs in the faeces of dogs or cats are frequently detected using only a concentration method (see p. 308), but this is not successful on every occasion. Repeated examination of faeces may be necessary.
The greatest care is required when dealing with infected dogs or cats (especially

[1] According to Mannweiler (1982), the indirect enzyme method is also suitable. Gottstein (1984a, b) was able to prepare a species-specific antigen from *E. multilocularis* for the ELISA technique; sera did not react with *E. granulosus*.

with the faeces). Faeces and gut contents containing eggs represent a unique and extremely hazardous source of infection for man.
If lung echinococcosis is suspected, protoscolices and isolated hooks may be found in the sputum.

Differences between *E. multilocularis* and *E. granulosus* (after VOGEL, 1957, with additional information from RAUSCH, 1968)

	E. multilocularis	*E. granulosus*
Body length (after water-alcohol treatment)	1.11 – 2.71 mm, mean 2.13 mm	2.10 – 6.0 mm, mean 3.36 mm
Length of terminal proglottid	Less than half the total length (0.44 – 1.11 mm, mean 0.85 mm)	Mostly greater than half the total length (1.02 – 3.2 mm, mean 1.94 mm)
Number of proglottids	3 – 5 (in dog mostly 4)	3
Sexually mature proglottid	Third from last	Penultimate
Number of testes	14 – 31, mean 22	38 – 65, mean 44.2
Number of testes anterior to genital pore	0 – 5, mean 2.3	9 – 23, mean 15.8
Uterus in gravid terminal proglottid	Without lateral branches	Lateral branches usually evident
Position of genital pore	Near to mid-line	Behind mid-line

There are also minor differences in the dimensions of the scolex and hooks in the protoscolices (see VOGEL, 1957). The ratio of the height of the dome (between the root processes) to the length of the base is measured. This is 1 : 3.4 for *E. multilocularis* and 1 : 6 for *E. granulosus*.

Chemotherapy. To date, it has not been possible to kill the developmental stages of echinococci in man with any degree of certainty. However, work done by the Swiss research group led by AMMANN and ECKERT (AMMANN et al., 1979) has shown that high doses of mebendazole (1.2 g per day over about 100 days) can inhibit the growth of the cysts. The cellular epithelium in the cysts cannot be destroyed, but treatment with mebendazole (or flubendazole) leads in most cases of *E. granulosus* infection to subjective improvement and sometimes to substantial regression of the cysts (SCHANTZ et al., 1982).
With the larvae of *E. multilocularis*, there is growth inhibition, but the treatment does not kill the parasites. Side effects are mostly slight. There may be allergic reactions, hair loss, and reversible neuropathies. However, in general the benzimidazoles are well tolerated (AMMANN et al., 1979). Of these, albendazole with its metabolite albendazole sulphoxide has given substantially better results than

mebendazole according to initial results with laboratory animals (10 mg/kg daily by mouth for 1–3 months; Stallbaumer et al., 1983; Morris et al., 1983). It should be mentioned that level of specific IgG, IgA, and IgE antibodies decreases with mebendazole treatment. In untreated patients, on the other hand, the level increases (Gottstein et al., 1984).

Praziquantel is active against the adult worms in the dog. The drug very rapidly destroys the proliferative zone. However, the wall of the brood capsule containing the protoscolices inside the echinococcus cyst is not penetrated by praziquantel (Thomas and Gönnert, 1977).

Plate XXIII ⇨

Nematodes (Roundworms)

Trichinella spiralis

Nematodes

The nematodes (roundworms) that are human parasites are mainly gut worms, but a few are blood and tissue parasites. Some species occur world-wide, while others are restricted to tropical areas, where in general they occur very commonly and are of considerable significance in health.

The nematodes are predominantly unsegmented worms with separate sexes. The digestive tract runs the length of the body as an almost straight tube. The mouth opening is sometimes surrounded by papillae or lips, and sometimes has a buccal capsule. Growth is mainly by cell enlargement, with the number of cells remaining

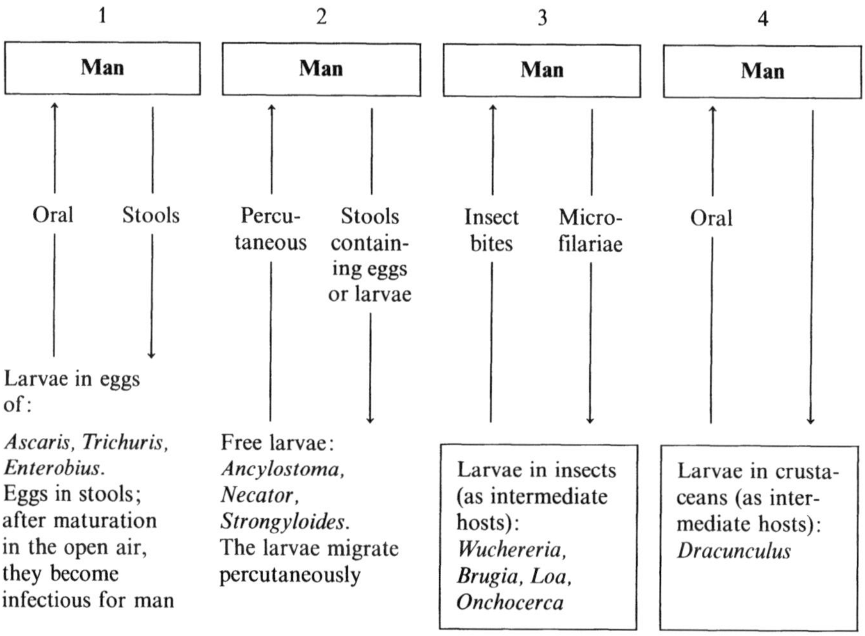

Comparison of the various modes of development of nematodes affecting man:

1 Without intermediate host but with host change (Plates XXIV, XXVIII)
2 Without intermediate host but with host change (Plates XXV, XXVI)
3 With intermediate host as active vector (Plates XXIX, XXX)
4 With intermediate host as passive vector (Plate XXXI)

approximately constant after the first larval stage. During growth, the nematodes shed their cuticles four times.
Nematodes mostly enter man directly (orally or percutaneously), but some do so via intermediate hosts (insects or crustaceans); they do not multiply in the intermediate host. With gut worms, the eggs pass out of the host with the faeces. Larvae develop within the eggs after a temperature-dependent latent period. The eggs then become infectious. In some species, the larvae mature in the open (e.g. hookworms) and enter man percutaneously. Others are released in the gastrointestinal tract only after the eggs have been ingested orally and begin to migrate (e.g. *Ascaris* species). Larvae of *Strongyloides* species are released in the human gut and are passed out with the faeces, then entering man again via the skin. With *Strongyloides* the parthenogenic parasitic generation can be followed by a free-living generation with separate sexes (generation change). The larvae can also develop directly in man into parthenogenic adult female worms and thus cause an infection to persist. *Trichinella* species and the filariae produce either viviparous or ovoviviparous larvae. *Trichinella* larvae taken in with food, invade the intestinal epithelium where they undergo four molts and become sexually mature. These adults produce a new generation of larvae, which become encapsulated in muscle cells of the same host.

Microfilariae are taken in by blood-sucking insects, in which they develop to the infectious stage (see pp. 270). In a new vertebrate host, they then become sexually mature and produce microfilariae (pp. 206, 279).

Plate XXIII

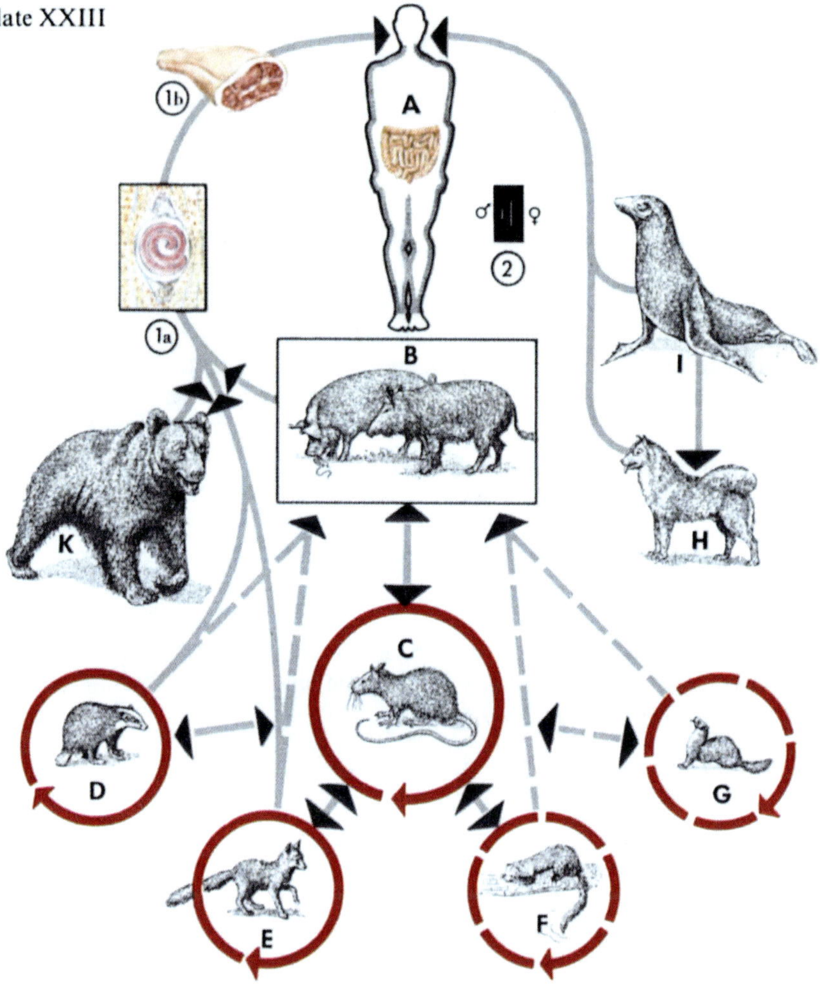

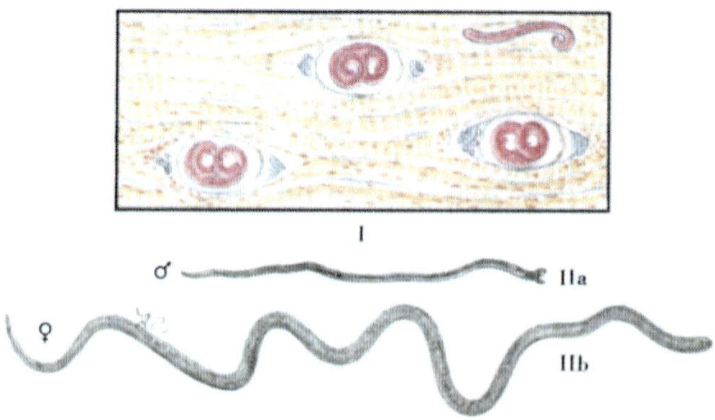

Trichinella spiralis (OWEN, 1835) RAILLIET, 1895

Etiologic agent of Trichinellosis (Trichinosis)

Epidemiology of trichinellosis

Ⓐ Man: Trichinella infection is produced by ingesting raw meat containing "muscle trichinae" (trichinella larvae (*1a, 1b, I*);
2 Males and females approximately life size

Ⓑ Domesticated and wild pigs: the main infection source for man

Infection sources for pigs

Ⓒ rat, Ⓓ badgers, Ⓔ fox, Ⓕ mink, and Ⓖ marten (animals reared for furs); these animals also become cross-infected by cannibalism

Other infection sources for man

Ⓓ badger, Ⓔ fox, and Ⓚ bear from wild areas; Ⓗ husky, and Ⓘ sea mammals in polar regions

I Muscle trichinellae

Microscopic preparation; muscle fibres containing larvae, some are encapsulated; Stain: borax-carmine with picric acid, approx. ×40

II Gut trichinellae

a males
b females, larvae emerging, approx. ×40

(See also Plate XXXIII, *g*)

Trichinella spiralis undergoes its entire development in the same host, i.e. larvae taken in by mouth by the host, for example with raw meat (muscle trichinae, Plate XXIII, *I*), develop in the small intestine to mature worms (gut trichinae, Plate XXIII, *II*). The females produce larvae in the same host (i.e. are viviparous), and these larvae become muscle trichinae. To continue their development, the larvae must enter a new host (host change, see routes of transmission, p. 211). This developmental cycle repeats each time the larvae reach a new host. As the trichinae at no time live outside the host, the distribution is independent of the climate. Consequently they are found throughout the world, from the poles to the equator.

The Arctic explorer ANDREE and his associates died on an expedition because, in an emergency, they consumed raw polar bear meat containing trichinae.

Trichinellosis is relatively common in man in the USA, in Canada, and in Eastern Europe. Australia remains free of the parasite, apart from patients who have acquired the infection outside the continent.

In addition to *T. spiralis*, it has recently been shown that there is at least one additional species, namely *T. pseudospiralis* GARKAVI, 1972. This is known in the North Caucasus as a distinct species, although the distinction is not without criticism (MADSEN, personal communication), partly because it is not clear whether or not this species infects man (PAWŁOWSKI and RUITENBERG, 1978). Different strains of *T. spiralis* have been identified from several geographic areas and different mammals, and it has been suggested that these be raised to the species level (*T. nativa, T. nelsoni* and *T. domestica*), but so far this has been rejected.

Morphology and Development. The mature trichinae (gut trichinae) are about 1.5 mm (males) and 3–4 mm (females) long and are thus amongst the smallest parasitic worms of man. The worm tapers towards its anterior end (Plate XXIII, *IIa, b*). The gut trichinae live in the mucosa of the small intestine (and sometimes intracellularly according to WRIGHT, 1979). A female produces about 1000–2000 larvae viviparously. Each larva is about 100 μm in length. These pass through the lymphatics and the thoracic duct to the bloodstream, where they can be detected from day 5 onwards (blood trichinae).
The larvae migrate through the heart and lungs and spread with the arterial blood into all parts of the body. They invade the skeletal muscles and actively penetrate the muscle fibres (from about day 6). Here they grow to a length of about 1 mm and then become surrounded by connective tissue capsules (trichina capsules, Plate XXIII, *1a, I*). From day 15, they are infectious for a new host. In this state, the larvae can remain viable for 20–30 years (possibly for the host's lifetime), but under some circumstances they die and become calcified. The larvae have a predilection for the muscles of the diaphragm, the intercostal muscles, the muscles of the larynx and tongue, and the extraocular muscles, although they also occur in the muscles of the extremities (muscle trichinae).

T. pseudospiralis does not produce capsules, in contrast to *T. spiralis*. The larvae migrate into the muscle fibres and persist there but are not strictly localized, since they perform rotary movements and tend to burrow and coil up. This species is found in small predators, rodents and birds. Many laboratory animals can be infected with *T. pseudospiralis,* including monkeys. Infection induces immunity so the gut worms are expelled more rapidly in any subsequent infection. So far, *T. pseudospiralis* infections have not been observed in man, although this should be possible (KOCIECKA et al., 1981).

Further development requires a change of host. Raw or inadequately cooked or preserved meat must be consumed by the new host, which can be of the same species. The capsules are digested in the stomach and duodenum and the larvae are freed. They enter the wall of the small intestine as far as the muscle, but some of them return to the gut lumen. In this migration they cause appreciable damage to the gut wall. Subsequently, the worms re-enter the mucosa and become mature after about 5–7 days. Immediately after mating, the females, which live only 25–30 days, produce the first larvae. At the same time, the young larvae can be detected in the blood.

Clinical Symptoms. The disease symptoms in trichinellosis are dependent on the number of trichina larvae ingested. Symptoms also vary with the incubation time, which is between 1 and 30 days. Other factors influencing the disease are the virulence of the strain of parasite and the age of the patient. Children, for example, are less severely affected than adults.

The symptoms in trichinellosis include both pains in the region of the small intestine plus gastroenteritis (gut trichinae) and muscle pains (muscle trichinae). Very often there is high fever, eosinophilia and also signs of severe toxicity and allergy, which may be associated with oedema (mainly of the eyelids). These signs occur from about the day 11 until 3–4 weeks after the onset of the attack. The length of the incubation period and the symptoms are generally related to the severity of the attack, but there is no relationship between the severity of the disease and the degree of eosinophilia. The latter reaches its peak after about 3–4 weeks, then falls slowly and persists at a low level for years. The pains can also persist for many years, since the muscle trichinae can survive for decades (electromyographic detection). Elevated levels of lactate dehydrogenase and other enzymes occur in the first few weeks following infection and this response is earlier than the peak antibody response. Consequently, biochemical tests are of considerable diagnostic value (BOCZON et al., 1980). According to FRÖSCHER et al. (1982), psychomotor epilepsy may also be a symptom of trichinosis. This feature can contribute significantly to the making of a firm diagnosis when reviewing sequelae of the disease (KASSUR et al., 1978). The muscle cells are probably the major source of the elevated serum enzyme activity in trichinellosis, although not the only one. This is due to increased permeability of the muscle cell membranes. After a period of greater permeability, enzyme synthesis in the muscle cells increases (GENTILINI et al., 1976; POZNAŃSKA et al., 1981). With very heavy infection, the patient may die as a result of myocarditis, lung complications, and complete collapse caused by massive toxaemia (after 4–5 weeks).
It should also be mentioned here that immunity develops following a primary infection. This leads in mice to the elimination of the gut worms within 24 hrs in any subsequent infection. There is an increase in mast cells in the small intestine which can be inhibited by prior immunosuppressive treatment, but can also be restored by immune-cell transfer (CRUM et al., 1977; DESPOMMIER et al., 1977; ALIZADEH and WAKELIN, 1981).
The clinical symptoms in *T. pseudospiralis* infection are very similar to those with *T. spiralis* according to observations on experimentally infected monkeys. However, in general symptoms are milder (TEPPEMA et al., 1981).

Transmission. The natural trichina reservoir is provided primarily by rats, foxes, badgers, and wild pigs. Consequently, domesticated pigs Ⓑ can be infected by various routes and then become the principal source of infection for man Ⓐ. *T. spiralis* is extraordinarily non-specific in its choice of host, and develops in almost all mammals. It is found, for example, even in the polar regions, where seals Ⓘ are frequently carriers of trichina and therefore Eskimos and huskies Ⓗ are infected (see also MADSEN, 1974).

The unique mode of transmission lies in the consumption of uncooked or insufficiently cooked (heated to less than 60 °C) muscle from animals containing trichina. For man, the main sources are raw pork eaten as mince and raw sausage. In the wild, bears Ⓚ, badgers Ⓓ, foxes Ⓔ, and possibly also dogs Ⓗ may be infected. Birds on the other hand are not infected by trichina.

As the meat from slaughtered animals is mostly consumed simultaneously by a fair number of people, there may be outbreaks of mass-trichinosis in which many people may be affected. If cases of trichinosis occur the source of infection must be identified at once (the outbreak may stem, for example, from uncontrolled distribution of pork, game, etc.) in order to eliminate it (PAWŁOWSKI, 1981).

A less commonly identified source of infection in the wild is the decomposition of infected carcasses. This way releases encapsulated trichinae, which by virtue of their resistant capsules, can be distributed (much like eggs). This may also be the source of infection for horses. According to ANCELLE et al. (1986), outbreaks occurred in Italy and France in 1985 (409 and 500–900 patients respectively) due to the consumption of raw horse meat which had been imported from the USA or the Federal Republic of Germany. Latent infection sources occur also in fur farms, where the stripped carcasses are fed to other animals (transmission of trichinosis by "human hand" Ⓕ, Ⓖ).

The larvae are killed by freezing the infected or suspect meat to –15 °C for at least 20 days, or to –35 °C (taking care to ensure the meat is frozen throughout) for 24 hrs.

Diagnosis by Microscopy. Trichinella larvae can be diagnosed following the digestion of a biopsy sample or can be detected by direct examination of the blood within the first 3–4 weeks (blood trichinae), or with more certainty in the muscles during the chronic rheumatic phase by biopsy. If the clinical picture suggests infection but the biopsy is negative, there may often be subacute interstitial myositis, which is a typical pathohistological finding in trichinellosis. As is done in official meat inspection, to examine muscle small samples of the fresh tissue are compressed between two glass slides until translucent. The specimen is examined under a miscroscope at medium magnification. The larvae can also be detected in venous blood using the following method: 10 ml blood is mixed with 200 ml 3% acetic acid followed by centrifugation. The larvae are found in the sediment. Adult worms and larvae can also sometimes be found in the stools. With dead animals it is possible to detect the trichina larvae by artificial digestion of larger muscle specimens.

In official meat inspection, hazelnut size pieces are taken from both ends of the diaphragm (so-called kidney-shaped parts), or else, if the second side is lacking, two equal samples from one side, or from the rib or chest section of the diaphragm or else from abdominal muscle. Each sample is used with seven barley-kernel-sized pieces of meat for trichina compression preparations (14 pieces in total) (perhaps after 10–20 min in potassium hydroxide solution). The specimens are examined under a miscroscope at a magnification of ×30–40. Meat inspection has greatly reduced the frequency of infection.

Diagnosis by Immunobiological Methods. In the natural course of the life cycle the parasite never enters the outside world and microscopic examination is not possible, apart from during the blood phase and by muscle biopsy. Consequently immunobiological detection is very important. It is, however, difficult. Larvae can be detected in the blood or cerebrospinal fluid at the start of the infection. The most sensitive and reliable test is indirect immunofluorescence (titre up to about 1:2000), together with indirect haemagglutination (titre up to about 1:1280). There is no correlation between the clinical course of the disease and the level of the positive titre. Other methods in use include the complement fixation test and microprecipitation using live larvae, or better still deep-frozen preserved ones (Constantinescu and Căprarŭ, 1980; positive from 2–3 weeks following infection for up to 45–65 days). The latex test is simple and rapid. This is based on agglutination of latex particles sensitized with trichinella antigen. The latex suspension is mixed with an equal volume of test serum. Agglutination occurs if the serum contains specific antibody. The test is about 90% reliable and is suitable for a preliminary examination instead of the skin test previously used. ELISA has also been found to have high sensitivity, having given positive results in known cases. However, not every infection can be identified serologically. On the other hand, results are independent of the duration of the infection. Positive reactions can occur even after 30 years. Cross-reactions with filariae are to be expected (Ruitenberg et al., 1975, 1978a, b). It is hoped that ELISA will prove highly specific for the detection of circulating antigen.

Chemotherapy. The clinical course of trichinosis can be influenced by chemotherapy if the treatment is given at the gut stage. The difficulty in initiating prompt treatment is that the correct diagnosis is mostly made after the gut attack, i.e. when the larvae are present in the muscles. Treatment is, therefore, mostly too late for gut trichinosis. Thiabendazole is recommended in a dosage of 25–50 mg/kg on 2–4 successive days, but not more than 3 g daily. Mebendazole is much better tolerated and has given equally good results (100–300 mg twice a day or 20–40 mg/kg body weight (≅1200–2400 mg) daily over 2–3 weeks) especially when given before the occurrence of clinical symptoms, or as a prophylactic (Klein et al., 1978). This provides extensive protection against muscle invasion within the first 2–3 weeks. The muscle trichinae can be attacked by the administration of 100 mg Mebendazole daily for 6–10 days (Mittermayer and Spaldonova, 1981). Alternatively, symptomatic treatment may be given, e.g. with steroids. Treatment can produce a sudden reduction in the eosinophil count. Taylor and Pearson (1981) refer to the use of the drug ivermectin, which appears to offer a new mode of chemotherapy for nematodes (see also Schulz-Key et al., 1984; and p. 284).

Plate XXIV ⇨

Enterobius vermicularis
Trichuris trichiura

Plate XXIV

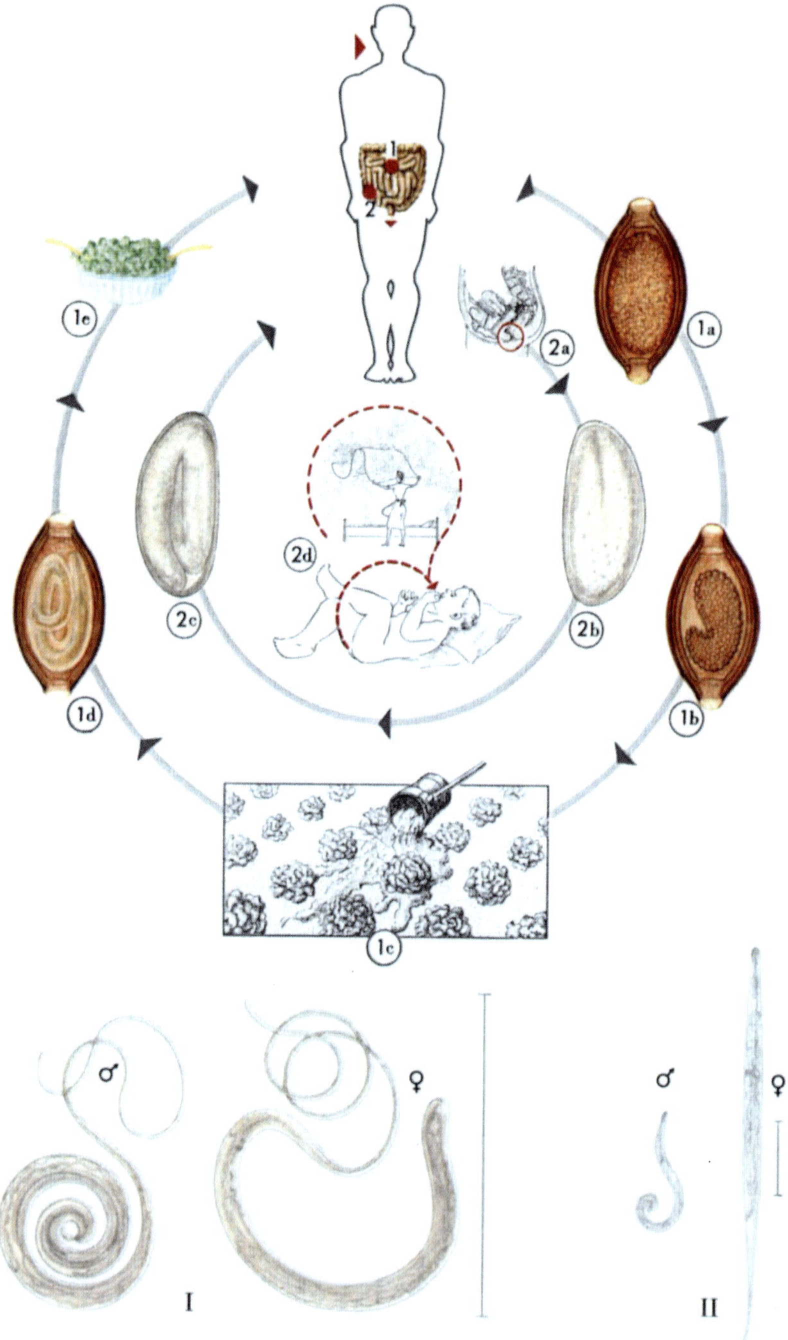

Enterobius vermicularis (LINNÉ, 1758) LEACH, 1853

Pinworm

Trichuris trichiura (LINNÉ, 1771) STILES, 1901

Whipworm

Man is the only host for *Enterobius vermicularis* (pinworm) and *Trichuris trichiura* (whipworm)

1 Occurrence of worms in the lower small intestine, appendix and large intestine

- *a* Egg from whipworm in freshly excreted stools
- *b* Onset of embryonic development
- *c* The spreading of human faeces (used as fertilizer) containing material from worm carriers causes eggs to contaminate salad plants
- *d* Infectious egg, containing larva
- *e* Intake of the very resistant infectious eggs can occur from the consumption of fresh salad vegetables (see *1*)

2
- *a* Pinworm eggs deposited externally around the anus
- *b* Freshly deposited pinworm egg
- *c* Infectious egg about 6 hrs old containing larval stage
- *d* Routes of infection: in children, disturbance produced mainly by irritation around the anus leads to scratching and causes eggs to be transferred from the anus to the mouth during sleep. In adults, transmission is mainly by eggs in dust from bedding, which is disturbed and enters the mouth

I Mature male and female whipworms

II Mature male and female pinworms

The lines show the natural length of the worms.

(See also Plates XXXII, *l*, *m* and XXXIII, *e*, *f*)

The pinworm, *Enterobius vermicularis* (also known under the name *Oxyuris*) is distributed world-wide and is one of the commonest worm parasites of man, particularly amongst school-children. The reason for the parasite's frequent occurrence is that the eggs contain potentially invasive larvae within a few hours of being laid. Consequently, carriers can not only reinfect themselves but also infect their associates. Infections frequently persist for years. *E. vermicularis* develops only in man.

Morphology and Development. The mature females are 10–12 mm long, and each has a long pointed tail ("pinworm", Plate XXIV, *11*). The much smaller males (2–6 mm long) die immediately after mating and are thus scarcely seen. The mature females migrate out of the anus in the evening and at night to lay their eggs on the perianal skin. Within a few minutes, they can lay 5000–10,000 eggs. The eggs are almost colourless, ovoid and flattened on one side (about 55–25 μm; Plate XXIV, *2a*). Within 5–6 h, infective larvae develop within the eggs (Plate XXIV, *2c*). After oral ingestion, the larvae enter the small intestine and become mature worms in the upper part of the large intestine and in the caecum.

Clinical Symptoms. Direct gut injury from the pinworm has not been observed, but there can be general symptoms, especially in children. The worms are found in the lower part of the small intestine, in the caecum, in the upper large intestine, mainly in the region of the ileocaecal valve, and sometimes in the appendix. In *E. vermicularis* infection, the nocturnal migration of the female worms to the anus for laying eggs frequently leads to severe irritation there. This causes sleeplessness and skin irritation due to scratching and bleeding. Schoolchildren may often become restless and inattentive. Such symptoms should lead to appropriate examination for pinworms. In female patients, the worms sometimes also migrate to the genital tract (vulvovaginitis), from where they can gain access to the body cavity. In this manner, they may rarely cause salpingitis and even peritonitis.

Transmission. Eggs are transmitted to man by: (1) auto-infection, e.g. in children who scratch because of anal irritation, infect their fingers with eggs, and then suck their fingers during sleep, (2) by direct contact between people touching hands, particularly schoolchildren, and (3) by intake of eggs in dust, e.g. during respiration. Sometimes entire families can be infected by the latter route with the same applying to the dust in classrooms, where schoolchildren are permanently at risk. The combination of these various routes of transmission may lead to obstinate worm infection. According to Schüffner and Swellengrebel (1949), in adults there can also be retrograde migration (retrofection) of the larvae into the gut. Pinworms may also play a part in the transfer of *Dientamoeba fragilis* (see pp. 55, 57; Burrows and Swerdlow, 1956; Ockert, 1972).

Prophylaxis. Of particular importance in prophylaxis is hand washing (with soap and nailbrush) after passing faeces and before eating. Bedding and clothes should be changed frequently and boiled on a daily basis for 8 days after the start of the treatment. Floors should be cleaned with a vacuum cleaner every day to prevent the accumulation of dust (see the recommendations in the section on chemotherapy).

Diagnosis by Microscopy. Pinworm eggs are generally best detected, not by examining stool, but by means of an anal smear, which can be obtained in a simple fashion with strips of clear adhesive tape. The adhesive strip is pressed onto the anal region first thing in the morning and then examined at medium magnification, preferably after treatment with a drop of toluene. In the USA an anal swab is recommended, by which the eggs are taken up with cellophane from the anal skin. The cellophane is then examined in a drop of water, toluene, or dilute caustic soda for eggs. Quite often the stool is found to contain isolated or even numerous adult worms, which emerge spontaneously or after effective treatment (see p. 310).

Diagnosis by Immunobiological Methods. No specific antibodies of diagnostic value have been demonstrated in pinworm infection.

Chemotherapy. A pinworm infection must be treated by chemotherapy, since there is a considerable danger of reinfection (see above), in addition to the possibility of frequent recurrences in particularly sensitive people. It must be remembered that the development time from ingesting an egg to the production of mature egg-laying worms is at least 35 days. The indicator of therapeutic success in pinworm infection is the absence of eggs in the anal region for at least 5 weeks after the treatment. Any pinworm attack occurring after that represents a new infection. The most reliable drugs for dealing with pinworms are pyrvinium embonate (50 mg/10 kg; note that it colours the stool a bright red, which may be confused with blood); pyrantel embonate (10 mg/kg), and mebendazole (100–200 mg)[1], which are effective in single doses. Repeated microscopic checks are necessary, and the treatment must be repeated as required (mebendazole has also been found active against *Ascaris, Trichuris,* and hookworms, as well as *Enterobius,* see pp. 221, 229, and 265).

Symptomless pinworm carriers represent a permanent source of infection, particularly for family members. It is therefore recommended that all members of a group living together should be examined and treated simultaneously as required.

[1] Mebendazole; should not be given during pregnancy and to children under 2 years of age.

Trichuris trichiura

The whipworm *Trichuris trichiura* develops directly in the host, without migration or an intermediate host. It occurs mainly in the large intestine and the caecum, and only results in substantial pain if there is a heavy infection. Although it is found throughout the world, it is more common in hot countries than in temperate zones. Current estimates indicate that about 500 million people are carriers. In some tropical areas the infection rate is over 90%, while in many it is 30%–60%.

Morphology and Development. The mature worm is about 3–5 cm long and owes its name to its whip-like shape. For a long time the thread-like anterior part was erroneously considered a tail, and this gave rise to the name *Trichuris,* meaning thread tail. The posterior part is mainly filled by the sex organs and is quite distinct from the slender front part. The eggs (about 50 × 25 µm) have a very typical appearance on account of the two pale polar prominences and the pale to dark brown colour. They are unlikely to be confused with the eggs of other nematode genera that affect man (see Plates XXIV and XXXII, *m*).
Larval development in the eggs takes about 4–6 months at 15 °C, about 3–4 weeks at 26 °C, and about 11 days at 35 °C. The eggs can survive for years. Following oral ingestion of the eggs, the larvae are released and pass into the small intestine, but they then leave it and end up in the large intestine. There they become sexually mature within 2–3 months. The slender front part penetrates into the mucosa. The worms live for about 1.5 years. The prepatent period lasts about 90 days.

Clinical Symptoms. In general, there are no symptoms associated with mild whipworm infection, apart from eosinophilia. If numerous worms are present, there may be loose stools, anaemia, weakness, weight loss and abdominal pain. These features are not, however, specific for *Trichuris* and are rarely serious. There is inflammatory change and necrosis around the worms and a pronounced cellular infiltration. Additional symptoms include anal eczema, pruritis and urticaria. Rectal prolapse is a complication which arises from severe colitis and proctitis, particularly in undernourished children (Cooper and Bundy, 1988).

Comparative studies of schoolchildren with and without heavy *T. trichiura* infections, have clearly shown that numerous worms can lead to a state of impaired nutrition. Bacterial and protozoan infections are associated with a high worm burden, as well as a higher incidence of invasive amoebiasis. All these features improve after specific treatment (Gilman et al., 1983).

Transmission. This is mainly by the consumption of raw vegetables that have not been properly cleaned (e.g. lettuce or carrots), particularly those that have been contaminated with the faeces of worm carriers. Children are generally infected more frequently than adults. They do not represent any direct hazard to their companions, because the eggs are not immediately infective upon being laid (see above). Precautionary measures include removing the eggs by proper cleaning of raw vegetables and killing the eggs by processing the faeces (e.g. by heat treatment). Whipworms have also been found in pigs and monkeys, but it has not yet been confirmed whether these are identical to the one found in man. *T. trichiura* eggs have been transmitted from macaque monkeys to volunteers with a prepatent period of 127 days (Horii and Usui, 1985).

Diagnosis by Microscopy. Whipworm infection can be detected relatively easily by examination of the stools for the typical eggs. Concentration techniques are to be preferred to direct stool examination, since the number of eggs is often relatively low (see p. 308).

Chemotherapy. The treatment of trichuriasis was difficult for a long period, because the location of the worms in the gut mucosa restricted the activity of antihelminthic drugs. A reliable agent is mebendazole in a dose of 100 mg twice a day for 3–4 days (children up to 2 years should take half the dose for 3–4 days). Pyrvinium embonate is also effective in a dose of 5–10 mg/kg given on a single day or 0.5–1.0 mg/kg over 6–7 days, with a reliability of about 92%. In obstinate cases, it is recommended that thiabendazole should be added (2 × 25 mg/kg daily for 2–7 days). Thiabendazole acts in the eggs of the whipworm, but not on the worm itself.

Plate XXV ⇨

Ancylostoma duodenale
Necator americanus
Trichostrongylus species
Oesophagostomum species

Plate XXV

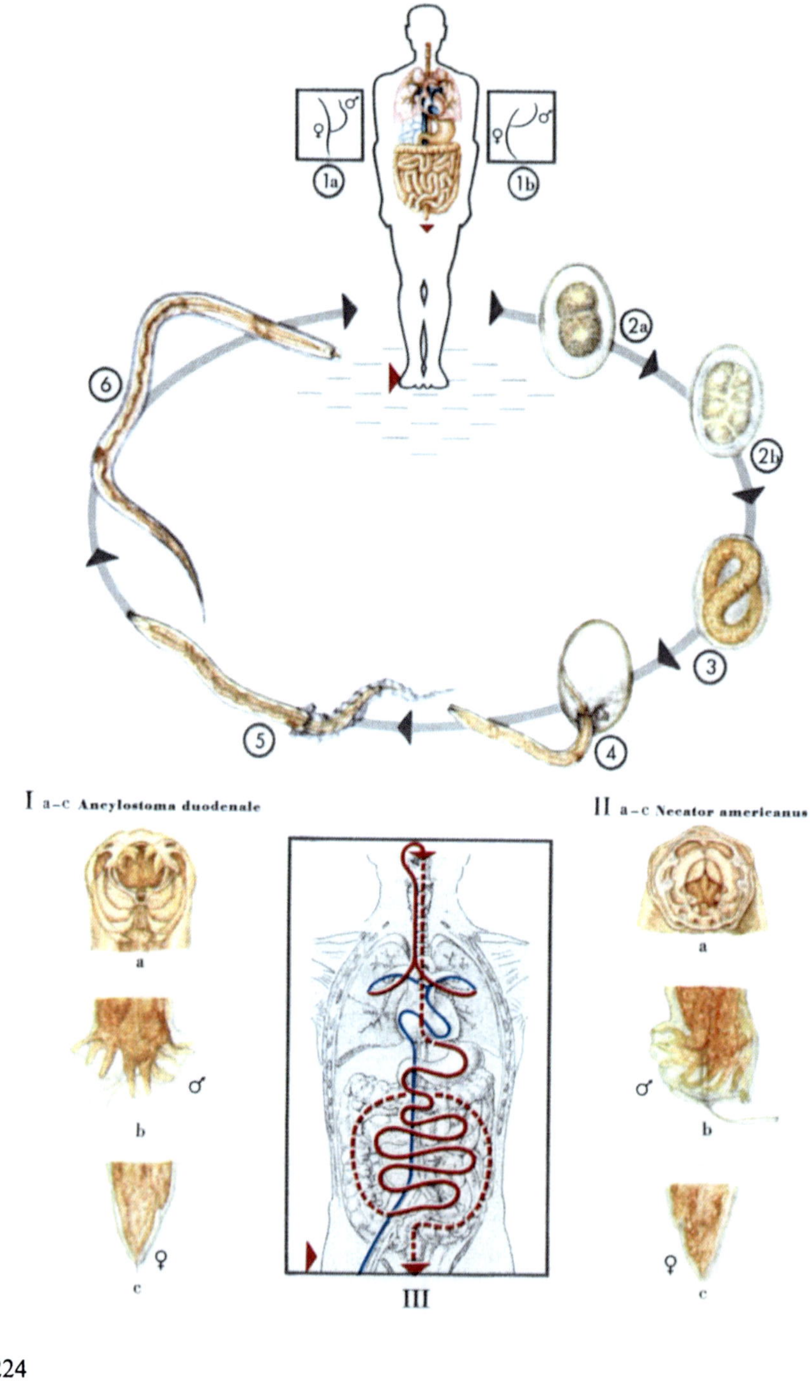
1a
1b
2a
2b
3
4
5
6
I a–c Ancylostoma duodenale
II a–c Necator americanus
a
b
c
III

Ancylostoma duodenale (DUBINI, 1843) CREPLIN, 1845
Necator americanus (STILES, 1902) STILES, 1903

Hookworms

1 a	*Ancylostoma duodenale*	in characteristic copulation position
b	*Necator americanus*	(about natural size)

Development outside the body

- *2 a* Freshly laid egg, two-cell stage
- *b* Egg in six-cell stage
- *3* Egg containing larva
- *4* Emerging rhabditiform larva
- *5* First moult
- *6* Filariform larva after second moult; the larva retains the old cuticle (sheathed larva).

Some characteristic differences between species (see table on p. 227)

I *Ancylostoma duodenale* *II* *Necator americanus*

- *a* Buccal capsule with mouth
- *b* Caudal bursa in male
- *c* Tail in female

III Migration route of hookworm larvae

blue After active percutaneous entry of the filariform larvae into man, the larvae migrate via the veins to the heart, and from there to the lungs, where they enter the vessels and penetrate the alveoli

red From the alveoli, the young worms ascend the bronchioli, bronchi, and trachea and thence to the pharynx. Here they are swallowed and pass via the stomach into the small intestine, which is the location of the sexually mature hookworm

(See also Plates XXXII, *n* and XXXIII, *d*)

The name of hookworm is given to two species parasitic in man which are frequent in the tropics and which are partly restricted to separate geographical areas: *Ancylostoma duodenale,* the so-called Old World species, and *Necator americanus,* the so-called New World species. The latter has been transferred from Africa to America. The main distribution area for hookworms lies roughly between 30 °S and 40 °N, with *N. americanus* mainly in Central- and northern South America, Equatorial Africa, South and Southeast Asia, Polynesia and Australia, and *A. duodenale* mainly in North Africa and in North and South Asia.
Carriers of the worm have also brought the parasites to more northerly areas with suitable climatic conditions, as for example in mines and in construction sites for major mountain tunnels and so on, whence the name tunnel disease or miners' disease. A third species, *A. ceylanicum,* which affects cats and dogs as well as man, occurs only in Taiwan, Southeast Asia and Surinam. These worms resemble certain related species in the dog and cat in that they usually enter the body through the skin, but the hookworms of domestic animals do not become sexually mature in man (see p. 245). Recent estimates indicate that the proportion of hookworm carriers is about 20%–25% of the Earth's population, i.e., more than 1 billion people (e.g. 60% in the Dominican Republic, 25% in Puerto Rico and Mexico). About 60,000 people die each year from hookworm disease.
In heavy or long-lasting infections considerable physical stress is placed on the individual concerned because of the hypochromic, microcytic, iron-deficiency anaemia that develops. The social and economic significance of this parasite is therefore considerable.

Morphology and Development. *Ancylostoma duodenale* (Plate XXV, *1a*; female about 10–12 mm long) has two pairs of two fused curved teeth in a rigid mouth capsule (Plate XXV, *1a*), while the somewhat smaller species *Necator americanus* has two sickle-shaped cutting plates (Plate XXV, *IIa*). The hookworm uses these

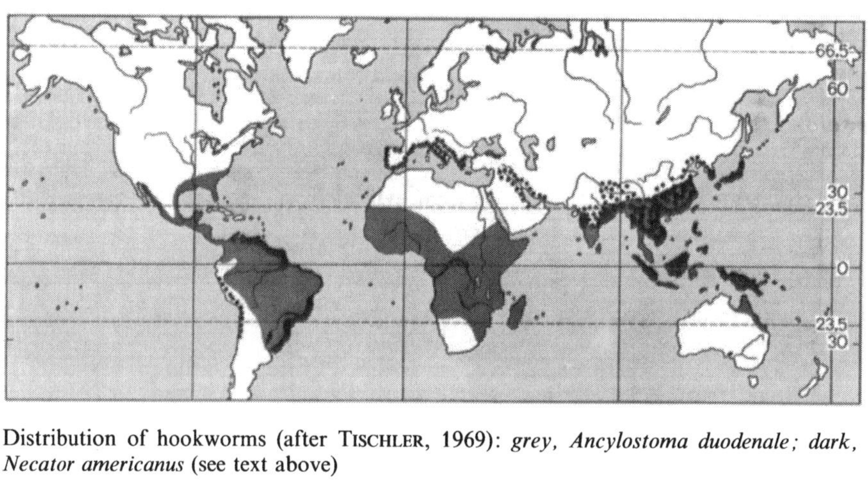

Distribution of hookworms (after Tischler, 1969): *grey, Ancylostoma duodenale; dark, Necator americanus* (see text above)

to attach itself to the mucosa of the small intestine and this damages blood vessels. Freshly expelled worms are frequently reddish in colour due to the ingestion of pieces of mucosa, mainly from the jejunum, and blood. The body of the female ends in a sharp point (Plate XXV, *Ic, IIc*). The male has a bursa copulatrix at the end of the body equipped with a pair of spicules (Plate XXV, *Ib, IIb*). The female is encompassed by the bursa during mating at the level of her sexual orifice. The mating between pairs gives rise to the characteristic copulation pictures for each hookworm species, which aids in identification (Plate XXV, *1a, b*).

Characteristics of different species (see Plate XXV, *I, II*)

	Ancylostoma duodenale	*Necator americanus*
Buccal capsule with mouth	2 pairs of two fused curved teeth	2 crescent-shaped cutting plates
Spicules of caudal bursa	Divergent, ending in a point	Lying close together, and ends barbed
Tail in female	With sharp spine	Without spine

The hookworm develops without entering an intermediate host. The eggs are laid (Plate XXV, *2;* above 60 × 40 μm) mainly in the two-cell to 16-cell stages. They appear glass-clear to pale yellowish with thin outer capsules. Consequently, the eggs can be seen under the microscope only if the light source is properly stopped down. They withstand external temperatures between 10° and 45 °C. If the environmental conditions are favourable, i.e., availability of oxygen, appropriate humidity, and a minimum temperature of 18 °C, the eggs develop into larvae which emerge after 1–2 days at an optimal temperature of 28°–30 °C (Plate XXV, *3, 4*). Within about 5–6 days, the rhabditiform larvae moult twice and then become the sheathed filariform larvae retaining the second cuticle (Plate XXV, *5, 6*). These alone are capable of penetrating man (*A. duodenale* can also be acquired by mouth). The larvae migrate (Plate XXV, *III*) via the veins to the right side of the heart, from which they pass to the vessels of the lung (heart-lung passage). They pass out through the vessel walls and enter the alveoli (pneumonia; see *Ascaris,* p. 263). While *N. americanus* mainly stays in the lungs, *A. duodenale* immediately enters the air passages. The larvae are then carried by the air flow outwards via the bronchioli and bronchi and migrate via the trachea, epiglottis, and oesophagus into the gastrointestinal tract. They reach the small intestine about 3–5 days after percutaneous penetration. After shedding the filariform cuticula and after a further 4–5 weeks, the worms become mature and begin to lay eggs (prepatent period 5–6 weeks). *A. duodenale* lays about 30,000 eggs daily, *N. americanus* about 9000. The lifespan can be as much as 20 years, but seldom exceeds 2 years for *A. duodenale,* although *N. americanus* often survives for many years.

Clinical Symptoms. Symptoms can be produced directly by the entering larvae, especially on repeated infection, in the form of dermatitis (ground itch). There is marked irritation, oedema, and erythema, mostly as a consequence of sensitization, followed later by papulovesicular eruptions, which persist for about 2 weeks.

While a few adult hookworms do not produce notable injury (a maximum of 100 for *N. americanus* or about 20 for *A. duodenale*), a massive infection leads to severe iron-deficiency anaemia and hypoproteinaemia. There may also be lack of working capacity, fatigue, shortness of breath, swelling in the legs, anorexia and impotence. (About 0.02 ml blood is lost per worm per day with *N. americanus* and about 0.1 ml with *A. duodenale.*) The degree of anaemia is dependent on the iron content of the diet, the iron reserve, and the extent and duration of the hookworm infection. These factors vary substantially in endemic areas. Children are particularly at risk from the chronic protein loss. They become very retarded in development and maturation, and are particularly liable to die from the disease. Another characteristic feature is the generally high level of eosinophilia in the white blood picture (33%–40%, or even higher). A sudden fall in eosinophils is a sinister sign, which is often followed by a fatal outcome due to circulatory collapse. Coloured Africans acquire a grey or ashen skin because of the anaemia, with the connective tissue appearing pale or yellow, or even of an ivory colour in extreme cases. The anaemia is due to blood loss caused directly by the worms, which ingest blood, and also due to their changing positions in the small intestine. This leads to severe bleeding from the points of attachment (MILLER, 1979). Inflammatory symptoms also occur in the skin when dog or cat hookworms enter the body, but they do not mature and migrate in a random fashion under the epidermis producing injurious burrows (creeping eruption, see p. 245). Hookworms acquired during mining is recognized as an occupational disease (see also *Strongyloides,* p. 235).

Epidemiology. Hookworm eggs require a minimum temperature of about 18 °C with suitable humidity for their development. This is the main reason why these worms are restricted to hot countries. The eggs die when the temperature falls below 10 °C. The larvae do not remain permanently in the soil but can also migrate upwards, for example on moist plants (up to 1 m). Over the years, the number of worm carriers has decreased in certain areas (for example, in Brazil, from 77% to 28.8% over 45 years), but the total number of those affected has remained constant on account of population growth.

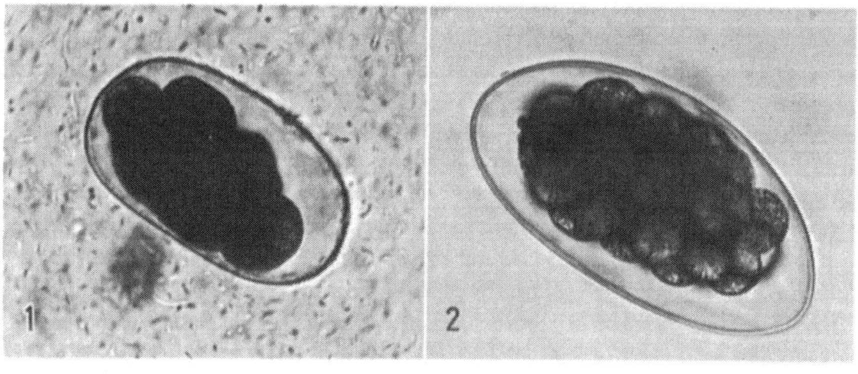

1 Egg of *Ancylostoma duodenale* or *Necator americanus*. *2* Egg of *Trichostrongylus species.* Magnifications approx. ×600

Hookworm infection is a major health problem in the affected countries, since it can lead to severe anaemia and thus to inability to work. Hookworm infection is more common in rural areas than in towns, because in many developing countries the fields and streets are contaminated by excrement (a disease of rural workers caused by having bare feet).

Prophylaxis and Control. Three aspects have to be considered here: improving public health, building latrines, and chemotherapy (including treatment of anaemia). Protection from hookworm attack rests primarily on the use of hard shoes and general health education! Care must also be taken that moist sandy soils are not contaminated by the faeces of hookworm carriers. Pets and other domestic animals (cats and dogs) act as hosts e.g. for *A. ceylanicum* (however, see also p. 245).

Diagnosis by Microscopy. Hookworm infection can be diagnosed by microscopic examination of stools. The eggs are usually quite numerous (about 10,000–20,000 daily) and can frequently be identified without special concentration techniques. In microscopic examination it must be remembered, however, that the almost colourless eggs with their thin egg shells can be overlooked (for culture method see p. 238, 240/241). As there is a danger of infection (the larvae may have already emerged if the air temperature is high), the same care should be exercised as in handling *Strongyloides* (see p. 238).
A particular note should be made of the similar appearance of the eggs of *Trichostrongylus* species (about 75 μm with 32 embryonic cells as against 60 μm with 6–8 embryonic cells in hookworm eggs; see figure on p. 228) and of *Ternidens deminutus* (see also faecal culture for *Strongyloides*, p. 238). These species can occur simultaneously in certain areas.

The species *Ternidens deminutus* (Railliet and Henry, 1909) is a rare parasite of man (occurring, for example, in East Africa, Zaire, Zimbabwe and South Africa), but it is notable that the eggs laid by this worm can readily be confused with hookworm eggs. The distinction can be made by calculating the egg volume. Eggs with volumes over 170,000 μm^3 belong to *T. deminutus*, while those with volumes under 150,000 μm^3 are hookworm eggs (see Goldsmid, 1968).

Chemotherapy. The disease symptoms can often be ameliorated before the worms are eliminated simply by giving protein and iron (e.g. by the administration of iron sulphate, 200 mg two or three times daily, by mouth, for 2–3 months), or blood transfusion if necessary. Hookworm iron deficiency anaemia is basically different from the pernicious anaemia such as occurs in broad fish tapeworm infection (see p. 178). Effective agents are mebendazole (two doses of 50–100 mg daily, for 3 days), bephenium (one dose 5 g for adults, 2–4 g for children) and pyrantel embonate (10 mg/kg for 2–3 days). The treatment depends on the hookworm species present in the patient. *Ancylostoma duodenale* and *Necator americanus* show somewhat different responses to the same drugs. For example, bephenium has a good effect on *A. duodenale*, but is less effective against *N. americanus*. *A. duodenale* can also be eliminated with a single dose of albendazole (400 mg).

Trichostrongylus orientalis Jimbo, 1914
T. colubriformis Giles, 1892
T. axei Cobbold, 1879

The representatives of the nematode genus *Trichostrongylus* are predominantly stomach parasites of ruminants. Amongst the numerous species frequently found in warm countries and in man are *T. orientalis, T. colubriformis* and *T. axei,* common for example in the Near East (Egypt, Iran), Africa and Asia. Locality, distribution, and frequency vary considerably and depend on the prevailing climatic conditions in each case (frequency of infection is about 10% in Japan, approximately 80% in Korea; worms are often found in humans where livestock are kept in the Federal Republic of Germany). These worms are of relatively little pathogenic significance for man. However, a heavy infection of the intestine can cause severe general symptoms.

Morphology and Development. The adult worms (about 4.5–9 mm) develop in man without the lung migration typical of, for example, *Ascaris lumbricoides.* The third larval stages, which are capable of invasion when ingested orally, establish themselves directly in the small intestinal wall (duodenum and jejunum) and grow there to sexually mature worms. They survive for 5–8 years. In ruminants they live mainly in the stomach. The females lay about 100 embryonated eggs (approx. 75–90 μm) with 32 embryonal cells. These are present in the human stool about 25 days after invasion (prepatency). Under favourable conditions, i.e. with sufficient ground moisture and warmth, the larvae develop within 24 hrs, their tail ends showing the typical shape for this species (p. 240/241). Further development takes place in the fresh air, with two moults each of which requires a period of about 60 hrs. These "sheathed" third-stage larvae, which are capable of invasion, are still encased in the last larval cuticle. They are characterized by having a great capacity for resistance, withstanding, for example, dry heat for many days. The developmental pathway principally resembles that of *Trichuris trichiura* (see Plate XXIV). The infectious larvae mainly enter the next host by the oral route, but they can also actively penetrate human skin (see below). Then, according to Watson (1960), they migrate through the lungs, as with the hookworms. The relatively long lifespan and the possible heavy worm infections suggest that immunity does not develop.

Clinical Symptoms. These are usually relatively mild even with a severe infection. However, the worms, situated deep in the small intestinal epithelium, occasionally ingest blood. Inflammation and desquamation of the affected intestinal wall (enteritis) and mild secondary anaemia are the consequences. It is thought that the parasites also produce a toxin, but this only causes extensive damage to the intestinal wall and manifestations of general toxicity (abdominal pain, malaise and other non-specific symptoms) when the parasites are very numerous. A moderately elevated eosinophilia (10%–15%) is frequently present.

Epidemiology. Transmission to man usually takes place via the oral route with contaminated food, mainly by the consumption of raw green salads and uncooked vegetables which have been fertilized with faeces from worm carriers (man; for *T. orientalis*, cow). The infectious larvae adhere to the plants and can thus be eaten by man. They are evidently also able to actively penetrate human skin. Under some conditions the larva-containing eggs can survive dryness and cold and even periods of frost for a long time so they may over-winter in the open air.

Prophylaxis. It is recommended that in affected countries, with unhygienic living conditions, the consumption of raw vegetables be avoided. The gasing of pastures using e.g. methylbromide with severely infected soil, is possible. This kills the larvae to a depth of about 20 cm.

Diagnosis by Microscopy. *Trichostrongylus* infection can be detected by examining fresh stool samples for eggs. Faeces which have been kept for approximately 24 hrs before investigation may already contain hatched larvae. An accurate knowledge of the morphology of *Trichostrongylus* eggs and larval stages is therefore very important, because these are morphologically very similar to the eggs and larvae of hookworms and the larvae of *Strongyloides stercoralis*, and may easily be confused (see p. 240/241). The plump, oval, colourless eggs of *Ancylostoma duodenale* and *Necator americanus* measure approximately 60 µm. The equally colourless eggs of the *Trichostrongylus* species, with a size of approx. 75–90 × 40–43 µm, appear more slender, and on leaving the host already contain an advanced morula stage (see figure on p. 228). Faecal culture methods can also be used for detection (see p. 238).

Chemotherapy. *Trichostrongylus* infection can be eliminated in man by the use of bephenium (1 × 5 g for adults, 1 × 2.5 g for children) and thiabendazole (1 × 50 mg/kg body weight for adults, 1 × 30 mg/kg body weight for children, in each case divided into 2–3 doses). Pyrantelembonate is active in a single dose of 10 mg/kg body weight.

Oesophagostomum species

Oesophagostomum species are primarily parasites of ruminants and pigs, but they can also affect man. The infection is initiated by the sheathed third-stage larvae (see figure on p. 240/241), which, when taken orally, pass through the stomach and intestine, lose their sheaths and attack the walls of the caecum and large intestine. A granuloma forms, inside which the larva moults ("nodular worm"). It then leaves the nodule (2–10 mm in diameter) establishes itself in the mucosa, and develops into an adult worm. The sizes vary greatly between species, from 8.5 to 20 mm in the female and from 8 to 17 mm in the male.

The eggs measure, depending on the species, 60–120 µm × 30–50 µm; they are very similar to hookworm eggs and possess a thin colourless shell with a smooth surface. The number of blastomeres in eggs from fresh stool samples lies between 8 and 32 (in hookworms between 2 and 8). The prepatent period is between 5 and 8 weeks. The excretion of eggs (patency) can last for over 1 year.

Clinical Symptoms. The main clinical symptom is dysentery. The larvae can also perforate into the peritoneal cavity and induce peritonitis. The exact symptomatology of *Oesophagostomum* infection is not known, because the number of reported clinical cases is very small.

Prophylaxis. This consists of the thorough cleaning of raw vegetables before consumption (carrots, salad, vegetables, etc.).

Diagnosis by Microscopy. The eggs must be found to establish the diagnosis. Only stool culture, which leads to the development of third-stage larvae, permits a reliable determination of the species of worm eggs. The length of the oesophagus corresponds to about ¼ of the length of the parasite. The larva is sheathed, and its diameter amounts to more than 20 µm. The intestinal tube is not straight, but in a zig-zag pattern. The sheath is relatively thick compared to the cuticle of the larva, and the posterior tip is rounded. The sheath ends in a long filamentous structure (see figure on p. 240).

Chemotherapy. See *Trichostrongylus*, p. 231.

Plate XXVI ⇨

Strongyloides stercoralis

Plate XXVI

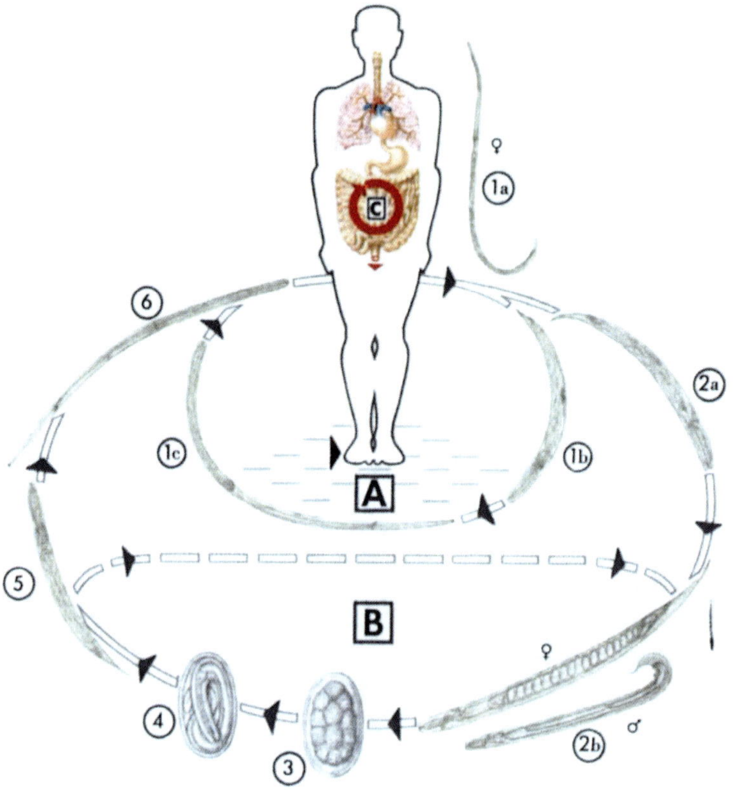

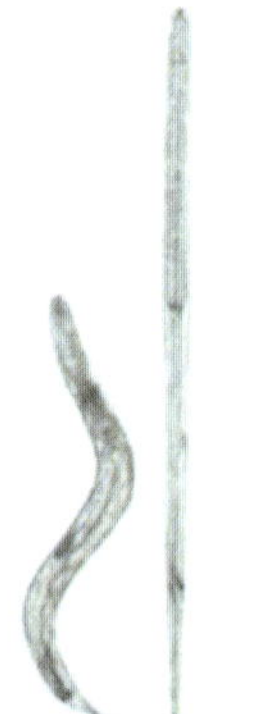

a b

II

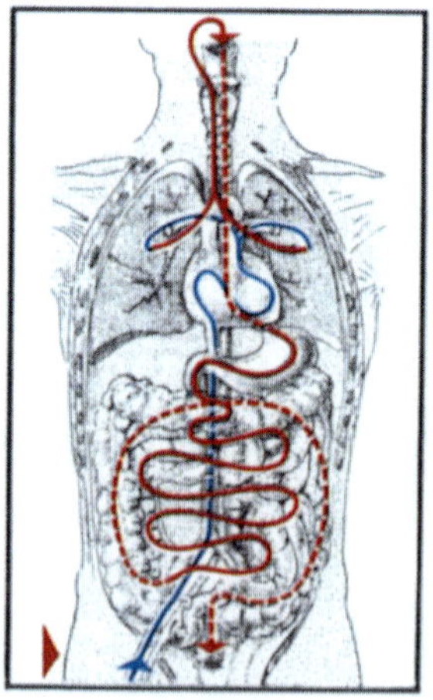

I

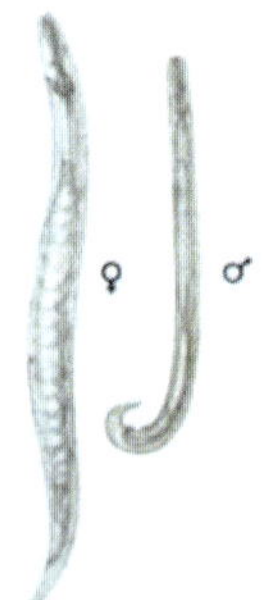

III

Strongyloides stercoralis (BAVAY, 1876) STILES and HASSALL, 1902

Dwarf threadworm

1 a Parasitic female; parthenogenetic generation:

Ⓐ Direct development in the soil

1 b Rhabditiform larva from fresh stool sample (see *II a* and Plate XXXII, *p*)
1 c Filariform larva (stage capable of invasion; see figure on p. 240)

Ⓑ Indirect development in the soil

2 a Rhabditiform larva from fresh stool sample
b Separate-sexed generation, males and females (see *III*)
3 Egg from separate-sexed generation
4 Egg with larva
5 Rhabditiform larva
6 Filariform larva

Ⓒ Autoinfection (*red line*)

Rhabditiform larvae of the parthenogenetic generation develops into filariform larva within the intestinal lumen; these again migrate into the veins of the intestinal wall.

I Migration route of larvae in man
blue After percutaneous invasion of the filariform larvae into the host, they pass in the veins to the heart and from there to the lungs, where they penetrate into the lungs and then into the alveoli.
red From the alveoli the larvae migrate through the airways upwards to the epiglottis, are swallowed and pass through the stomach to the small intestine, the site of the sexually mature worms.

II a Rhabditiform larva
b Filariform larva (approx. ×100)

III Sexually mature forms of free-living generation (see *2 b*; approx. ×50)

(See also Plate XXXIII, *p* and XXXIII, *h*)

Strongyloides stercoralis (dwarf threadworm) is a frequent intestinal parasite of man in warm countries. Its main area of distribution covers approximately the same regions as those endemic for hookworm (see p. 226). Characteristically these species of worm have free-living stages which require a mean temperature of at least 15 °C for their further development. Here-in lies one of the reasons for the limitation of the parasites to certain geographical regions. In Africa, Kenya, Mozambique, Ethiopia, Asia, Iran, South America, North Peru and Columbia 30%–60% of people are infected! On the other hand, in East Asia (Japan) and Southern Europe only a few cases of *S. stercoralis* infection are reported. Suitable climatic conditions prevail in temperate regions below ground in mines. Therefore the parasite occasionally occurs there (occupational disease; miner's disease).

Morphology and Development. The parasitic parthenogenetic females (about 2 mm long and 50 µm wide; Plate XXVI, *1 a*) live in the upper small intestine, bore into the epithelium of the crypts of LIEBERKÜHN and deposit eggs (55 × 30 µm). Within the eggs, the rhabditiform larvae develop (Plate XXVI, *1 b*), which hatch out into the intestinal lumen of the host and pass out in the faeces. Because they move actively, they are relatively easy to recognize microscopically in fresh, warm human faeces. The rhabditiform larvae (L_1 : 280–310 µm) develop in the soil after a moult (L_2) to the invasive filariform larva (L_3: approx. 550 µm), which are able to penetrate actively the intact skin (direct development) Ⓐ. Under conditions which are not yet precisely known (genetic influences depending on chromosome number and activities) the rhabditiform larvae, present in the intestine of an individual, can transform into filariform stages (without reaching the exterior). They may then penetrate the intestinal mucosa (endo-auto-infection) Ⓒ. Alternatively, outside the intestine on the perianal skin, they penetrate through the skin of the host (exo-auto-invasion; Plate XXVI).
Besides direct development Ⓐ, alternation of generations can also occur. From rhabditiform larvae in the soil, free-living female and male worms develop (Plate XXVI, *2b*), the females of which produce fertilized eggs (Plate XXVI, *3*). Within these, rhabditiform larvae again form (Plate XXVI, *4, 5*), which either give rise to filariform larvae (Plate XXVI, *6*) or again mature into free-living separate-sexed worms (Plate XXVI, *2b*; indirect development) Ⓑ.

After penetrating the skin, the filariform larvae begin a migration in the host. They pass through the venous system to the right side of the heart and from there to the lungs (Plate XXVI, *1* blue route), where the larvae leave the vessels and so reach the alveoli. From here they migrate up through the bronchioles, bronchi and trachea to the epiglottis, are then swallowed and reach the small intestine via the stomach. Within the small intestine they develop into parthenogenetic females (Plate XXVI, *1* red route). At least 17 days elapse between the percutaneous entry of the larvae into the host (which is followed by sexual maturation of the females) and the appearance of the larvae in the faeces (prepatent period).

The adult worms appear to have a long life-span, allegedly more than 20 years (possibly even 35 years; GROVE, 1982), although this may be due to continued infection through auto-infection.

Clinical Symptoms. The clinical symptoms of infection change according to where the parasite is located. On penetrating the skin, the larvae can cause dermatitis. As the larvae pass through the lungs, there are frequently an irritating cough and pneumonic signs, comparable to LÖFFLER's syndrome, which regularly occurs in ascariasis (see p. 263). At this stage it is possible to detect the young worms in the sputum. The worms lead to abdominal symptoms on entry into the mucosa of the small intestine. These vary according to the severity of the infection. Frequently epigastric pain occurs, combined with anaemia, nausea, loss of weight, diarrhoea or constipation, and general weakness; there may be a clinical picture of haemorrhagic gastroenteritis. Frequently there is a definite blood eosinophilia, which is, however, very variable in its degree, and can even be absent (in case of a immunosuppressed patient). In many regions (Columbia, Venezuela) fatal cases of *S. stercoralis* infection are not uncommon. If intractable infection persists for many years, then constant auto-infection by endogenous larvae is likely to be the cause of the problem. An infection with the immunosuppressive virus of AIDS (see p. V, 6, 91), treatment with immunosuppressive preparations (e.g. cortisone), or immunodeficiency from other causes is becoming responsible for intractable or even fatally progressive cases with general dissemination of the parasites. In endemic regions patients must be examined for *S. stercoralis* infection before the start of immunosuppressive therapy, and if necessary be treated (opportunistic parasite). This recommendation applies especially to travellers who have stayed in endemic regions. It is suspected that strains of *S. stercoralis* of different degrees of virulence are responsible for the variability in the clinical picture. An infection should always be taken very seriously, and the patient should be urged to get treatment.

If *S. stercoralis* larvae pass from animal hosts (e.g. dogs) to man, they frequently wander around under the skin for a long time, producing the so-called creeping eruption. The worms, however, do not become sexually mature and die off (see p. 244). In rare cases, the larvae of *S. stercoralis* "roam" around in the human body, and most become "captured" mainly in the region of the mesentery, and appear at post mortem examination as worm nodules (see also *Ascaris*, p. 263).

Epidemiology. For man there are the following routes of infection.

1. Endo- and exo-auto-infection by endogenous *S. stercoralis* larvae Ⓒ
2. Percutaneous larval invasion
   a) after direct development in the soil Ⓐ
   b) after indirect development with alternating generations Ⓑ. This developmental pathway means that damp soil, under certain conditions, may contain large numbers of worms.

The transmission of the parasite to the newborn via the mother's milk is thought possible and has been demonstrated for *S. fülleborni* in primates.

For further development in the soil the worms require a mean temperature of at least 15 °C with sufficient moisture; the optimum temperature lies between 23 °C and 30 °C. Such conditions are also found underground in mines. In this way infection in mining regions can occur, following the introduction of *S. stercoralis* by parasite carriers. The filariform larvae penetrate mainly through the foot skin

of man, who must be regarded as the sole host. Oral infection acquired from contaminated food or drinking water must be regarded as a rare event.
Through prophylactic measures, e.g. building of lavatories, wearing of sturdy footwear and also drainage of the soil, it is usually possible to largely control *S. stercoralis* both outside and in mines. In the first place careful disposal of human faeces is most important to ensure the worms cannot get into the soil. Dogs, foxes and cats are to be regarded as possible animal reservoirs. However, these hosts are only lightly and temporarily infected with *S. stercoralis*. It is thought that other worm species of the same genus are present in these animals.
People infected with *S. stercoralis* must be regarded as source of infection for other humans because the larvae excreted with the faeces develop within a few hours to the filariform stage, and can actively penetrate the human skin. Therefore, caution is to be observed in the laboratory when dealing with suspected positive stool samples (see also *Ancylostoma*, p. 229). In temperate latitudes there is no danger to the indigenous population from a *S. stercoralis* carrier with a hygienic lifestyle on account of the climatic conditions. Only under the "favourable" climatic conditions as in mining under ground, i.e. with warmth and moisture (see above), the larvae have any chance of survival.

Diagnosis by Microscopy. If a fresh stool sample contains free moving larvae, then it is highly probable that there is a *S. stercoralis* infection. The eggs, which are similar to those of hookworms, are very rarely found. However, the larvae are not excreted continuously, but often in batches, so that with high clinical suspicion but negative stool findings at least two further investigations (see below) on different days are necessary. Also, duodenal aspiration and small intestinal biopsy should be considered. In very severe infections and widely disseminated ones as in immunosuppressed patients, larvae can also be found in the sputum and the urine.

Culture and Concentration Techniques

1. Using a stool culture at about 28 °C eggs and larvae will further develop and multiply (see above; caution is required as there is danger of infection). Numerous infectious larvae develop after a heterosexual free-living generation.
Harada and Mori (1955) have modified the old charcoal faeces culture (also suitable for *Ancylostoma, Trichostrongylus*, and related genera), by using filter-paper strips (13 × 120 mm) instead of charcoal. The middle third of their paper strips is thickly (1–2 mm) covered with faeces and placed in a suitable centrifuge tube, which contains about 3 ml distilled water. The filter paper is placed with the clean side adjacent to the wall of the tube, so that the water can rise up by capillary action. The water must be replenished daily, thereby keeping the filter-paper and faeces moist. The culture is kept for 10 days in the dark at 24°–28 °C. Hatched larvae migrate downwards into the water, where their presence can easily be confirmed with a hand lens.

In the practical execution of the method (1) the water must be taken from the bottom of the tube with a pipette, and transferred to a clean tube. (2) Fresh water must again be added to the culture tube, in order to wash off the larvae adhering to the paper strip, which some species favour. The paper strip should be removed with forceps and laid aside for disinfection. The larvae are investigated microscopically for speciation. The filariform larva of *S. stercoralis* is differentiated from all other larval species by the long, attenuated oesophagus, which extends to almost half the length of the larva. The slender body has no sheath and the tip of the tail is not pointed, appearing notched (see figure on p. 240).

2. If a large stool sample in a cloth bag is suspended in a sedimentation glass filled with luke-warm water, the larvae migrate out, sink down, and collect within a few hours as a deposit in the bottom of the vessel. They can easily be taken from the deposit with a pipette for examination under a microscope. Basically, this technique can also be used for hookworms (see p. 229; method according to BAERMANN, 1917).

Diagnosis by Immunobiological Methods. In *Strongyloides stercoralis* infection immunological detection methods play no greater role than in ancylostomiasis. So far, no suitable antigens have become available for the tests. Serological results (from IIFT and ELISA) moreover, are only of limited usefulness because of the lack of specificity.

Chemotherapy. The drug of choice is thiabendazole and it is effective in a dosage of 25 mg/kg body weight, twice daily, on 3–5 successive days. It has a high ovicidal and larvicidal action, and so is equally suitable for the treatment of larva migrans. However, it has troublesome side-effects, such as nausea, confusion and dizziness (see p. 255).

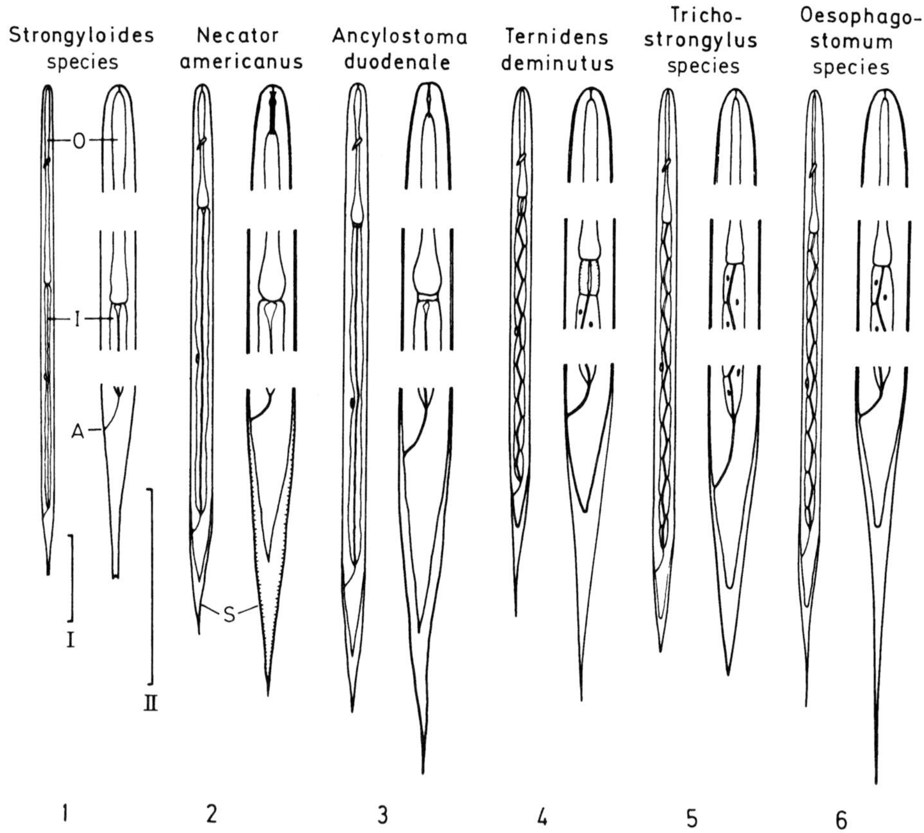

Diagnostic features of filariform larvae found in human coprocultures (from WHO, 1981). *A*, anus; *O*, oesophagus; *I*, intestine; *S*, sheath. (*I* = 100 μm for the illustrations of the intact larvae, ×150; *II* = 100 μm for the illustrations of portions of larvae, ×310)

Key for the Identification of Filariform Nematode Larvae (L_3) in Human Stool Samples or in Coprocultures (see p. 238) (from WHO Technical Report 666, 1981)

Parasite appears in stool: * as egg, ** as rhabditiform larva

1 a Oesophagus about ½ the length of the body; body slender (14–17 µm) without cuticular sheath; tip of tail not pointed, appearing notched
see p. 236 ** *Strongyloides stercoralis*
see p. 237 * *Strongyloides fülleborni*

1 b Oesophagus about ¼ length of body; cuticular sheath present, body thicker than 20 µm 2

2 a Intestinal lumen straight 3

2 b Intestinal lumen not straight, but zig-zagged (see figure on p. 240, 4–6) 4

3 a Body (without sheath) about 500–600 µm long; tail (i.e. anus to tip) less than 73 µm long (50–72 µm); intestine at oesophagus-intestine junction, as wide as oesophageal bulb; buccal "spears" conspicuous, parallel throughout the length, about 15 µm; clear conspicuous transverse striations present on sheath in the tail region (see p. 226)
* *Necator americanus*

3 b Body (without sheath) about 600–700 µm long; tail more than 73 µm (75–94 µm); intestine at oesophagus-intestine junction narrower than oesophageal bulb; buccal "spears" inconspicuous, about 10 µm long; transverse striation of the sheath in tail region inconspicuous (see p. 226) * *Ancylostoma duodenale*

4 a Sheath relatively thin (thinner than the cuticle of the larva); a pair of elongate sphincter cells between oesophagus and first pair of intestinal cells; tip of larva's tail pointed; posterior end of sheath elongate, tapering to thread-like tip; body about 630–730 µm long, 29–35 µm wide (see p. 229) * *Ternidens deminutus*

4 b Sheath relatively thick (thicker than the cuticle of the larva); no sphincter cells between oesophagus and intestine; tip of larval tail rounded-off or blunt.......... 5

5 a Posterior end of sheath relatively short; not drawn out to a fine point; distance between tip of larval tail to end of sheath smaller than the distance between anus and tip of larval tail (see p. 230) * *Trichostrongylus* species

5 b Posterior end of sheath relatively long, drawn out finely to a long point; distance between tip of the larval tail and end of sheath greater than the distance between the anus and tip of larval tail (see p. 232) * *Oesophagostomum* species

Nematode Larvae as Pathogens

Cutaneous Larva Migrans (Creeping Eruption)
Visceral Larva Migrans

Herring Worm Disease Due to *Anisakis* and Related Genera

Nematode Larvae as Infectious Agents

In Europe in recent years there has been an increasing number of cases of infection due to aberrant nematode larvae. These larvae are not able to reach sexual maturity in man and are not transmitted further. "Creeping eruption" has been known for a long time in the tropics and subtropics, the result, for example, of the larvae of the canine hookworm *Ancylostoma braziliense* or *A. caninum*. Other nematodes from dogs or cats, e.g. members of the genus *Toxocara*, can migrate in their larval form virtually anywhere in the human without becoming mature. They can affect liver, lung, central nervous system and the eye, amongst others. This type of condition is known as visceral larva migrans and the disease described as toxocariasis.

Larvae of herring worms of the genus *Anisakis*, which in their adult forms live in marine mammals, lead to stomach and intestinal disease. The larvae get into man through the consumption of raw fish and cause disease in the region of the gastrointestinal tract. They are incidentally found at surgery or post mortem – mainly – usually causing an eosinophilic granuloma in the stomach or intestines. The larvae of the nematode *Angiostrongylus cantonensis*, a rat lung worm from the Pacific Islands, can cause an eosinophilic meningitis or encephalomyelitis and damage to the eyes. Another primary parasite of rats is the species *Angiostrongylus costaricensis*. Its larvae can migrate into the arteries of the mesentery and of the intestine and can lead to severe intestinal granulomas and ileus. Many worms even become mature but the deposited eggs or larvae remain buried in the tissues of the intestine and are not excreted. Other nematode larvae can also cause similar clinical pictures. They are quite rare, however, and their occurrence is limited to small geographical areas (e.g. *Strongyloides ransomi*, a porcine parasite; canine-infecting species of *Dirofilaria; Gnathostoma spinigerum*, mainly in Thailand).

In this section the following aberrant nematode larvae will be dealt with:

1. Cutaneous larva migrans (creeping eruption)
2. Visceral larva migrans (toxocariasis)
3. Herring worm infection with *Anisakis marina* and related species
4. *Angiostrongylus cantonensis*, the cause of eosinophilic meningitis
5. *Angiostrongylus costaricensis*, the cause of intestinal disease.

Cutaneous Larva Migrans (Creeping Eruption)

Cutaneous larva migrans or creeping eruption is caused primarily by the third stage larva of the canine hookworms *Ancylostoma braziliense* and *A. caninum*. After entering man – a "foreign" host – through the skin, they wander around in the subcutaneous tissue without being able to reach the site where they would normally mature, i.e. the intestine. The disease usually starts in the skin with a mild irritation limited to one area, and the appearance of papules, usually on the feet and hands but possibly on any part of the body. During their wanderings through the stratum germinativum, the larval stages, which secrete a proteolytic enzyme, only advance 3–5 cm per day. The linear skin lesions disappear again as the larvae move on, but they can give rise to a purulent eczema due to secondary infection from scratching. The larvae remain active for several days to a few weeks and then die off. It is uncommon for a larva to get into the circulation.

The general damage to the host is relatively slight (irritation in the affected area). The illness is, however, associated with a relatively high eosinophilia. Infection with the larvae occurs mainly by lying on ground soiled with dog or cat excreta.

It should be mentioned that a picture similar to creeping eruption can also be produced by other "foreign" parasite larvae, e.g. by cercariae of *Schistosoma* species of water fowl, by sparganum stages of some fish tapeworm species from Asia and some species of fly larvae (cercarial dermatitis see p. 166; sparganosis see p. 178, cutaneous myiasis).

Chemotherapy. Local application of 15% thiabendazole ointment is effective for creeping eruption. Freezing with ethyl chloride spray is also recommended.

Visceral Larva Migrans

Toxocariasis

Visceral larva migrans is the name given to disease caused by those dog and cat nematode larvae which in man – and hence in a "false" host – become liberated within the gastrointestinal tract following the ingestion of eggs, mainly of the genus *Toxocara (T. canis, T. cati)*. They penetrate the intestinal wall and wander aimlessly in the foreign host until their migration is blocked by an inflammatory reaction with granuloma formation. In the end the larvae die. Localization is thus purely fortuitous.

Clinical Symptoms. After liberation in the gastrointestinal tract the larvae initially begin to wander as they would have in the "proper" host, i.e. they reach the liver and lungs via the veins. They do not, however, penetrate the alveoli and from here on they wander aimlessly around in the body, thus causing varied, though characteristic, symptoms. This results in conditions resembling asthma due to passage through the lungs, epilepsy from injury to the CNS, damage to the eye or even retinal detachment (ocular larva migrans) and blindness (GLICKMAN et al., 1979). A persistent, often very high, eosinophilia is typical of toxocariasis, and possibly a leucocytosis. Liver enlargement, hypergammaglobulinaemia and lung infiltration similar to LÖFFLER's syndrome (as in ascariasis) have been observed. Pyrexia of unknown origin and high eosinophilia in childhood should raise suspicion, necessitate a serological examination and lead to chemotherapy (see the experimental findings of PEPPERSACK, 1981).
In one British study it was claimed that each year there were 50 cases of blindness due to toxocariasis, other authors cite up to 200 cases. 30,000 people are said to suffer from asthma in Great Britain as a result of toxocariasis, and 12,000–15,000 from epilepsy (?).
Migrating nematode larvae can spread microorganisms in some ways.

Transmission. The source of infection is mainly young dogs and cats, which are more often infected with round worms than older animals. As there is often very close contact with pets, children between the ages of 1–5 years are at greater risk than adults. Children meet with dogs and cats in playgrounds, for example, in the sand box. The eggs excreted by the pets are not infective immediately, as the infectious larva forms within the egg days to weeks later (depending on the climatic conditions). The highly resistant eggs adhere for days to the coat (in the anal region) and are transmitted by mouth to man through hands contaminated by stroking the animal (see GILLESPIE, 1988).
As a **prevention measure** young dogs should be regularly dewormed and above all kept away from children's playgrounds. Soiled play sand must be renewed frequently.
The number of dogs and cats in towns has continued to rise in the last few years (e.g. 3 million dogs in the Federal Republic of Germany in 1982). Approximately 20% of the dogs are infected with *T. canis*. Hence toxocariasis has attracted increasing interest in the public health service.

Diagnosis. The diagnosis is often made only with difficulty, as the index of suspicion for toxocariasis usually is not high, in the face of what are really quite specific symptoms. The key sign, however, is usually an increase of blood eosinophils. Antibodies can be demonstrated by a variety of serological techniques (IIFT, microprecipitation test, OUCHTERLONY test and ELISA with approx. 78% sensitivity; GLICKMAN et al., 1979; SPEISER and GOTTSTEIN, 1984).

Chemotherapy. Mebendazole and thiabendazole are considered effective. TAYLOR and PEARSON (1981) recommend the antibiotic ivermectin (see p. 284).

Herring Worm Disease Due to *Anisakis marina* and Related Species

Herring worm disease is caused by nematode larvae of the genus *Anisakis* and its relatives. According to research by VAN THIEL (1966) mammals (dolphins, whales, seals and porpoises) in the North Sea and South Atlantic are the natural definitive hosts for herring worm larvae of the species *Anisakis marina* and *A. simplex* (LINNÉ, 1767, see also KIKUCHI et al., 1967). If by "mistake" they enter man (a false host) through ingestion of raw herring, the clinical picture which was described by VAN THIEL as anisakiasis is produced. This results in eosinophilic granulomata particularly in the antral region, or a small bowel obstruction due to stenosis.

The natural **life cycle** of these worms is only incompletely known. Presumably the larvae must be obligatorily in sea water and ingested by isopods (tiny crustaceans; Euphausiids krill; first intermediate host). The larva bores through the intestinal wall of the copepod and continues its development in the haemocoel or in the surrounding tissues. If the infected host is eaten by a second intermediate host (transport host), a fish or squid, the larva again bores through the wall of the intestine and lodges in the peritoneal cavity or other tissues of that host, but it does not become sexually mature. The liver is most often involved, resulting in considerable injury from which the host may die. The infective third-stage larva (up to 3 cm long) is found in various marine fish (e.g. herring, mackerel) (OSHIMA, 1987). The consumption of such fish or squid uncooked leads to infection of the definitive host, and of man as well, although in man the worms do not become sexually mature. Instead infection results in the eosinophilic granulomata in the gastrointestinal tract described above (SMITH and WOOTTEN, 1978).

The first case of an eosinophilic granuloma of the stomach due to an *Anisakis* larva was described in Japan following an operation on a 35-year-old man. Subsequently, in a literature review ASHBY et al. (1964) reported 89 cases of eosinophilic granuloma (including the UK). Over half of the patients had lesions in the stomach, particularly in the antrum; they exhibited corresponding gastric or pyloric symptomatology. In a third of the cases small bowel obstruction had occurred; intestinal resection was necessary in many patients. The number of cases has since increased substantially, ISHIKURA et al. (1983) summarize 1523 cases of stomach anisakiasis and 223 intestinal cases in Japan in 1983.

Even though at first species of the genus *Anisakis* were seen as causal, WILLIAMS (1965) cautioned people not to overlook the fact that several species of nematode from different genera (which are in part still poorly defined) might also be suspected of causing the same clinical picture. Considering our still very circumscribed knowledge, he does not consider it constructive to talk simply of one herring worm disease. ASAMI et al. (1965) also reported two cases of stomach granulomata with "*Anisakis*-like larval nematodes" causing the illness, which was histologically confirmed (see also YOKOGAWA and YOSHIMURA, 1965).

Reaction in the tissues appears as an old foreign body reaction, but as shown by the experimental studies of OYANAGI (1967) the eosinophilic granuloma in the intestinal tract should be considered an allergic reaction. Rabbits sensitized by intraperitoneal injections of *Anisakis* larvae show changes in the intestine following oral reinfection such as are known to occur in man. Similar lesions arise when homogenized larvae are injected directly into the stomach lining of sensitised animals. Eosinophilic skin abscesses form when a suspension derived from an in vitro culture of *Anisakis* larvae is injected intradermally into sensitized animals. The observation of KUIPERS (1964) argues for this sensitization and subsequent development of granulomata. He suspected that in three out of four cases of herring worm infection an old foreign body reaction was present, which arose following recent invasion by *Anisakis* larvae. The local reaction is presumably produced by an earlier infection of the same kind.

According to observations by POLAK (1966) and BIJKERK (1969) the frequency of herring worm disease had noticably been increasing in the Netherlands. The number of cases of illness always risen there after the start of the herring season. Laws introduced in 1965 regulating the salting of herrings had a favourable effect locally. Following this, freezing of herrings for 24 hrs was introduced, and this procedure killed the herring worms. VAN THIEL and BAKKER (1981) reported that herring worm disease due to *Anisakis simplex* was no longer found in the Netherlands, whereas, for example, in Japan it is not uncommon. The custom there of eating raw marine fish leads to the gastric form of the illness, occasionally even among tourists. This is the main form in Japan, whereas in the Netherlands mainly intestinal anisakiasis occurs. OSHIMA (1987) reported about the increase in cases of anisakiasis and codworm anisakiasis, infection with the codworm, *Pseudoterranova decipiens* in Japan and USA, related to advance in diagnosis.

Diagnosis by Immunobiological Methods. The parasitic infection can be diagnosed serologically by means of a complement fixation test and an indirect immunofluorescence test using an *Anisakis* antigen (MERKELBACH, 1964). A skin test developed by MORISITA et al. (1965) is based on an allergic response. An antigen derived from *Anisakis* larvae is injected intradermally. As yet, however, no great confidence has been gained with these methods.

Chemotherapy. Thiabendazole is considered a reliable agent against larva migrans (25 mg/kg body weight approx., twice daily on two successive days). When the diagnosis has been confirmed, surgical removal of the lesions caused by *Anisakis* larvae is often necessary. But if the diagnosis is correctly established from the patients history of ingested raw fish just before the onset of disease, conservative treatment without surgery (endoscopy of the duodenum) usually results in complete cure.

Plate XXVII ⇨

Angiostrongylus cantonensis

Plate XXVII

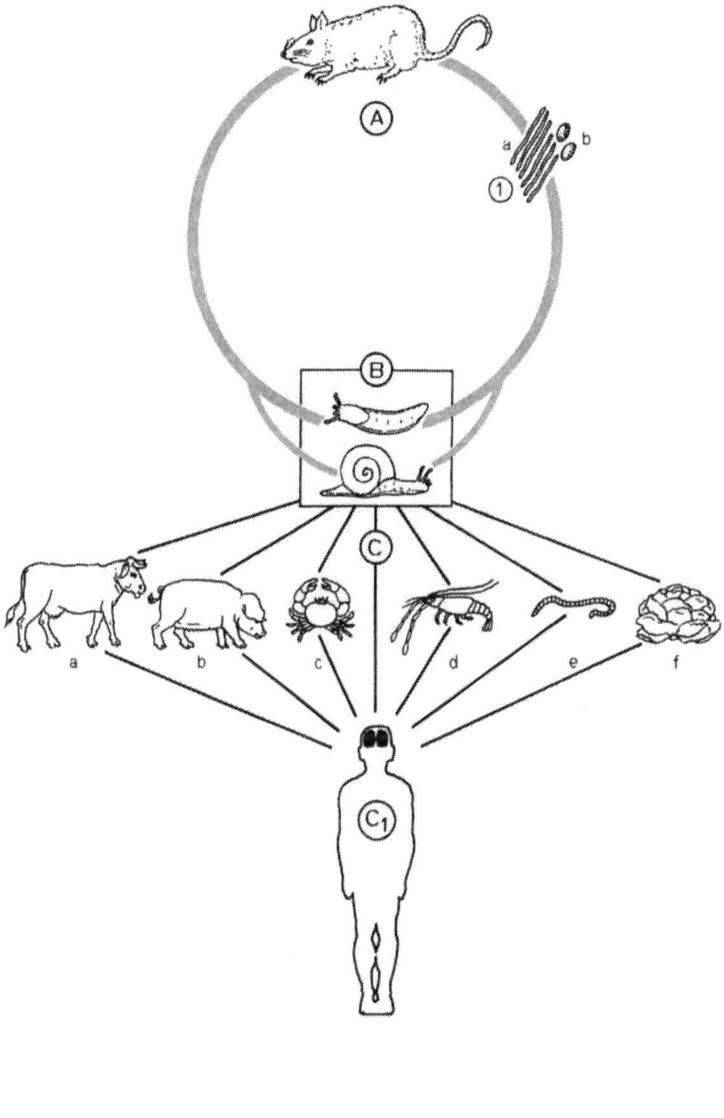

Angiostrongylus cantonensis (CHEN, 1935) DOUGHERTY, 1946 (= *Parastrongylus cantonensis* DROZDZ, 1970)

Rat lungworm
A cause of eosinophilic meningitis

Ⓐ Specific host: rat

1 a Adult worms
b Eggs

For migratory pathway of the larvae in the definitive host, see figure on p. 254

Ⓑ Intermediate hosts: snails

The larva moults twice in the snail. The infective third-stage larva has to be ingested by a rat

Ⓒ Non-specific hosts: *a* cow, *b* pig, *c* crab, *d* crayfish, shrimp, *e* planarian; for *f* see text p. 255

(C_1) Man as an "accidental host" becomes ill with meningo-encephalitis, because of the predilection of the parasite for the CNS.

(From ALICATA, 1965)

It was in 1944 in Taiwan that NOMURA and LIN demonstrated for the first time in man a particular type of eosinophilic meningitis, which was due to infection with larvae of the nematode species *Angiostrongylus cantonensis* (Metastrongylidae). The worm lives as an adult in the pulmonary arteries of rats. Snails act as intermediate hosts. The number of infections seen in man has increased substantially (see ALICATA and JINDRAK, 1970).

Currently the area of distribution of *Angiostrongylus cantonensis* includes the Pacific Islands of New Caledonia, the Cook Islands, Tahiti, Hawaii, Guam, Micronesia and Taiwan, and as far away as Thailand, Cambodia, Vietnam, Malaysia, Indonesia (Sumatra) and the Philippines. The area also extends from the North Japanese islands to Queensland in Australia. The incidence of the parasite in rats is very variable and reaches more than 50% in some localities. The appearance of the illness in man coincides with the geographic range of the rodent lung worm. MACKERRAS and SANDARS (1955) unravelled the life cycle of *Angiostrongylus cantonensis* in Queensland; the species was described for the first time by CHEN (1935). In addition, BHAIBULAYA (1968) found a new species of *Angiostrongylus* in the Australian rat *Rattus fuscipes* (*Angiostrongylus mackerrasae*). Altogether about 16 species have been identified thus far. However, of these, apart from *A. cantonensis*, only *A. costaricensis* MORERA and CÉSPEDES, 1971 is important in human medicine (see p. 257).

Morphology and Development. The adult worms (approx. 20–25 mm) live in the two main branches of the pulmonary artery of the rat Ⓐ or, in a severe infection, in the right ventricle (see Plate XXVII). Characteristic of the female are the milky white uterine tubes which are arranged in a spiral around the blood red coloured intestine. Once laid, the eggs pass into the alveolar capillaries, where, within 6 days, the first larval stages develop and hatche. The first stage larvae bore through the vessel walls (*8*) and enter the alveoli and migrate through the respiratory passages, the bronchioles and bronchi, and up the trachea. They are then swallowed, and when they have passed through the stomach and intestine, appear in the rat's faeces 42–48 days from the start of the infection (prepatent period).

The larvae (approx. 0.3×0.015 mm) survive on damp ground for up to 2 weeks. They must then, however, either be eaten by an intermediate host, e.g. land living snails Ⓑ, or must actively penetrate these. The larvae moult twice. Only the third larval stage, which appears after about 28 days, is infectious. Initially, it remains inside the two shed larval skins (approx. 0.5 mm long). If the infected snails (more than 20 species) are eaten by r a t s, the larvae then migrate out of the rat's stomach (*1*) into the lumen of the small intestine and enter the blood stream via the veins of the intestinal wall and of the mesentery (see below, migratory pathway of the larvae). Within 2 hrs the larvae have already reached the right ventricle (*6*), by means of the portal vessels (*3*), hepatic veins, inferior vena cava (*4*) and right atrium (*5*). After 2 days they have passed further via the pulmonary arteries (*7*), capillaries (*8*) and veins (*9*) through to the left atrium (*10*) and left ventricle (*11*),

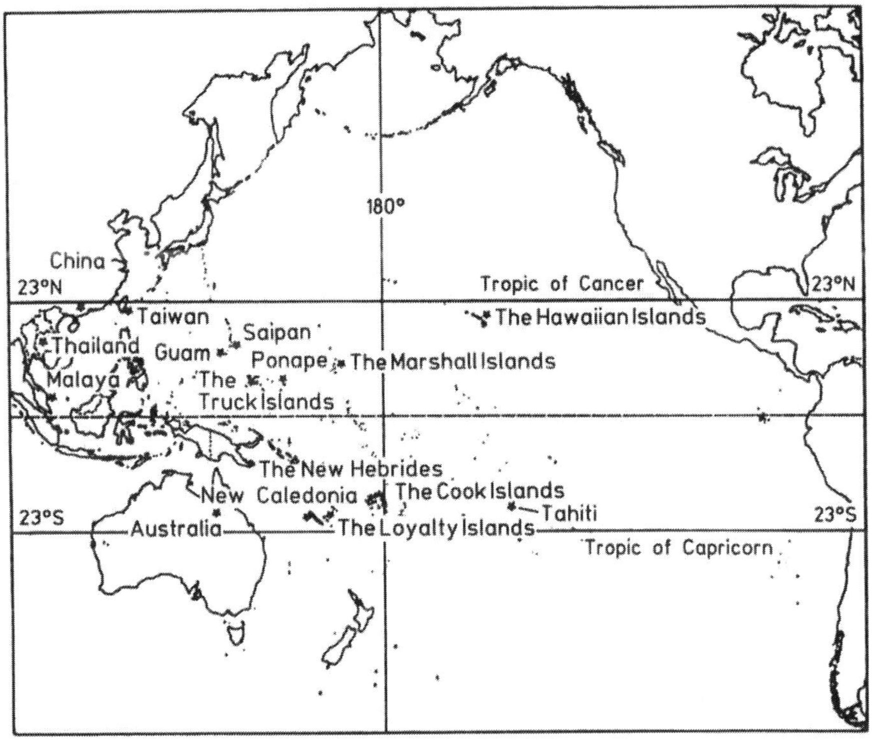

Distribution of *Angiostrongylus cantonensis* in the Pacific region (from Alicata, 1965).

through the arch of the aorta (*12*) and common carotid artery (*13*), to the capillaries of the brain (*14*).

After two moults, the larvae develop into adult worms (approx. 12–13 mm long) within the rat's brain. These then migrate from the brain, probably via the subarachnoid space where they remain for about 2 weeks, to the cerebral veins, the jugular vein (*15*) and via the superior vena cava (*16*) once more to the right ventricle. After about 29–32 days they reach their final destination by means of the pulmonary artery, i.e. the two main branches of the pulmonary artery. Here they reach maturity (Alicata, 1965). The total development takes 42 days (prepatent period).

The larvae may occasionally use the lymphatics of the digestive tract to reach the right atrium, passing through the hepatic circulation and reaching the heart by means of the thoracic duct. The route of migration of infectious larvae to the spinal cord has not yet been determined.

Depending on the site of the worms a visceral, intracranial or pulmonary reaction may be produced in the rat. If the larvae do not reach the rat (the specific host) but enter a "non-specific" host Ⓒ (such as cows, pigs, crabs, crayfishes, shrimps, planarians), or either directly or indirectly via an intermediate or transport host

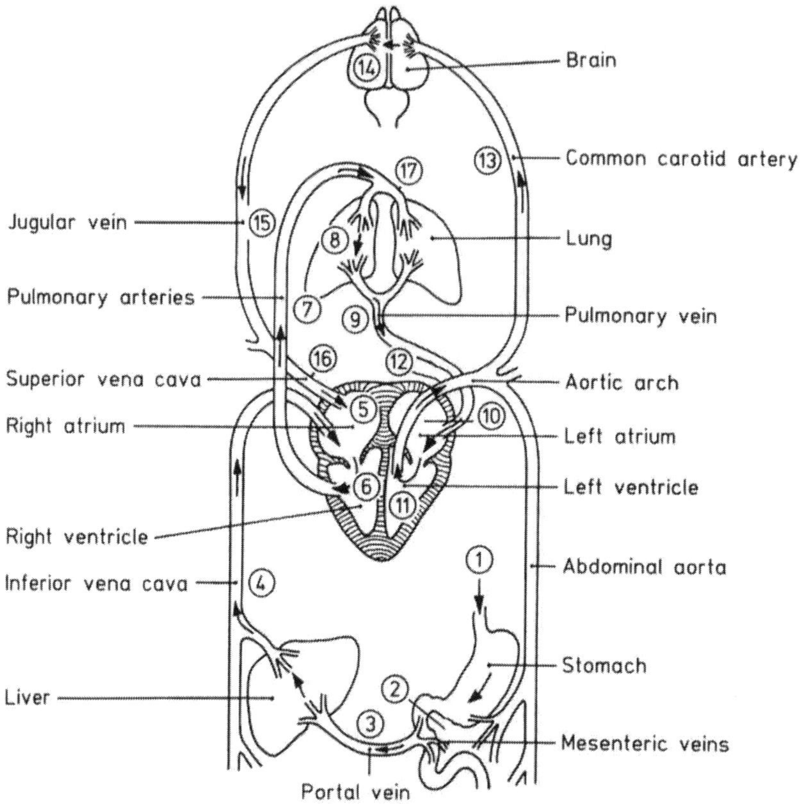

Migratory pathway of the larvae of *Angiostrongylus cantonensis* (from ALICATA, 1965)

into man, then they wander aimlessly without developing further. In man ©, they may proceed, among other places, into the CNS, where they then cause an eosinophilic meningoencephalitis.

Clinical Symptoms. Various clinical symptoms have been described in man, depending on the geographical location. However, not every infection leads to a severe illness. In Thailand, cases with partial loss of sight are said to predominate, with worms present in the anterior chamber of the eye. In the Pacific Islands patients with eosinophilic meningoencephalitis (incubation period 2–3 weeks) complain of local paresthesiae of the skin in various parts of the body, associated with burning sensations, pain and marked sensitivity to touch. On Tahiti 5% of patients are said to have facial palsies, which often remain for several weeks. Longer lasting paralyses have also been observed. In the leucocytosis, which is often present, the eosinophil count may rise to above 50%.

Epidemiology. Infection in man is caused by the consumption of infected intermediaries (Plate XXVII, ©). The infectious stages can also actively enter when they come into contact with injured skin. The portals of entry are largely determined by local eating customs. In some regions of Thailand and on Samoa, the infection can be traced to the ingestion of raw infected snails of the species *Pila ampullacea* and *Achatina fulica*. The fleshy foot of the relatively large snails is prepared in a salad, mixed with various juices and vegetables, and then eaten raw. On Tahiti raw shrimps are often eaten. According to ALICATA and JINDRAK (1970) up to 4% of these are infected (© *d*). As these shrimps are eaten in particularly large quantities on social occasions and festivals, mass infection occurs. This is thought to be the cause of the often quite numerous cases of eosinophilic meningitis occurring locally. Consumption of raw lettuce (© *f*) can also lead to worm infections in many areas. Tiny infected slugs, which can also be intermediate hosts, and land-living flatworms may be found, for example, in ripe strawberries, fallen ripe figs and mango fruits. These intermediate hosts can then be ingested unnoticed with the fruit. Since pigs and calves can be infected, the consumption of raw liver, much liked by Europeans, leads to infection. Water can also be contaminated by liberated larvae, but this is a much less dangerous route. Presumably other routes of invasion exist which are as yet unidentified.

Numerous land-living species of snail have in the meantime been found to be naturally occurring intermediate hosts. On the Pacific Islands these include the garden snails *Bradybaena similaris*, *Opeas javanicum* and *Subulina octona*, and also the smooth snails *Deroceras laeve*, *Vaginulus plebeius* and *Veronicella alte*. In Southeast Asia other species of snails serve as intermediate hosts, e.g. *Girasia peguensis*, *Macrochlamys resplendens* in Malaysia and the amphibious snail *Pila ampullacea* in Thailand. Under experimental conditions even more snails can be infected with *A. cantonensis*. Many other species of animal can serve as paratenic hosts (i.e. transport hosts, passive hosts), for example, predatory planarians and some snail-eating species of crayfish.

Along with several rat species of the genera *Rattus*, *Melomys* and *Bandicota* (infection rates in Micronesia are up to 22%, in Thailand up to 35%, in New Caledonia 45.5%–94.1%, and on the Cook Islands up to 61.9%), the larvae can attack man and also the CNS of monkeys, cats, mice and many other mammals, although not calves, pigs or hens (see ALICATA, 1965). *A. cantonensis* has been found in the following five species of rat amongst others: *R. norvegicus*, *R. rattus*, *R. conatus*, *R. fuscipes* and *M. littoralis*.

Diagnosis by Microscopy or by Serological Methods. Such methods of any practical significance do not exist. Attempts have been made to demonstrate the parasites in rats using an indirect haemagglutination test. A skin test is available for man.

Chemotherapy. Thiabendazole has been shown to be a highly effective agent in rats infected with *Angiostrongylus* species. The larvae are particularly sensitive at the time of the migration to and within the brain. From this CUCKLER et al. (1965) deduced that thiabendazole should also be effective in eosinophilic *Angiostrongylus* infection of man, since the same stages reach the human brain (25 mg/kg body weight, 1–2 times daily, for 2–3 days; see p. 239). The antibiotic ivermectin has been recommended by SANO et al. (1983) for *A. cantonensis* (quoted by CAMPBELL, 1985; see also p. 284).

Angiostrongylus costaricensis MORERA and CÉSPEDES, 1971 (= *Morerastrongylus costaricensis* (MORERA and CÉSPEDES, 1971) CHABAUD, 1972

The second species of *Angiostrongylus* which can attack man is *A. costaricensis*. It gives rise to an intestinal angiostrongyliasis, a condition noted for the first time by MORERA (1970) in Costa Rica. Subsequently, cases have been recognized in Texas, Mexico, Middle America, Venezuela, Columbia and Brazil. The complete extent of its spread is as yet unknown (MALEK, 1981).

Angiostrongylus costaricensis lives primarily in domestic rats (*Rattus rattus*) and cotton rats (*Sigmodon hispidus;* definitive hosts). These become infected mainly by eating affected smooth snails (predominantly *Vaginulus plebeius*). In the rats, the parasites are found in the mesenteric arteries in the caecal region, and their eggs and larvae in the wall of the intestine; inflammatory reactions are, however, absent. The youngest larvae are passed out in the faeces (ECKERT and LAEMMLER, 1972).

In man too, the adult worm of *A. costaricensis* parasitizes the arteries of the mesentery and the intestinal wall, particularly those of the ileocaecal region. This causes inflammation, thrombosis of vessels and necrosis. In some cases the bowel wall becomes so grossly distorted that a partial or complete ileus can occur. In the bowel wall and in the regional lymph nodes eggs can be found which cause a granulomatous eosinophilic inflammation. Hence larvae are rarely liberated. All the signs indicate that development is not taking place in the "right" host, even though the worms become sexually mature.

In man the tissue reaction in the bowel wall stops the larvae getting free, but leads to the pathological changes described above, which can then cause an ileus.

Surgical treatment consists in some cases of appendectomy, right and left hemicolectomy, colostomy or removal of intestinal lymph nodes, depending on the site affected. Eggs, larvae and mature worms can be found histologically. Usually there is a high eosinophil count and a marked leucocytosis.

LORÍA-CORTÉS and LOBO-SANAHUJA (1980) examined 116 children (64% boys and 36% girls) between the ages of 6 and 13 years. In these patients the signs resembled those in an acute abdomen, including the patients complained of pains in the right iliac fossa, and pain on palpation and rectal examination. A tumour-like mass could be palpated, and there was abdominal rigidity usually associated with a moderate pyrexia. The surgical procedures used in 90 patients were appendectomy, or right or left hemicolectomy. Sites affected were mainly the caecum, the ascending colon, appendix and small intestine. On histological examination the tissue was found to be oedematous with yellowy granulations in the subserosa with eggs, larvae and adult parasites present.

The first larval stages can also appear in human faeces. These are much smaller (approx. 260–290 μm long, 14–15 μm wide) than all the filariform larvae mentioned in the key above (see p. 240/241). Care must be taken not to confuse them with adult or larval stages of free-living nematodes (genera *Rhabditis, Rhabditoides* etc.).

Chemotherapy. Although diethylcarbamazine and thiabendazole are used, these drugs are not always effective, since it is not usually possible to treat successfully without surgery.

Plate XXVIII ⇨

Ascaris lumbricoides

Plate XXVIII

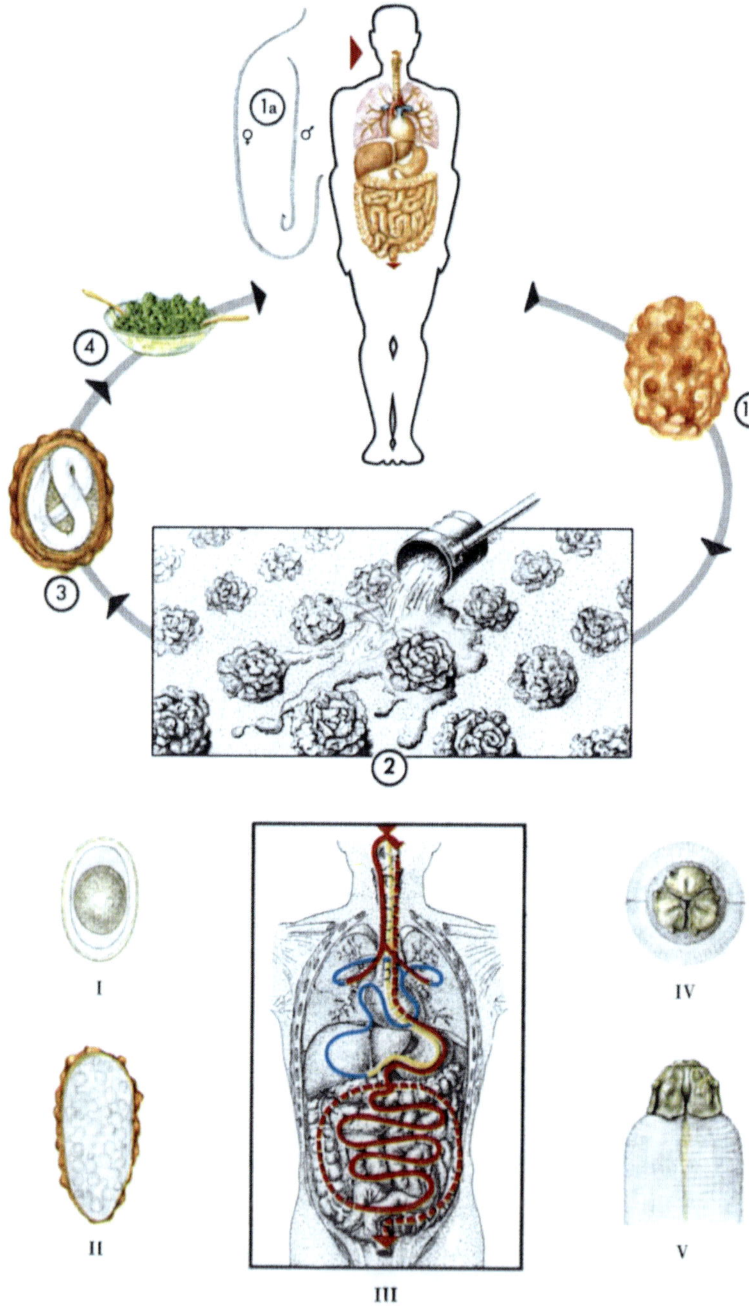

Ascaris lumbricoides Linné, 1758

The large intestinal roundworm (ascarid roundworm)

Development and mode of transmission

1 a Adult worms: female and male
b Typical egg from faeces (see *I* and *II*)
2 Use of manure from roundworm carriers as fertilizer leaves eggs on, for example, lettuce
3 Egg contain infective larva
4 Eating lettuce leaves contaminated with eggs containing infective larvae leads to roundworm infection (see *2*).

I Fertile egg without its outer shell in cross-section

II Infertile roundworm egg

III Pathway followed by the roundworm larvae following ingestion of eggs containing larvae (see *3*):

yellow Egg with larva passes into the gastrointestinal tract with food

blue Having made its way into the small intestine the larva bores through the intestinal wall and gets into the liver via the venous drainage. From here it passes to the right ventricle and through the pulmonary artery into the lungs, where it leaves the blood vessels and moves into the alveoli

red The larva migrates from the alveoli upwards via the bronchi and trachea to the epiglottis. Here it is swallowed and enters the gastrointestinal tract once more via the oesophagus. The small intestine is the definitive site of the adult roundworm

IV View of anterior end ("mouth"): three lips, the upper with dorsal and the lower with ventral papillae

V Ventral view of anterior end of the female

(See also Plates XXXII, *o* and XXXIII, *i*)

The ascarid roundworm, one of the biggest and most common intestinal parasites of man, is found world-wide, although it becomes less common nearer to the North Pole. On the whole children are infected more frequently than adults (see Plate XXVI). The WHO estimates the number of carriers of the worm to be 1 300 million (with approx. 12,000 deaths per year). In many parts of Asia and Latin America, 50%–95% of the inhabitants are affected (Philippines, Malaysia, Indonesia, Brazil etc.) (CROMPTON, 1988). Man is the only host. The closely related species *Ascaris l. suum* generally remains at the larval stage in man and only rarely reaches full maturity.

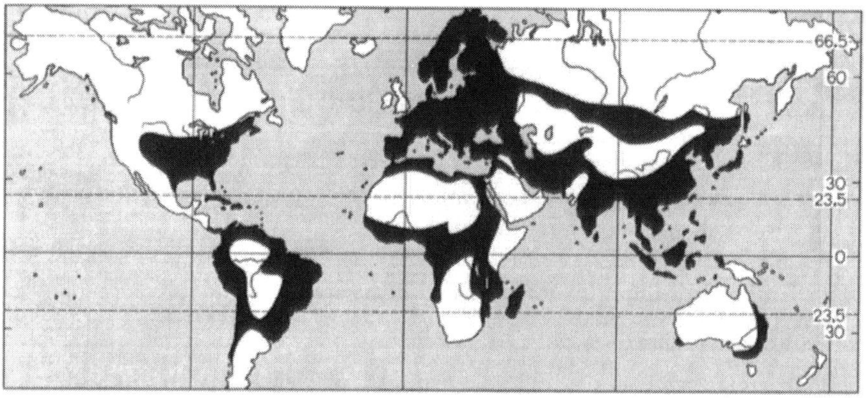

Distribution of *Ascaris lumbricoides*

Morphology and Development. *A. lumbricoides* adults inhabit the small intestine. They are yellowish-white or light pink in colour and usually reach 20–30 cm in length. The development of this roundworm proceeds without any intermediate host. A female can lay up to 240,000 eggs per day. The plump, oval or spherical eggs are approximately 60–70 × 50 μm; their shape can vary slightly. A typical egg (Plate XXVIII, *1b*) is yellow-brown in colour with a rough outer shell. It is very resistant to unfavourable conditions and can even survive for several months in weak solutions of formalin. It has even been claimed that egg shells have been identified in the sewage in prehistoric finds from salt mines (ASPÖCK et al., 1973; FERREIRA et al., 1980).

The ovum can be identified in a cross-section of a freshly laid egg (Plate XXVIII, *I*). Infective eggs contain a larva (Plate XXVIII, *3*), and occasionally the rather thinner and longer unfertilized eggs are found (Plate XXVIII, *II*).

The larva only forms within the egg casing in the open air after exposure to oxygen. Depending upon the external temperature, the embryonation can take from 8 to 50 days (Plate XXVIII, *3*); it can however also take much longer than this (see *Trichuris*, p. 220). The first and second moults actually take place within the "egg" (second larval stage). After ingestion of embryonated eggs, the young worms are liberated in the small intestine and begin migration (see Plate XXVIII, *III*, yellow pathway), passing through the small bowel wall and the portal venous system to the liver, where the third stage larvae develop, following another moult. They move to the right heart in the circulation and from there, via the pulmonary artery, to the lung within 1 week. The young worms leave the circulation through the alveolar capillary net and thus reach the lumen of the alveoli (heart-lung passage; blue pathway). Here the larvae moult once more (fourth larval stage) and pass into the gastrointestinal tract via the bronchi, trachea and throat are swallowed and then settle predominantly (approx. 87%) in the jejunum (red pathway). After a total of approximately 1.5–2 months (and one further moult) the worms become mature (prepatent period). The adult worms survive for 1–2 years. The laying of eggs is limited on average to 9–12 months, during which 50–60 million eggs are produced.

Clinical Symptoms. After ascarid roundworm infection in man, coincident with the passage through the lung, there are often symptoms such as fever, cough – often with blood-stained sputum – and transient eosinophilic lung infiltration demonstrable on X-ray (LÖFFLER's syndrome). If there is secondary infection, particularly in small children, the damage to the lungs proves very harmful and in some places this is associated with high child mortality (WHO, 1964). The worms inhabiting the intestine cause a variety of non-specific symptoms such as abdominal pain, vomiting, general restlessness and insomnia. Individual worms may sometimes wander into the pancreatic and biliary ducts, the appendix, bronchi or other organs. This can lead to corresponding illness, sometimes severe, such as acute obstructive pancreatitis, suppurative cholangitis, liver abscess or appendicitis. In severe cases an ileus verminosus may occur, which can even be provoked as a reactive phenomenon when only a few worms are present. There are often high blood eosinophil counts. As a result of small bowel perforation, adult worms can cause secondary bacterial infection resulting in peritonitis.

The death rate from *A. lumbricoides* infection is estimated at 6 per 100,000 carriers. The majority of the carriers (approx. 85%) remain asymptomatic. This does not, however, mean that they are entirely unaffected. *A. lumbricoides* infection leads to malnourishment through diminished appetite and impaired absorption of nutrients.

Generally speaking a roundworm infection has an immunosuppressive effect on the host and makes more susceptible to other infections. The *A. lumbricoides* antigen acts as one of the most potent of all parasitic allergens. It can lead to hypersensitivity, producing allergic reactions in the lungs, skin, conjunctiva and in the gastrointestinal tract. Each stage of development has its own immunizing activity. The host reacts to *A. lumbricoides* infection with high IgE production. A certain protective immunity develops, which can be ascribed to the activity of the migrating larvae, and which produces a cellular immunity with formation of IgM antibodies.

The protective role of humoral and cellular immunity can be shown experimentally in animals. For example, it has been shown that during reinfection the migrant larvae die rapidly because of precipitates blocking the excretory pores and anus, and they are then walled off in granulomas. In man, the so-called "knot of worms" is not infrequently found at post mortem.

According to observations on pigs infected with *A. suum*, the infection leads to hypertrophy of the muscle layers of the intestine, the small bowel mucosa takes on a wrinkled appearance, the crypts become shallower and the production of mucus falls. The result is an impaired absorption of nutrients probably also in man. Decreased growth has been observed in children and adolescents infected with roundworms, particularly when associated with malnutrition (see STEPHENSON, 1980). After deworming, growth and efficiency are considerably improved.

Transmission. Ascarid roundworm infection is acquired by eating raw lettuce and vegetables, when these plants have been fertilized with human excreta from carriers of *A. lumbricoides* (Plate XXVIII, *2, 4*). As the eggs must first mature for several days in the soil, Carriers cannot be a direct source of infection for their fellow humans (cf. *Enterobius vermicularis,* Plate XXIV and p. 218). However, eggs containing larvae (Plate XXVIII, *3*) can remain "dormant" for a long time and survive adverse environmental conditions well (they can even remain viable in a 5% solution of formalin for several months), so that contaminated ground can harbour viable eggs for months or up to 6 years (Samarkand, 14 years; PAWŁOWSKI, 1982). On the other hand, in tropical conditions they can die after only a few hours. The presence of ascarids is largely related to the socio-economic conditions of the people (population density, agrarian economy, use of human faeces as fertilizer). Thus, people in developing countries are not only frequently but also severely infected. Cockroaches have been suspected of having a role in the spreading of roundworm eggs but it is probably of only small significance in their distribution. Intrauterine infection has also been observed, with the newborn baby already being a carrier of roundworms.

Prophylactic Measures. Vegetables which are to be eaten raw must be thoroughly cleaned so that all eggs are removed, as it is possible that human excreta have been used as fertilizer.

Killing *A. lumbricoides* eggs in faeces is relatively difficult. Therefore, it is recommended that only purified sewage be used as a fertilizer. Chemical preparations are in any case not very effective, but generally speaking the eggs do become non-viable after about 6 months in temperate latitudes. In endemic countries the building of sanitary facilities to prevent ground contamination and the practice of personal hygiene are of paramount importance in reducing infections. Where possible in these countries, systematic deworming of the population with drugs should be practised.

Diagnosis by Microscopy. A diagnosis of ascariasis can be confirmed by examining a stool sample. The eggs, which are usually quite numerous (see Plate XXVIII, *1b, I, II*) are relatively easy to identify even under moderate magnification (×100–200) due to their characteristic shape and colour. Through careful microscopic examination of stools they can usually be found without concentration techniques (see p. 309). Also, the presence of worms can often be recognized by radiological examination using contrast media.

Diagnosis by Immunobiological Methods. Immunological techniques do not play a key role in the diagnosis of ascarid roundworm infections, as, on the whole, the eggs are easy to find on microscopic examination of the stool. Microscopy is also aided by the relatively high number of eggs even in a mild worm infection (see above). On the other hand, in the presence of aberrant ascaris larvae immunological detection affords the possibility of confirming the infection (see Larva migrans, p. 245). In these cases LEJKINA (1965) and LAMINA (1970, 1980) recommend a microprecipitation reaction using live second-stage larvae. This technique is considered suitable for the identification of visceral larva migrans, because it permits definite, genus-specific diagnosis. The first *Ascaris* antibodies appear 5–6 days after infection and reach their peak on the 25th–30th day, falling again after 90–100 days. At first IgM antibodies appear, and later IgG. During reinfection the antibodies influence the migration of the larvae by the formation of precipitates which interfere with their metabolism (see above).

Chemotherapy. Mebendazole (100 mg both morning and evening for 3 consecutive days) is considered the treatment of choice, although not in the first 6 months of pregnancy. Similarly, children under 2 years should be excluded from treatment with mebendazole. For them pyrantel embonate in a single dose is recommended (adults 10 mg base/kg body weight; children, 2.5–10 mg/kg body weight). However, medical supervision is necessary, since complications (ileus verminosus) can set in with a severe worm infection and drastic therapy. Similarly, a single dose of albendazole should be adequate (1 × 400 mg in adults; AYAD EL-MASRY et al., 1983).

Plate XXIX ⇨

Filariae

Wuchereria bancrofti

Brugia malayi

Loa loa

Plate XXIX

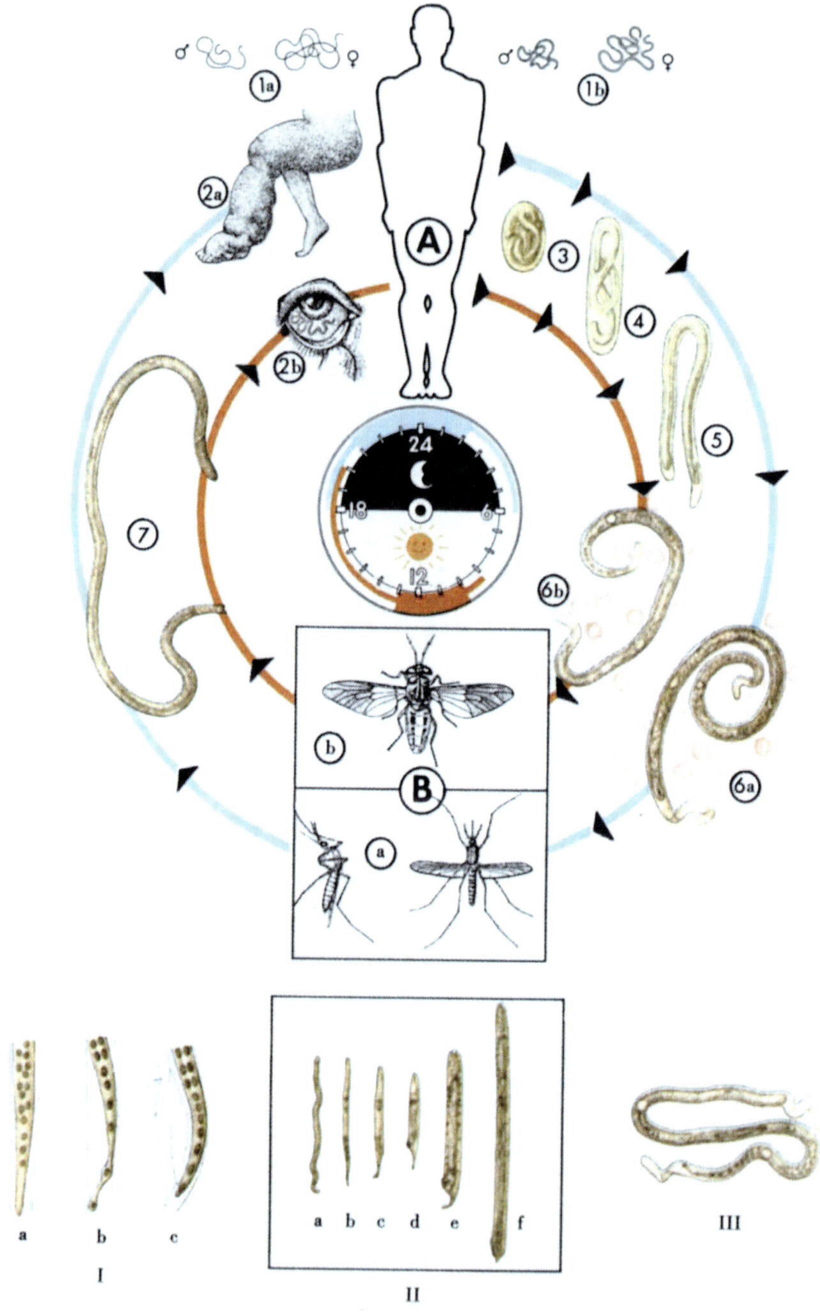

Wuchereria bancrofti (COBBOLD, 1877) SEURAT, 1921
Brugia malayi (BRUG, 1927) BUCKLEY, 1960
Loa loa (COBBOLD, 1864) CASTELLANI and CHALMERS, 1913

Ⓐ Definitive host: Man

1 Adult stages, male and female
a *Wuchereria bancrofti*
b *Loa loa*
2a Characteristic symptom in a *W. bancrofti* infection: elephantiasis
2b *Loa loa* passing through the conjunctiva
3 Egg containing microfilaria
4 Egg elongation
5 Free microfilaria (sheathed)
(3–5: development of the filaria in the uterus)
6a Microfilaria bancrofti (M. nocturna)
6b Microfilaria loa (M. diurna)
(Note the larval periodicity and date in the centre of the Plate)

Ⓑ Intermediate hosts and vectors: dipterans

a *Aedes* species and *Culex* species
b *Chrysops* species
7 Metacyclic infective microfilaria

Blue pathway, *Wuchereria bancrofti;* yellow pathway, *Loa loa*

I Posterior ends of some sheathed microfilariae (see p. 285)
a *Wuchereria bancrofti*
b *Brugia malayi*
c *Loa loa*

II Development of microfilaria to the metacyclic form in the vector:
u Microfilaria from blood
a–e Development period to the first moult
f Growth by elongation to the second moult

III Microfilaria malayi (M. nocturna)

(See also Plate XXXIII, *c*)

Filariae

Wuchereria bancrofti, Brugia malayi, Brugia timori and *Loa loa* are four species of filarial worm, the larvae (microfilariae) of which possess the remarkable property of periodicity. The microfilariae are found in the blood of man Ⓐ at night in the case of *Wuchereria bancrofti* and *Brugia* species and during the day in case of *Loa loa*. However, the subperiodic subspecies *W. bancrofti* var. *pacifica* does not have this characteristic. Transmission to man takes place by insect vectors (dipterans) Ⓑ (arthropod borne diseases). Individuals can be protected from bites of vector species using repellents, chemical compounds, which reject arthropods from the skin and clothing. All species are confined exclusively to tropical and subtropical countries.

The periodicity of the microfilariae is still an unexplained phenomenon. Many hypotheses have been formulated (circadian rhythms, CO_2 or O_2 tension of the peripheral blood) but so far it has not been possible to prove any of these theories experimentally. One hypothesis considers the periodicity an adaptation to the habits of the diptera which transmit these parasites. This view is also supported by the observation that the species of fly which live in close proximity to man, *Chrysops silacea* and *C. dimidiata*, are the essential intermediate hosts for *Loa loa* with diurnal periodicity. By contrast, monkey-infecting *Loa loa* which exhibit nocturnal periodicity are transmitted by the species *Chrysops centurionis* and *C. lenyi;* both species bite in the evening and the early part of the night. Moreover, these insects live in the undergrowth of the forest and bite humans extremely rarely. The monkey reservoir (many anthropoid species of monkey) nevertheless seems to be of considerable importance for humans. Another hypothesis considers the oxygen pressure inside the vertebrate host as the most important factor for the periodicity.

Apart from these pathogenic species, there are at least three others which are non-pathogenic and whose microfilariae may be found in the peripheral blood: *Dipetalonema perstans, D. streptocerca* and *Mansonella ozzardi*. They are widely distributed in Africa and America, are transmitted by insects of the genus *Culicoides* (small midges with a painful bite), but do not lead to any serious disease (see pp. 288–290). Morphologically these microfilariae can be easily distinguished from those of the pathogenic species as they are unsheathed (see figure on p. 285). A fourth non-pathogenic species, *Microfilaria bolivarensis,* has been found in human in Venezuela; it is larger than *M. ozzardi* or *D. perstans*. Presumably it occurs more frequently than has previously been recognised. It might also have been confused with *M. ozzardi* (Godoy et al., 1980 and according to Botto et al. (1984) it may belong to the genus *Onchocerca*).

Wuchereria bancrofti; Brugia malayi, B. timori

Parasitic agents of lymphatic filariasis

Wuchereria bancrofti endangers man in most tropical countries, especially Central Africa, South America, India and South China (about 900 million people exposed with about 80 million infected; WHO, 1982). The parasite is also to be found in South Spain, Tanger, the Nile delta and Turkey, and in the southern hemisphere to Uruguay, the Transvaal and Brisbane (Australia). On the other hand, *Brugia malayi* occurs in South Asia up to China and Korea (8.6 million infections).
The subperiodic subspecies *W. bancrofti* var. *pacifica* is also endemic in New Guinea, as well as in the islands of the Central and South Pacific, and especially in the Eastern Pacific. *B. malayi* on the other hand is limited to Indonesia, Malaysia, Indochina, South China, Central India and Sri Lanka. *B. timori,* with nocturnal periodicity, is found in a small focus in Indonesia.

Morphology and Development. The whitish, filamentous, sexually mature worms (Plate XXIX, *1 a*) reach a length of up to 10 cm in the females, and about 4 cm in the males (*W. b*). Both sexes live in the connective tissue, the lymph vessels and lymph nodes. The females are ovoviviparous producing juveniles known as microfilariae (i.e. Plate XXIX, *3*). While still in the proximal part of the uterus the larvae are coiled within the thin, pale egg shell, but as they progress toward the uterine pore, the egg shell elongates to accommodate the uncoiling of the larva and is retained as a delicate sheath of the microfilariae (Plate XXIX, *3–6*). In this stage they are released from the female. They appear in the peripheral blood nocturnally, mainly between 21.00 and 02.00 hours. (Microfilaria nocturna; see blue pathway in Plate XXIX). The arrangement of the nuclei at the end of the tail is characteristic of the individual microfilarial species (see Figure on page 285), and becomes recognisable in a stained preparation.
The adult stage of *B. malayi* closely resembles *W. bancrofti* in form and development but it is somewhat smaller. Their microfilariae show nocturnal periodicity (see Plate XXIX, *III*; pp 286/287).
For further development the microfilariae must be taken up by certain mosquitoes (genera *Culex, Aedes, Mansonia* and *Anopheles*). The larva penetrates the stomach wall of the mosquito and migrates into the thorax muscles where it develops through a relatively short, sausage-shaped stage into the infectious form (Plate XXIX, *B, II, a–f, 7*), moulting twice. The minimum temperature allowing development is 22 °C. With temperatures between 27° and 30 °C the parasite requires about 9.5–12.5 days to reach the infectious stage. Finally, the parasite enters the labium

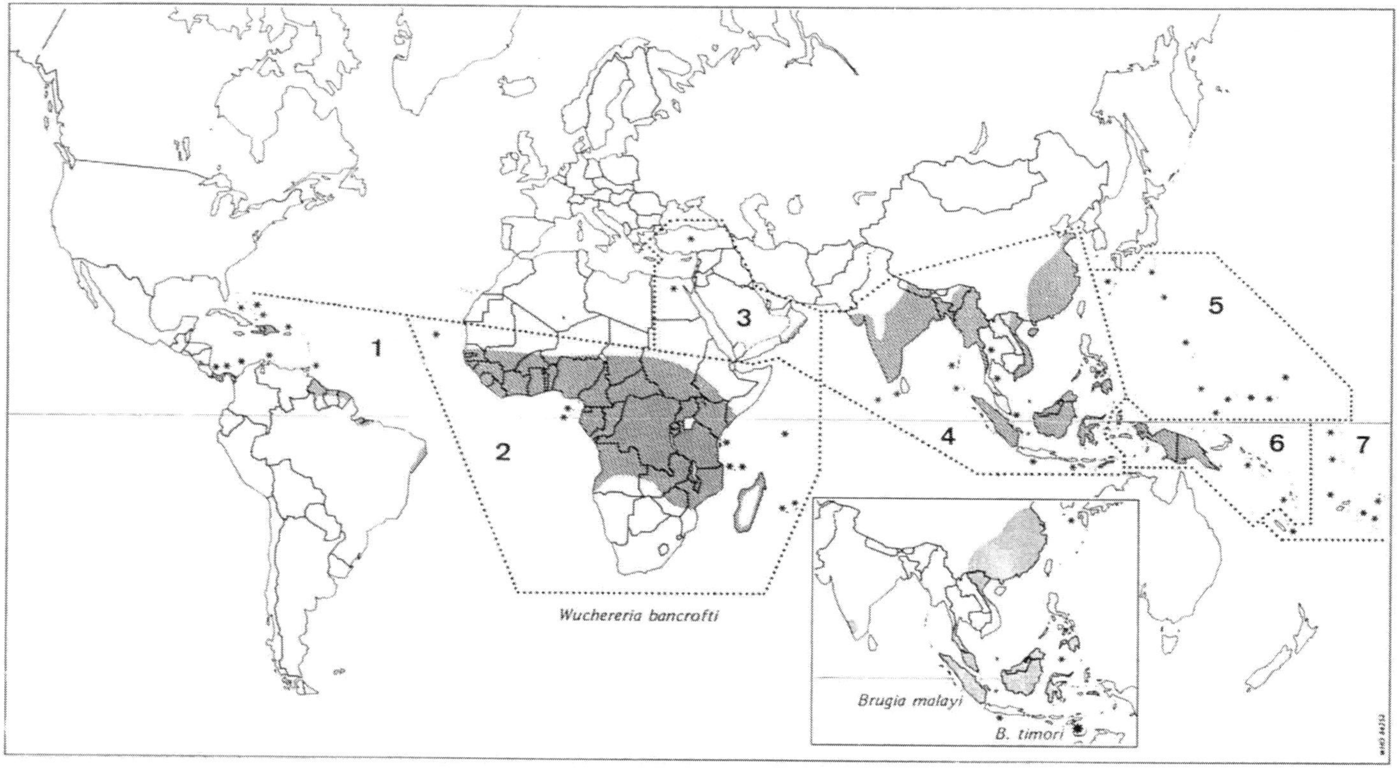

Distribution of *W. bancrofti (hatched)*, *B. malayi (grey)*, and *B. timori* * (i.e. of lymphatic filariasis; WHO, 1984 b).
1–5, Areas with region-specific vectors: *1* Tropical America, *2* Tropical Africa, *3* Middle East, *4* South Asia, *5* Far East, *6* New Guinea, *7* Polynesia. *1–6*, *Wuchereria bancrofti,* periodic form; *4, 7, W. bancrofti,* subperiodic form.
Inset, Distribution of *Brugia malayi* in South Asia and *B. timori* on Flores and Timor (WHO, 1984 b).

of the mosquito. When the insect bites a person, the parasite bursts out of the labium and enters the human skin. The moment of release coincides with the bending of the labium. Only after months do the filariae become sexually mature and deposit the first microfilariae (prepatent period). For *B. malayi* the prepatent period is about 2 months and for *W. bancrofti*, 7–8 months. In an infection with *B. timori* it is stated to be 3 months.

Clinical Symptoms. The clinical manifestations of lymphatic filariasis include a broad range of symptoms; these can differ markedly from one endemic region to another, so only general data can be given here. The reaction in the lymphatic system induced by the adult worm is conspicuous and leads to lymphadenitis and lymphangitis. Within 10–15 years a chronic picture develops, the chief characteristics of which are hydrocele, elephantiasis and chyluria as a consequence of severe lymphatic obstruction.
As part of a hyperergic reaction, diffuse lung infiltrations with a high eosinophilia may occur. The reaction continues with enlargement of the spleen and lymph nodes. Microfilariae are then found in the lungs and lymph nodes, but not in the peripheral blood (tropical eosinophilic lung syndrome). The incubation period is usually between 8 and 16 months, but it can also be much shorter.
In *Brugia* filariasis, elephantiasis of the lower leg and lower arm are typical. The oedematous swelling can double the size of the normal limb. In chronic cases the microfilariae are frequently absent, though this finding is less common in the Pacific region and in Indonesia (40% of elephantiasis patients have microfilariae in the blood).

The clinical signs and symptoms begin with the penetration of the young adult worms into the lymph nodes. There is an inflammatory reaction in the inguinal and/or axillary lymph nodes and following this the deep abdominal lymph nodes can also be affected. Fever and malaise occur. The acute attack lasts about 3–15 days in *W. bancrofti* infection and can recur many times in the course of a year. The lymphadenitis due to *B. malayi* occurs at intervals with shivering, fever, and other general symptoms. After suitable rest the manifestations disappear spontaneously. Induration develops in the region of the lymph nodes and spreads to involve the whole limb. The microfilariae can remain for years in the blood without serious disease manifestations.

Epidemiology. Transmission occurs, because of the nocturnal cycle of the microfilariae, through night-flying mosquitoes predominantly of the genus *Anopheles* (e.g. *Anopheles gambiae; Culex quinquefasciatus* is also a vector). The subperiodic species *W. bancrofti* var. *pacifica*, which occurs in the Pacific Islands, shows no periodicity and is predominantly transmitted by day-flying *Aedes* species (*A. scutellaris*) with several subspecies. *B. malayi* is mainly transmitted to man by mosquitoes of the genera *Mansonia* and *Anopheles*, and *B. timori* by *Anopheles barbirostris*. Up to 60 larvae are transferred with a single mosquito bite (at temperatures of 30°–37 °C).
The probability of becoming infected with filariae in endemic regions is not very high with temporary visits to the tropics. However, with longer visits to severely infected regions, where there is a real possibility of repeated infection, there is a severe danger of disease. Congenital transmission may also occur.
For the control of mosquitoes, synthetic insecticides and biological preparations (e.g. from *Bacillus thuringiensis* H 14 and the larvicidal *Bacillus sphaericus*) can be used (see pp. 103 and 282). – In Malaysia the subperiodic species *B. malayi*

also develops in cats, which often live in close contact with man. Little is yet known of the actual extent of the animal parasite reservoir.

Prophylaxis. Prophylaxis consists, in the first instance, of the control of the vectors and the elimination of its breeding places. Through the use of modern long-acting insecticides it is possible to destroy the adults. The use of repellents and mechanical preventive measures offers further worthwhile protection from flying mosquitoes (for insecticide resistance see p. 102).

Diagnosis by Microscopy. This is possible by the examination of a fresh blood sample (taken at night), in which the vigorously moving microfilariae (see Plate XXIX, *1 a–c*) are relatively easy to detect (see also p. 276).
A provocation test performed by administering 100 mg diethylcarbamazine citrate (DEC-C) by mouth often leads to increased microfilaraemia after 30–60 min. Care must be taken in areas with mixed filariasis lest a severe MAZZOTTI-reaction occurs (p. 284).

Concentration Techniques. Venous blood (3–5 ml) may be treated with about 10–15 ml of the following mixture and then centrifuged: 95 ml 5% formalin, 5 ml acetic acid and 2 ml of a concentrated alcoholic gentian violet solution (4 g to 100 ml 96% alcohol). In the deposit the stained microfilariae and leucocytes are found. This technique permits the detection of microfilariae in blood regardless of species. The "thick film" technique, or biopsy of palpable lymph nodes, in which worms can be located, are recommended.
Membrane filter methods are very useful (e.g. Nucleopore filter, pore size 3 μm). For these, 5–20 ml fresh blood (citrated or heparinized blood may also be used) is immediately put through the filter using a syringe. After careful washing of the syringe and filter, the latter is fixed in ethanol for 15 min and stained with DELAFIELD's haematoxylin. The filter can then be mounted and studied microscopically (see PIEKARSKI and SEITZ, 1987).

Diagnosis by Immunobiological Methods. Methods in use include the complement fixation test, the indirect haemagglutination test, and the indirect immunofluorescence test. The ELISA can also be used, but with *W. bancrofti*-filariasis it serves more for sero-epidemiological studies than for individual diagnosis (KALIRAJ et al. 1980). *Litomosoides carinii*, a rat filarial species, is suitable as an antigen for this purpose. HINZ et al. (1981) have indicated a possible cross-reaction with *Echinococcus* species. These indirect detection methods are especially useful in cases where there is a well-founded suspicion of filariasis but microfilariae are absent (e.g. in tropical pulmonary eosinophilia or single sex infection; see p. 273).

Chemotherapy. Diethylcarbamazine citrate (DEC-C) has proven to be a useful drug (6 mg/kg body weight, daily, by mouth, over 12 days; with *Brugia* species 3–6 mg/kg body weight, daily, over 6–12 days; regional differences occur and repeated treatment is necessary under some circumstances). The drug kills pre-

dominantly the microfilariae but the sexually mature worms are also considerably damaged. For mass treatment of the population in severely affected areas, where treatment on an individual basis is difficult, household cooking salt containing 0.2% DEC-C is reliable. However, ethical scruples have been voiced about this practice, although in China it has produced satisfactory results.
About 40% of the DEC-C is normally excreted unchanged in the urine. However, in acid urine up to 60% (the therapeutic effect is questionable here) and in alkaline urine only in 5% DEC is excreted, i.e. under some circumstances an accumulation of the drug with possible toxicity is to be expected. The dosage can be adjusted individually according to the pH of the urine (BRECKENBRIDGE, 1981). Other drugs available are flubendazole (5 × 750 mg i.m. weekly) and ivermectin (single dose of 30–50 μg/kg body weight, orally). Ivermectin is also thought to have a chemoprophylactic action (see SCHULZ-KEY et al. 1984, 1985).

Loa loa

Migrating filaria

Loa loa, a filarial species limited to Western Central Africa, occurs chiefly in the region from Sierra Leone to Angola, and predominantly in the region of the great rivers, the Congo, Niger, Wellé and Ogowé. The number of *Loa loa* carriers is estimated as 20–40 million (10 years ago 2–13 million, PINDER 1988).

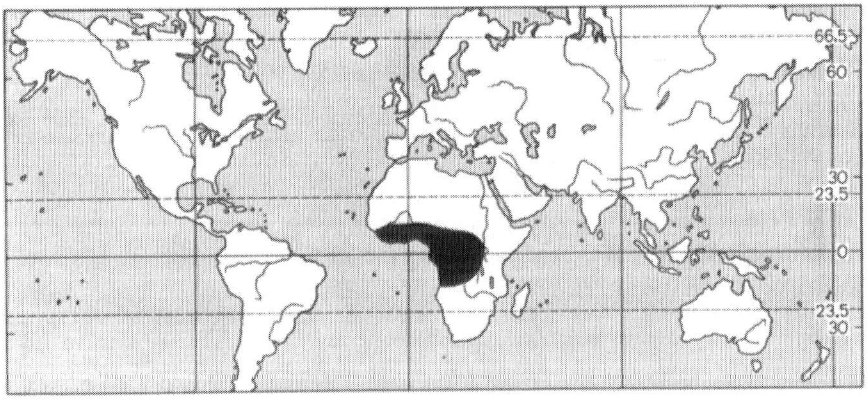

Distribution of *Loa loa*

Morphology and Development. The sexually mature worm (Plate XXIX, *1 b;* female up to 70 mm long and 0.5 mm wide; males up to 35 mm long) lives predominantly in the subcutaneous connective tissue. The sheathed microfilariae (about 275 μm long and 5 μm wide; microfilaria loa; Plate XXIX, *6 b, 1 c*) remain in the peripheral blood during the day (microfilaria diurna) between 08.00 and 20.00 hours (maxi-

mum between 10.00 and 13.00 hours) and are transmitted by flies of the genus *Chrysops,* which take them up in a blood meal. The microfilariae migrate to the thoracic muscles of the fly and develop within 8–10 days into metacyclic microfilariae. Finally, they pass into the head and proboscis of the fly. During biting they escape from the fly by bursting out of the proboscis and entering the human skin. The incubation period is usually several months long. The development to sexually maturity takes from 1–4 years (prepatent period); the lifespan of the females is quoted as 15 or more years (see pp. 286/287).
This filarial species is distinguished by the migratory drive of the sexually mature worms (migrating filariae); they are found in all possible sites of the subcutaneous connective tissue.

Clinical Symptoms. At sites where the adult worms are located, often on the forearm but also at any other place in the body, an allergic temporary swelling the size of a hen's egg appears (Calabar or Cameroon swelling). These swellings usually disappear within a few days but recur under some circumstances. Frequently there is a marked blood eosinophilia (30%–60%). Further sequelae include itching, general neuralgia and localized pain. The parasites are visible to the naked eye when they migrate through the ocular connective tissue (hence the name eye worm; Plate XXIX, *2b*); they cause severe itching and increased lacrimation. PINDER (1988) called *Loa loa* a neglected filaria.

Transmission. *Loa loa* is transmitted predominantly by flies of the species *Chrysops silacea, C. dimidiata* and *C. longicornis* (Ⓑ *b*). These insects fly during the day and can pick up the microfilaria loa with a blood meal. The breeding sites of the flies are widely distributed in damp meadows and in wooded areas and hence access for active control is difficult.

Diagnosis by Microscopy. During microscopic examination of the blood it must be borne in mind that the microfilariae only occur in the peripheral blood during the daytime and are not always very numerous. Therefore, it is necessary to search for microfilariae in fresh drops of blood where, because of their vigorous movements, it is easier to detect them, rather than in a "thick film" preparation (see p. 306). Microfilariae are usually first detected 1–4 years after infection.
The microfilariae may be concentrated with a Percoll density gradient of 1.09 g/ml (KIMMIG and BRAUN, 1980; FELDMEIER et al., 1981).

Diagnosis by Immunobiological Methods. A filarial infection can be detected by serological methods (see p. 274) and skin tests. The antigen is prepared from the adult worm of the dog filarial species *Dirofilaria immitis.* The results are only group specific but can be of value because the number of microscopically detectable microfilariae in the peripheral blood is often very small.

Chemotherapy. See *Wuchereria bancrofti,* p. 274.

Plate XXX ⇨

Onchocerca volvulus

Plate XXX

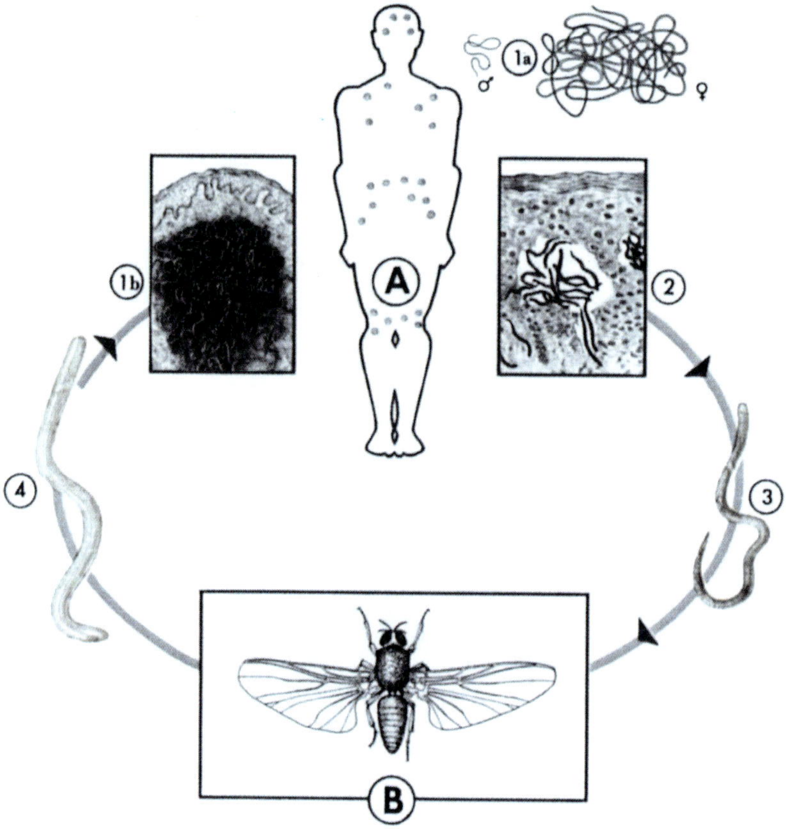

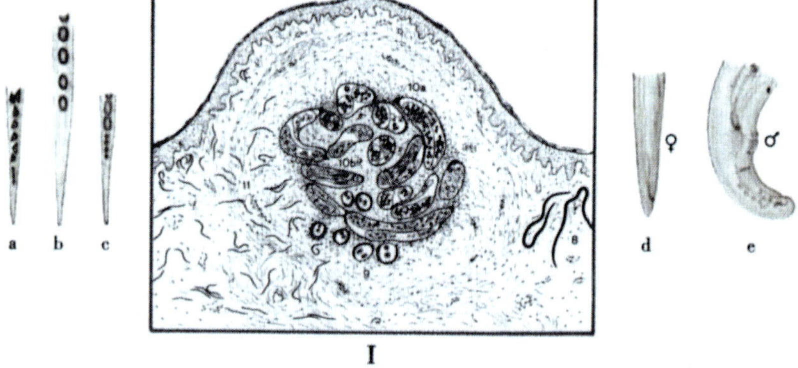

I

Onchocerca volvulus (Leuckart, 1893) Railliet and Henry, 1910 (= *O. caecutiens* Brumpt, 1919)

Pathogen of river blindness

Ⓐ Definitive host: man

1 a Male and female sexually mature worms
b Section through an *Onchocerca* nodule, approximately natural size
2 Microfilariae migrating into subcutaneous connective tissue
3 "Unsheathed" microfilaria

Ⓑ Intermediate host: *Simulium damnosum*

4 Metacyclic microfilaria from the proboscis of the intermediate host

I Section through *Onchocerca* nodule with numerous sections through adult worms and microfilariae (from Geigy and Herbig, 1953)

a–c Posterior ends of microfilariae of different species
a *Onchocerca reticulata* from the horse
b *O. volvulus* from man
c *O. gutturosa* from cow
d, e Posterior ends of sexually mature *Onchocerca* filariae

(See also Plate XXXIII, *b*)

Onchocerca volvulus is a very thin worm, up to 70 cm long, which grows in and under the skin in tumour-like nodules. The larvae, so-called microfilariae, migrate out of the nodules into the surrounding skin and often lead (in more than 10% of cases in certain areas) to blindness (river blindness). The microfilariae are taken up by blood-sucking flies of the genus *Simulium,* which transfer the worms from man to man. An animal reservoir apparently does not exist amongst house and domestic animals but does amongst some species of apes.
Onchocerca nodules are found predominantly on the trunk in a large proportion of the population of West and Central Africa, between Senegal and Tanzania, in the Nile Valley and in the Yemen. The nodules occur on the head and shoulder regions in parts of Central America (especially in Mexico and Guatemala as well as Ecuador; locally up to 83% of the population are infected). Altogether more than 50 million people are infected. In the region of the Volta basin in West Africa alone, more than 10 million people, with 70,000–100,000 of them blinded, are infected. In certain parts of this area up to 50% of the adult population may be affected.
The parasite was probably introduced from West and Central Africa to Mexico, Guatemala, Columbia and the northeast of Venezuela, although this may be questioned in view of the relatively large biological variations observed in the different geographic areas.
Onchocerciasis represents a significant social and economic problem in severely affected regions.

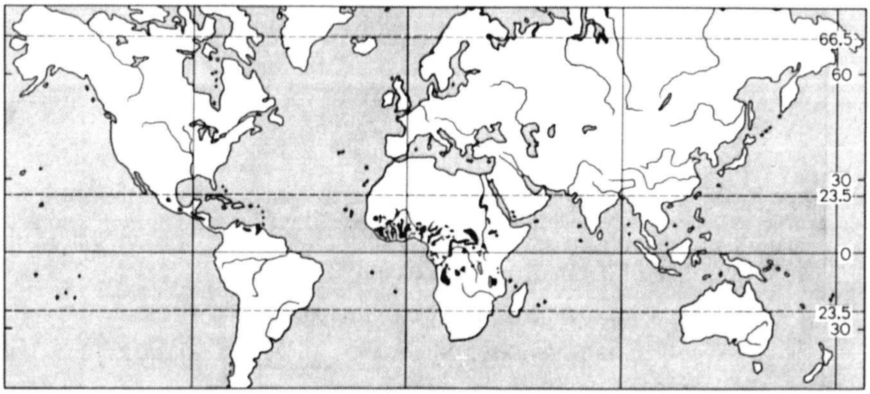

Distribution of *Onchocerca volvulus* (WHO, 1969)

Morphology and Development. The white, thin, filamentous female worms are 0.4 mm wide and have a mean length of 35–40 cm, but can be up to 60–70 cm long (Plate XXX, *1 a, b*); the males on the other hand are only 2–4 cm long. Several worms are usually grouped in a nodule (up to 60, on average 3–4) which is formed about 3–4 months after the initial infection, but they can also wander around freely without leading to nodule formation; it is predominantly the males which move to go to the females. The lifespan is assumed to be up to 10–16 years but is usually under 10 years. The surfaces of the worms have characteristic annular thickenings (FRANZ, 1982; FRANZ and BÜTTNER, 1983). The viviparous females produce about 1,000 "unsheathed" microfilariae daily (length approx. 300 µm; up to 1 million microfilariae in 1 year according to SCHULZ-KEY and KARAM, 1984). About 1 year after the start of infection the microfilariae appear in the subcutaneous tissue surrounding the nodules but they are only occasionally found in the peripheral blood. They live for between 6 and 30 months and show no periodicity as they are continually present in the skin (see *Wuchereria* and others pp. 271, 275).

Further development of the microfilariae takes place in buffalo gnats of the genus *Simulium* (black flies). In the thorax muscles of the fly the larvae develop (within 6–8 days at 21°–24°C and a relative humidity of 75%) into the infectious form (metacyclic larvae; Plate XXX, *4*). Morphologically these are quite similar to those of *Wuchereria* but are without a sheath (see Plate XXIX, *11*). At temperatures of less than 18°C larval development is arrested.

Simulium flies which take in more than about 20 microfilariae with a blood meal perish due to the infection. There are no morphological criteria for species differentiation. However *O. volvulus* has separate, geographically distinct variants which are differentiated from one another by their infectivity for the *Simulium* species, general epidemiology and clinical manifestations. Development of the filariae from the initial infection of the human host up to the appearance of microfilariae takes 9–20 months (usually 12–15, prepatent period; from the egg to the microfilarial stage is about 20 days).

According to the investigations of WEISSBRODT et al. (1984) there is a relationship between microfilaria density and age in patients from the hyperendemic regions of Liberia; in children there is a low density plateau, which is followed by an increase in density during adolescence. After this the density becomes constant again but at a higher level.

Clinical Symptoms. The clinical manifestations are mainly generalized but also may be confined locally and begin with dermatitis, itching and transitory papular erythema (rash). Secondary infection frequently occurs because of scratching. Following this stage there is skin thickening, oedematous swelling and lymphadenopathy, and finally atrophy of the skin. The skin nodules (up to 50) considered typical of this disease may, however, be absent (see above). ALBIEZ and BÜTTNER (1984) observed with regard to this that the nodes (onchocercomas), which are sometimes mobile and sometimes fixed, can grow together with the corium (e.g. Liberia, Burkina Faso). Many perforate externally, so that although rare, loops of worm may be visible (this is important when the differential diagnosis includes the possibility of infection with *Dracunculus medinensis*, p. 293).

Pathogenically the adult worms (in contrast to *Wuchereria bancrofti* and *Brugia* species, see p. 271) are of less significance than the microfilariae. Amongst other things the latter eventually lead to severe eye damage when they invade the cornea and finally die. Consequently, they can cause corneal opacities and inflammation of the iris, which finally lead to blindness but not to death (the blindness rate in parasite carriers is about 10%). The microfilariae reach the inner eye via the lymphatics, intercellular clefts and perivascular spaces, as well as the connective tissue cleavage spaces of the conjunctiva (GRÜNTZIG, 1984). Eye damage may even occur before the appearance of nodules. Five or more microfilariae in a skin biopsy from the region of the outer canthus indicates an increased risk to the eye. In Africans with long-standing onchocerciasis a characteristic skin picture develops (leopard skin, frequent in hyperendemic areas of Africa). These depigmented areas are seen particularly on the legs but also on the knees, shoulders, hands and feet. The picture resembles, amongst other conditions, that of leprosy or treponemal infection. The actual cause is still unknown (*Simulium* saliva?, a toxin of the microfilariae or adult worms?, anaphylactic reaction?; EDUNGBOLA et al., 1983). It is noteworthy that savannah onchocerciasis is accompanied by severe damage of eyes and rainforest onchocerciasis by less severe eye damage (ANDERSON et al., 1974). Evidently there exists a relationship between the location of the nodules and the density of the microfilariae in the skin. Thus, in Guatemala it is recommended that one should search for microfilariae in the region of the iliac crest in men and in the shoulder region in women. The severe complications of onchocerciasis are related to the long lifespan of the worms, which is on average 8 years but can extend up to 16 years. Microfilariae can still survive in the skin 30 months after the death of the adult worms. With increasing knowledge of the clinical manifestations, femoral hernias, elephantiasis and dwarfism are now recognized as possible signs of onchocerciasis (NELSON, 1970).

A useful **therapeutic measure** consists of the surgical removal of skin nodules (nodulectomy), in particular in the head region. This leads to a decrease in the density of microfilariae and a reduction in eye damage (particularly recommended in Middle America). However, by this technique only the adult worms of palpable nodules are removed and these frequently represent only a small fraction of the total adult worm population. Many adult worms lie in deeper tissues, for example between the muscle fibres as loosely encapsulated worm bundles which are operatively inaccessible. Ultrasound may assist in the location of non-palpable nodules (HOMEIDA et al., 1986).

Transmission. The only method of transmission is through the bite of an infected *Simulium* fly. In Africa the species particularly involved are those of the *S. damnosum* complex (about 10 transmitting species) and in Guatemala and Mexico it is the species *S. ochraceum*, *S. callidum* and *S. metallicum* amongst others. Repellents therefore offer effective protection from onchocerciasis. The elimination of *Simulium* flies – the goal of all control measures – presents a problem because they breed in rapidly flowing streams and have a generation time of 15–20 days. The preferred breeding sites in Africa are, depending on the species, the tropical rain forest or the savannah, but smaller bodies of water can also be used. In East Africa the larvae and pupae of *S. neavei* and *S. woodi* are frequently attached to the shell of a freshwater crab of the genus *Potamonautes*. These carry the larvae and pupae around with them, thus providing constant water changes (phoretic association). Although man is practically the only definitive host, isolated natural infections are also observed in the spider monkey (*Ateles geoffroyi*) and gorillas in the Congo region. No natural infection of the chimpanzee is known but it has been found to be a suitable experimental animal for the study of onchocerciasis.

In order to treat the most extensive breeding sites with insecticides, helicopters have been introduced and have proved most valuable. For this purpose low concentrations of the larvicides are usually used so as to avoid unintentional dosing of domestic animals (WHO; 1970). The results obtained so far are promising (WHO, 1985), but sanitation of the affected regions is difficult. It must be added that chemoresistance in *S. damnosum* has been observed in Africa (discovered in two species of the *S. damnosum* complex on the Ivory Coast). The strength of the resistance varies from region to region and is related to the intensity of the local agriculture.

A very effective, quasi-biological control of the larvae is possible with a toxic product from *Bacillus thuringiensis* H 14 (var. *israelensis*). Treatment of the breeding places permits selective elimination of the *Simulium* larvae (as determined by the standardized bioassay of the WHO) (Franz and Krieg, 1982). This method has proved valuable in, for example, cases of chemoresistance to the most often used insecticide temephos (Lacey et al., 1982).

Differences in the skin sites that have the maximum density of microfilariae occur between patients from Central America and patients from Africa because of the different localizations of the nodules. These sites correspond to the behaviour of the different vector species of the genus *Simulium*. For example, in Guatemala the microfilariae occur in the upper regions of the body where *S. ochraceum* prefers to feed, whilst in Africa they occur in the lower regions, which species in the *S. damnosum* complex seek out. The investigations of Renz and Wenk (1983) have clearly shown that the *Simulium* species of Africa bite predominantly on the foot joints and calves of the legs. In fact the greatest density of microfilariae is in the pelvic region and is relatively small in the calf and joint region. However, the transmission index is highest with the "calf-biters", because the frequency of blood meals from that site considerably increases the chance of the flies becoming infected. Suitable clothing offers a certain amount of protection from the bites, which therefore reduces the danger to man.

Diagnosis by Microscopy. Detection of the microfilariae is possible by examining one or more excised skin "snips". These should be bloodless if possible and taken from the area surrounding an *Onchocerca* nodule (with bleeding; care must be taken to avoid confusion with *Loa loa* or *Dipetalonema (Acanthocheilonema) perstans* blood microfilaria, and *D. streptocerca* skin microfilaria). The microfilariae are not evenly distributed throughout the skin but are present in particular regions according to strain (see above). A piece of tissue (2–3 mm in diameter) taken, for example, from the buttock region is placed in physiological saline. Some of the microfilariae emerge within about 30 min at a temperature of 37 °C. By centrifuging the saline solution the microfilariae can easily be concentrated and detected in the sediment by microscopy. There can be up to 1000 microfilariae per square millimeter, and they may be counted directly on a slide (partly hollow ground) after 30 min. Microfilariae are also detectable in urine by microscopy after provoking microfilaruria with 25–100 mg DEC-C (according to Roux and Picq, 1973). They occur in peripheral blood only in very severe attacks but in such cases are always present in the subcutaneous connective tissue. A more sensitive diagnostic method involves using collagenase for breaking up the tissue and thereby liberating all microfilariae (Schulz-Key and Karam, 1984).

Diagnosis by Immunobiological Methods. Previous experience has shown that reliable immunological detection methods still do not exist, because of the problem of cross-reactions with other nematode species (in particular with *Loa loa* but also with the cestode *Echinococcus granulosus*!). The direct haemagglutination test, the indirect immunofluorescence test, and the intradermal test using antigens from the adult worms are available. KLENK et al. (1984) used a purified extract of adult filariae of the species *Litomosoides carinii* in the ELISA and the latex agglutination test. The MAZZOTTI test has a certain diagnostic value. In this test, following the orally administration of 50 or 100 mg diethylcarbamazine an acute erythema and itching develops (as a consequence of an allergic reaction to killed microfilariae) accompanied by swelling of the regional lymph nodes (see RACZ et al., 1984). (Contraindicated in heavily infected patients!)

Chemotherapy. The microfilariae can be killed with diethylcarbamazine citrate (DEC-C; in adults 25 mg on the 1st day, 25 mg morning and evening on the 2nd day, 50 mg morning and evening on the 3rd day, 2 × 100 mg on the 4th day, 200 mg once daily from the 7th to the 14th day). However, severe allergic reactions (MAZZOTTI reactions) can occur, because the killing of the microfilariae causes a massive release of antigen. These reactions are reduced when the dose is decreased. An early reduction in the number of microfilariae prevents ocular onchocerciasis. For blind patients who have no symptoms other than blindness, DEC-C treatment brings no additional relief.

Suramin is as active against sexually mature filariae as against microfilariae (treatment regime as for trypanosomes see p. 15). Administration of corticosteroids inhibits the development of fatal progressive cases of exfoliative dermatitis. RIVAS-ALCALA et al. (1984) have shown that mebendazole exerts an obvious therapeutic effect on the embryogenesis of female worms.

A new drug for the treatment of onchocerciasis is ivermectin. It belongs to the avermectin group, a product of the actinomycete *Streptomyces avermitilis,* with broad-spectrum activity against nematodes and ectoparasites (according to experience in veterinary medicine). The rapid activity in human onchocerciasis does not extend to macrofilariae, but only to the free microfilariae and the developmental stages in utero. After initial stimulation of embryogenesis no more larvae are deposited; they degenerate and finally become absorbed. A single oral dose of 0.1–0.2 mg/kg body weight leads over 6–12 months to almost complete elimination of the microfilariae from the skin and eye with only the minor side-effects of DEC-C treatment (BENNETT et al. 1988). According to previous experience, ivermectin does not lead to the MAZZOTTI reaction. The microfilariae die more slowly following ivermectin treatment than after DEC-C, which apparently leads to fewer side effects. Ivermectin is apparently suitable for mass treatment (see AZIZ et al., 1982; AZIZ, 1986; SCHULZ-KEY et al., 1984, 1986; AWADZI et al., 1984; GREENE et al., 1985). According to LARIVIERE et al. (1986) a single dose of 0.150 mg/kg after 6 months suffices for the systematic control of onchocerciasis.

Distinguishing features of microfilariae in terms of the position of the nuclei in the cephalic and caudal ends (schematic; numbers in the illustration give the average size of the microfilariae after FAUST et al., 1970; PURNOMO et al., 1977).

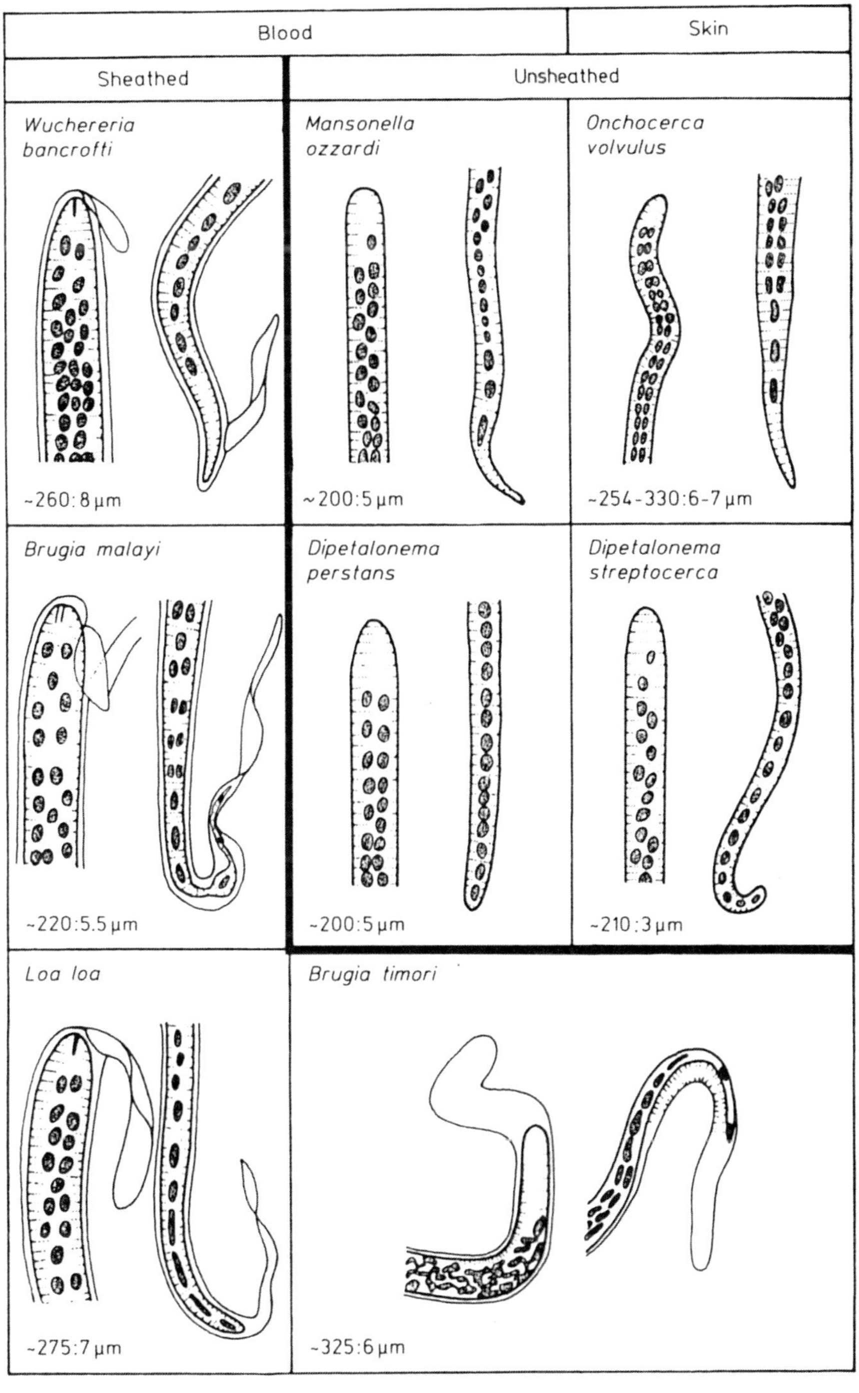
Blood
Skin
Sheathed
Unsheathed
Wuchereria bancrofti
~260:8 µm
Mansonella ozzardi
~200:5 µm
Onchocerca volvulus
~254-330:6-7 µm
Brugia malayi
~220:5.5 µm
Dipetalonema perstans
~200:5 µm
Dipetalonema streptocerca
~210:3 µm
Loa loa
~275:7 µm
Brugia timori
~325:6 µm

Review of pathogenic and non-pathogenic filariae (see figure on p. 285)

	Wuchereria bancrofti	*Brugia malayi*	*Loa loa*
Distribution	tropical and subtropical regions of Africa, Asia, S America	SE Asia, E Indies	Tropical Africa
Adult worms			
Site in definitive host	lymphatics	lymph nodes	subcutaneous tissue
Size (mm)	♂ 40 × 0.1 ♀ 85 × 0.25	♂ 18 × 0.08 ♀ 50 × 0.15	♂ 33 × 0.35 ♀ 60 × 0.5
Microfilariae			
length (μm)	230 – 290 × 6 – 8	170 – 260 × 5.5	250 – 300 × 7.5 μm
specific characteristics			
site of occurrence in definitive host	peripheral blood	peripheral blood	peripheral blood
detection method	fresh blood film thick film	fresh blood film thick film	fresh blood film thick film
lifespan			4 – 17 years
prepatent period	7 months	3.5 months	> 1 year
periodicity	nocturnal	nocturnal	diurnal
sheath	present	present, heavily stained	present
cephalic	blunt-round, 1 stylet	blunt-round, 2 stylets	blunt-round, 1 stylet
tail	tapering to delicate point	constricted between two	tapering gradually
nuclei (caudal tip)	non-terminal small, round	discrete nuclei in the tip small, ovoid	terminal large, oval
carrier (vector)	mosquitoes (*Culex, Aedes*)	mosquitoes (*Mansonia, Anopheles*)	flies (*Chrysops*)
duration of development in vector	6 – 20 days (according to species)	6 – 7 days (at 29 – 32 °C)	10 – 12 days
Chief host	man	man	man, ape baboons?
Clinical symptoms	elephantiasis, lymphangitis	elephantiasis, lymphangitis	skin nodes (Calabar swellings)
incubation time	8 – 16 months	8 – 16 months	8 – 16 months
drug therapy	DEC-C, flubendazole, ivermectin?	DEC-C, flubendazole	DEC-C

Onchocerca volvulus	*Mansonella ozzardi*	*Dipetalonema perstans*	*Dipetalonema streptocerca*
Africa, Central America	Central and S America, Caribbean Islands	Tropical Africa (Zaire), S and Central America (Guyana)	Central and W Africa
subcutaneous tissue, nodules	peritoneal connective tissue	abdominal, perirenal, pericardial tissue	subcutaneous tissue
♂ 30 × 0.16 ♀ 410 × 0.35	♂ 26 × 0.07 ♀ 49–81 × 0.15–0.25	♂ 45, diam. 0.06 ♀ 70–80 × 0.12	♂ 17 × 0.05 ♀ 27 × 0.08
254–332 × 6–8	190–230 × 5	190–200 × 4–5	180–240 × 3; hind part body enrolled
always in skin	peripheral blood	peripheral blood	subcutaneous tissue
skin snips	fresh blood film, thick film, skin snips	fresh blood film, thick film (confusion with *W. bancrofti, Loa loa*)	skin snips, scarified skin
15 years max,			
12–15 months		about 36 weeks	about 1 year (?)
none	none; many strains with cryptoperiodicity	none, but more frequent at night	none
none	none	none	none
blunt-round, no papillae	very short (3 µm), rounded off, blunt,	very short (3 µm), rounded off, no papillae	blunt, rounded off, papillae no stylet
tail tip free of nuclei	tail tip free of nuclei	nuclei up to tip of tail	nuclei up to tip of tail
large, oval	tapers to a filament		
Simulium	mainly *Culicoides, Simulium*	*Culicoides*	*Culicoides*
6–9 days	7 days (at 23–30 °C)	7–9 days	7–10 days (at 25–28 °C)
man	man, anthropoid primates	man, anthropoid primates	man, anthropoid primates
skin nodules, eye disease, blindness	none; hydrocoele?; enlarged lymph nodes?	non-pathogenic (?), Calabar swellings	pruritis, elephantiasis (?)
8–16 months		about 36 weeks	3–4 months
DEC-C, ivermectin Suramin	DEC-C?	mebendazole (side-effects), DEC-C	DEC-C

Mansonella ozzardi (Manson, 1897) Faust, 1929

Mansonella ozzardi occurs in Central and South America from Mexico to North Argentina as far as the Caribbean Islands.
The sexually mature worms (females about 49–81 mm, diameter 0.15–0.25 mm; males 26 × 0.07 mm) live in the retroperitoneal connective tissue, in the mesentery or in the visceral fatty tissue. The unsheathed microfilariae (190–230 μm × 5 μm; without periodicity) are found in the peripheral blood (see p. 285).

Transmission. This is by blood-sucking midges of the genus *Culicoides* (about 0.5–4 mm) of the family Ceratopogonidae as well as by *Simulium* species.

Development. The development of the microfilariae in the fly lasts about 5–7 days. The midges usually cause a very painful bite which in sensitized individuals can lead to strong allergic reactions. These insects frequently occur in large numbers and then become a plague for man and mammals; their larvae live in moist soil. *Simulium* larvae live in fast flowing water (see p. 282).

Clinical Symptoms. The clinical manifestations are variably interpreted. In general the parasites rank as non-pathogenic. However, small skin reactions of an allergic nature, enlarged lymph nodes and hydrocele occur in association with a *Mansonella ozzardi* infection.

Prophylactic Measures. These consist of the control of the vector with insecticides.

Diagnosis by Microscopy. Species differentiation is only possible in stained blood films (see p. 274). Moreover, other simultaneously occurring species (*Wuchereria bancrofti, Dipetalonema perstans*) must be considered. In *Mansonella ozzardi* microfilariae nuclei are absent from the rounded cephalic end and the pointed posterior tip (cf. *Dipetalonema perstans;* see figure on p. 285).

Chemotherapy. For drug therapy, as far as is necessary from the clinical point of view, DEC-C is recommended, but a successful outcome cannot be guaranteed.

Dipetalonema perstans (MANSON, 1891) YORKE and MAPLESTONE, 1926 (= *Mansonella perstans*)

Dipetalonema perstans is not uncommon in tropical Africa including the west coast from Senegal to Angola, in Algeria and Tunisia as well as Central America (Venezuela, Trinidad). These filariae may sometimes be found alone in man and sometimes associated with *Wuchereria bancrofti* and *Brugia* species or *Loa loa*.

The adult worms (females 70–80 mm, males 45 mm) live in the connective and fatty tissue of the mesentery, in the pleural cavity, in the pericardium and in the perirenal and retroperitoneal tissue. The unsheathed microfilariae (about 200 μm × 4.5 μm; see figure on p. 285) occur in the peripheral blood (without clear periodicity – subperiodic) but are more numerous in the heart, the lungs and large arteries. A high blood eosinophilia regularly occurs (allergic reaction). The infection is generally considered not to need treatment, but can be eliminated with DEC-C (see p. 274). Infection may, however, lead to troublesome generalized itching, abdominal pain, Calabar swellings, pleurisy and also possibly asthma. The prepatent period is assumed to be about 8–12 months.

Transmission. Vectors are blood-sucking midges of the genus *Culicoides* (e.g. *C. austeni, C. grahami*) in which development of the microfilariae lasts for 7–10 days up to the metacyclic stage. Several species of monkeys are considered to be animal reservoirs (Chimpanzees, Gorillas).

Control Measures. Against these filariae, control measures consist of reduction in the number of vectors, which breed in woods, jungles and swamps.

Diagnosis by Microscopy. Differentiation of the species is only possible in a stained blood film. The series of nuclei in the tail of the microfilaria extends to the tip (staining with haematoxylin see p. 285).

Chemotherapy. In an infection with *Dipetalonema perstans*, WAHLGREN and FROLOV (1983) reported good results with mebendazole (100 mg twice daily for 30 days; see footnote p. 219), whereas diethylcarbamazine is inactive in doses low enough to be well tolerated.

Dipetalonema streptocerca (MACFIE and CORSON, 1922) PEEL and CHARDOME, 1946 (=*Mansonella streptocerca*)

Dipetalonema streptocerca lives in the subcutaneous tissues of man and is considered to be non-pathogenic. It affects large parts of the population in West Africa and the Congo region, without causing serious disease, although itching, oedema and elephantiasis of the skin are attributed to this filarial species. It also occurs in some species of monkeys (genus *Pan*).

The microfilariae (length 180–240 µm, diameter 3 µm) are unsheathed. In fixed microfilariae the tail is curved inwards like a hook, and the anterior end is bluntly rounded. There is *no* oral *stylet* (in contrast, for example, with *Wuchereria bancrofti* and *Loa loa*). The tail is similarly blunt and contains oval nuclei up to within about 1 µm of the end. As in *Onchocerca,* the microfilariae live in the lymphatics of the subcutaneous connective tissue (not in the peripheral blood). The prepatent period is about 1 year; more precise data are not available.

Transmission occurs by insects of the genus *Culicoides* (see p. 288).

Diagnosis by Microscopy. Stained microfilariae are required for differentiation of the species and can be easily demonstrated in the scarified skin (snips). The row of nuclei in the tail of the microfilaria ends just before the tip of the tail, which is curved in the shape of a hook (see p. 285).

Chemotherapy. Diethylcarbamazine (6 mg/kg body weight daily for 8 days) is suitable for drug treatment. The pruritus which occurs with this can be controlled with corticosteroids.

Plate XXXI ⇨

Dracunculus medinensis

Plate XXXI

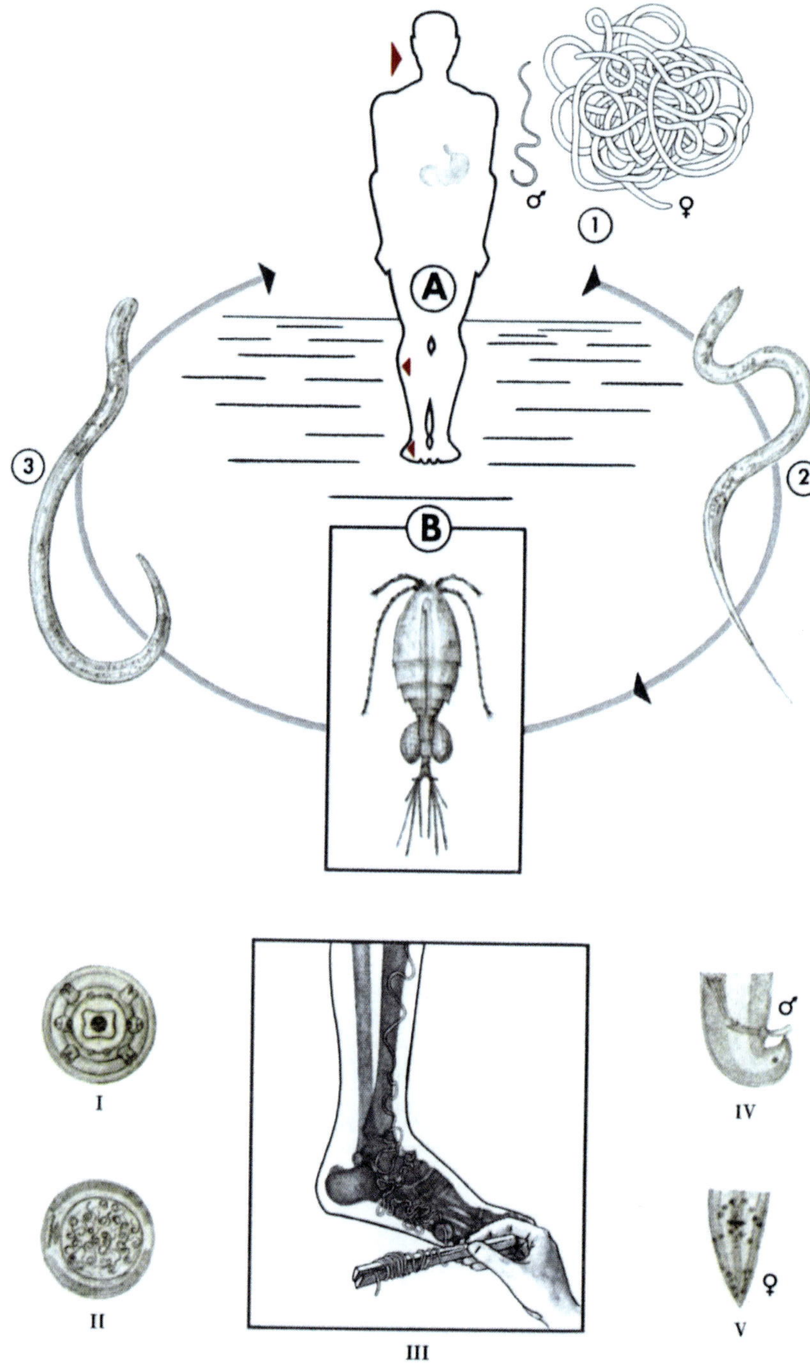

Dracunculus medinensis (LINNÉ, 1758) GALLANDANT, 1773

Medina worm (guinea worm)

Ⓐ Definitive host: man

1 Sexually mature medina worm, males and females
2 Free-living larva

Ⓑ Intermediate host: *Cyclops* species

3 Metacyclic larva from *Cyclops* species

I En face of the cephalic end of the female; characteristic arrangement of the papillae

II Cross-section through sexually mature female medina worm; uterus filled with larvae

III Leg infected with medina worm; the worm is drawn slowly out of a burst blister on the foot and is wound onto a stick (schematic X-ray picture).

IV Tail end of the male

V Tail end of the female

(See also Plate XXXIII, *a*)

A characteristic parasite of the rural regions of the Near East, Africa and Asia is the medina worm *Dracunculus medinensis* (also called dragon or guinea worm), the longest threadworm which lives in man. It occurs in Arabia, Iraq, Iran, East and West Africa and also in India and Pakistan, wherever the people collect their drinking water from ponds or open wells. The worm is acquired by drinking water containing small crustaceans (copepods). There are very marked variations locally in the frequency of infection (e.g. 54.7% of the population in Babanna/Nigeria). A total of about 5 million people are thought to be infected. The general incidence of dracunculiasis has declined in recent years (WHO).

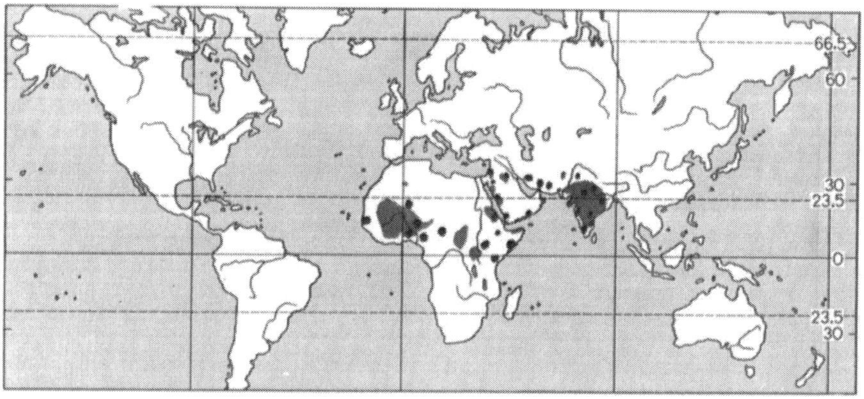

Distribution of *Dracunculus medinensis*

Morphology and Development. The sexually mature female worm, which is up to 1 m long and 1–2 mm wide (on average 690 × 2 mm; the males are about 3–4 cm long, Plate XXXI) lives in the subcutaneous tissue. It has species-specific arrangement of cephalic papillae (Plate XXXI, *1*). Whereas the male has a pair of spicules at the slightly hooked end of the body (length 0.4–0.5 mm), the female's body ends in a short straight point (Plate XXI, *IV, V*).

In adult females the narrow oesophagus is no longer functional and the intestine has no connection with the outside because the two uterine branches packed with larvae occupy almost the entire body space of the parasite (Plate XXXI, *II*). The vulva lies about 1 cm from the anterior end. The cephalic end of the worm ruptures to release the larvae, which may number more than 1 million per female (each larva is about 600 µm long and 15–30 µm wide). Development into the sexually mature worm, which takes place in the definitive host Ⓐ, takes 10–14 months (prepatent period).

The larvae (Plate XXXI, *2*) are dependent for their further development on small crustaceans Ⓑ of the genus *Cyclops* and related genera which live in ponds and wells and which ingest the larvae with their food. In the body cavity of these crustaceans, the larvae moult twice and become infective within 12–14 days at

25 °C (Plate XXXI, *3*). The larvae do not increase in size during this process. The minimum temperature for their further development is 19 °C. The larvae have a complete digestive tract.
If a definitive host (man and also dogs) swallows such a crustacean, the released larvae commence a migration which leads through the intestinal wall into the abdominal cavity. The worms mostly stay in the deep subcutaneous connective tissue and apparently pass via the lymphatic system into the region of the axillary and inguinal lymph nodes. An additional moult takes place during development into sexual maturity. Once sexual maturity is reached, and following copulation, the female medina worm moves to the body surface of the host and (10–14 months after the onset of infection) breaks through the epidermis. This occurs when the skin is in contact with water (stimulus of cooling), for example, the hands of washerwomen or the back of water carriers. The emerging worms are found most often on the legs and feet. Once in contact with water the female deposits larvae by expelling them from the prolapsed uterus protruding through the mouth or ruptured body wall near the anterior end.

Clinical Symptoms. Symptoms of disease are frequently absent or are non-specific before the emergence of the worm onto the body surface. Various toxic or allergic symptoms may occur, such as vomiting, diarrhoea, erythema, or exanthema. Prior to the breakthrough of the female, violent burning and itching occurs at the affected skin site. The skin changes, which develop as a result of the worm emerging, extend over an area 4 cm in diameter. A blister 3–5 cm in diameter forms, and after it has burst the head of the worm appears. The parts of the body mostly affected are the feet or lower leg, the arms and back (depending on the individual's lifestyle) and must frequently come into contact with water, because the water stimulates the release of larvae. The number of worms per patient is between one and four, rarely more. Worms which do not reach the surface die off and calcify. The greatest danger for the patients – they are mostly aged between 10 and 20 years – is the risk of secondary bacterial infection developing during the attempt to draw out the worm. A medina worm infection is generally accompanied by an eosinophilia (13%–36%). To date immunity has not been observed.

The mechanical removal of the worm may be attempted in the following way. If the worm emerges at the body surface, it can be cautiously withdrawn, even in the live state, by slowly winding it onto a stick, a few centimeters per day (Plate XXXI, *III*). This is the method used by the indigenous population. If the worm breaks, the remaining portion retreats to deeper tissues carrying pyogenic bacteria which can lead to severe illness with sepsis. It is therefore recommended that the worm should first be encouraged to expel all its larvae by constantly irrigating the breakthrough site with water, since it can then be removed more easily. Another method is to kill the worm prior to extraction by injecting it with phenothiazine emulsions or sublimate solutions, and only then slowly drawing it out or surgically removing it.

Transmission. *D. medinensis* infection is transmitted by drinking water containing infected copepods (mostly cosmopolitan species such as *Cyclops leuckarti*). During the dry period in particular, infected crustaceans are often concentrated at water sites which the indigenous population of affected countries use during a water shortage. As a result there is often a high incidence of infection related to a particular locality and season of the year. The disease is typical of poor rural regions. The larvae are about 0.5–1 mm long and remain alive in water for about 3–7 days, but for up to 6 weeks in moist soil. Even when slightly dehydrated they revive as soon as they contact water. Besides man, dogs, wolves and cats act as parasite reservoirs; they can also be infected relatively easily experimentally, but do not play an important role as natural animal reservoirs (possible confusion with *D. insignis*).

Prophylaxis. Worm carriers should be prevented from coming into contact with water for drinking. Chlorination of suspect water sites is very important. Insecticides (temephos, an organic phosphorus compound; 0.5–1.0 mg/l, effective for 5–7 weeks) and many molluscicides kill off the copepods, although they must be used with caution. If transmission can be successfully interrupted for a season in a community, new infections will not occur unless they are transported from other regions (migration of infected population groups, also individual nomads). For whole communities eradication can only be achieved through improved water supplies or water treatment, so closed water conduits are of particular value.
A simple filter made of dense material reliably traps the copepods, will provide individual protection against worm infection. The medina worm is therefore the most easily avoided parasite. As far as possible only boiled water should drunk in dangerous areas.

Diagnosis by Microscopy. There is no point in attempting diagnosis by microscopy because when they penetrate the skin, the adult worms are macroscopically visible and the larvae are released alive into the water (see section on transmission).

Diagnosis by Immunological Methods. An infection with this worm is usually recognized when the parasite breaks through the surface of the body (prepatent period 10–14 months). Prior to this both an indirect fluorescence test, using cryo-stored larvae or sections of *D. medinensis* as antigen (positive for up to 6 months after removal of the worms), and a skin test can be used.

Chemotherapy. Niridazole (25–30 mg/kg body weight, daily for 7–10 days) and thiabendazole (50–75 mg/kg body weight as a single dose or 2 × 25 mg for several days) have relieved symptoms. Metronidazole orally is also recommended as an effective drug (2 × 200 mg, 3 × daily for 10 days, elimination of pain and itching within 48 hrs). There is no effect on immature worms, however (SASTRY et al., 1978).

Plate XXXII ⇨

Protozoa – Helminths

Plate XXXII

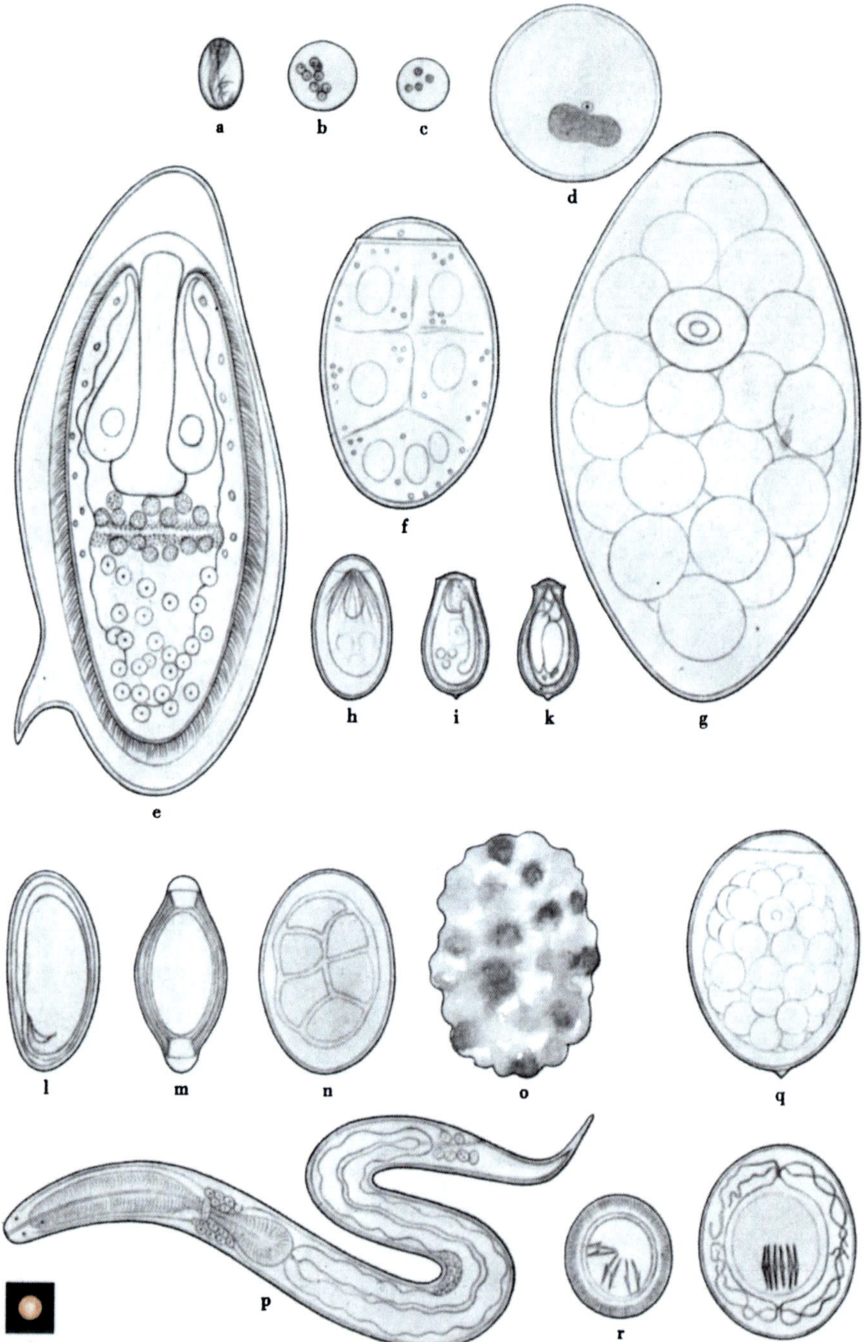

Protozoa – Helminths

(Scale: about 500:1)

	Parasite species	Fig.	cf. Plate
a	*Giardia lamblia*, cyst	(*1b*)	IV
b	*Entamoeba coli*, eight-nucleate cyst	(*4c*)	IV
c	*Entamoeba histolytica*, four-nucleate cyst	(*7b*)	IV and V
d	*Balantidium coli*, cyst	(*2*)	XII
e	*Schistosoma mansoni*, egg with miracidium	(*1b, 2*)	XVIII
f	*Paragonimus westermani*, egg	(*2a*)	XVII
g	*Fasciolopsis buski*, egg	(*1b*)	XIII
h	*Dicrocoelium dendriticum*, egg	(*1a*)	XVI
i	*Opisthorchis felineus*, egg	(*IIa*)	XV
k	*Clonorchis sinensis*, egg	(*2*)	XV
l	*Enterobius vermicularis*, egg	(*2b*)	XXIV
m	*Trichuris trichiura*, egg	(*1a*)	XXIV
n	*Ancylostoma duodenale*, egg	(*2a*)	XXV
o	*Ascaris lumbricoides*, egg	(*1b, I, II*)	XXVIII
p	*Strongyloides stercoralis*, larva	(*1b, c*)	XXVI
q	*Diphyllobothrium latum*, egg	(*1*)	XIX
r	*Taenia saginata*, embryophore	(*2*)	XXI
s	*Hymenolepis nana*, egg with oncosphere	(*1a*)	XX
t	Human erythrocyte (diameter about 7 μm) for comparison		XXXII

Plate XXXIII ⇨

Trematoda – Cestoda – Nematoda

Summary

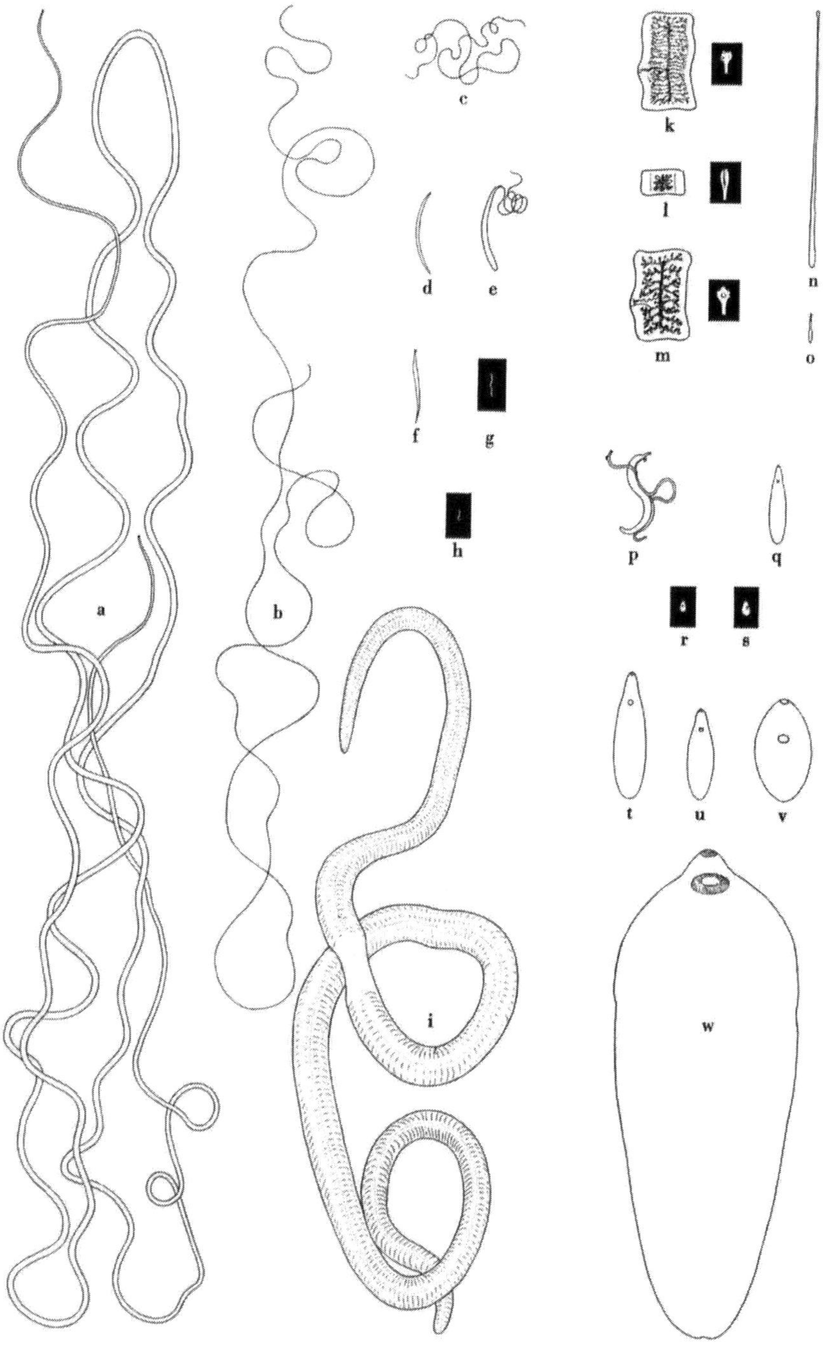
a
b
c
d
e
f
g
h
i
k
l
m
n
o
p
q
r
s
t
u
v
w

Trematoda – Cestoda – Nematoda

(approximately life size)

	Parasite species	Plate or page
a	*Dracunculus medinensis*	XXXI
b	*Onchocerca volvulus*	XXX
c	*Wuchereria bancrofti*	XXIX
d	*Ancylostoma duodenale, Necator americanus*	XXV
e	*Trichurus trichiura*	XXIV
f	*Enterobius vermicularis*	XXIV
g	*Trichinella spiralis*	XXIII
h	*Strongyloides stercoralis*	XXVI
i	*Ascaris lumbricoides*	XXVIII
k	*Taenia saginata*, gravid proglottid and scolex	XXI
l	*Diphyllobothrium latum*, gravid proglottid and scolex	XIX
m	*Taenia solium*, gravid proglottid and scolex	XXI
n	*Hymenolepis nana*	XX
o	*Echinococcus granulosus, E. multilocularis*	XXII
p	*Schistosoma mansoni, S. haematobium, S. japonicum*	XVIII
q	*Dicrocoelium dendriticum*	XVI
r	*Heterophyes heterophyes*	132
s	*Metagonimus yokogawai*	132
t	*Clonorchis sinensis*	XV
u	*Opisthorchis felineus*	XV
v	*Paragonimus westermani, P. kellicotti*	XVII
w	*Fasciolopsis buski*	XIII

The Most Important Methods of Microscopic Investigation [1]

General Preliminary Remarks

Under suitable conditions all parasites which affect man can be demonstrated relatively easily – presupposing relevant experience – with blood or tissues and in the excreta (urine, stools) with the aid of the light microscope. In everyday practice only a few species require immunobiological detection methods (e.g. species of the genera *Toxoplasma* and *Trichinella*). In microscopic investigations of worm infections it must always be remembered that the prepatent period (the time between infection and the formation of demonstrable development stages, e.g. eggs, larvae) varies depending on the species of parasite. Prior to the elapse of this period (see Table 1, p. 313) there is no prospect of obtaining a positive result, even if specific symptoms of disease are present.

Caution should always be exercised when handling material suspected of containing parasites. Samples of blood and stools should **basically be considered as infectious** with regard to fungi, bacteria and viruses. Cysts of amoebae and *Giardia lamblia*, sometimes larvae of *Strongyloides* species and some tapeworm eggs are capable of direct infection. Samples of blood, urine and stools not used in the investigation must be adequately disinfected or sterilized and then disposed of (for *Ancylostoma* and *Necator* see p. 229; for *Strongyloides* see p. 238).

Investigators must also familiarize themselves with artifacts of a non-parasitic nature present in specimens, so as not to confuse these with parasites. For example, leucocytes in a fresh stool preparation are often taken for amoebic cysts or if observed in the urinary sediment are mistaken for non-flagellated cells of *Trichomonas* species (see p. 42). Another source of confusion is ciliated cells, which in pathogenic processes are sometimes shed in increased numbers (e.g. from the nasal mucosa or the female genital tract) and which attract attention in fresh preparations because their cilia are still beating. They then give the impression of being ciliates. Pollen grains are sometimes taken for protozoan cysts or worm eggs, and plant and synthetic fibres for worms or worm larvae (see figure on p. 305).

Examination by microscopy requires much patience and care. When using concentration techniques, for example, at least 100 visual fields, or if possible an entire

[1] Details in Mehlhorn and Peters (1983); Mehlhorn (1988); see also Piekarski and Seitz (1987).

preparation, should be systematically scanned. This applies to both blood and stool investigations.
A special characteristic of the blood microfilariae is that, because of their periodicity, which depends on the species, they only appear in the peripheral blood during the daytime or at night (see p. 270). In the event of inadequate morphological knowledge and experience the advice of a specialist laboratory should be sought.

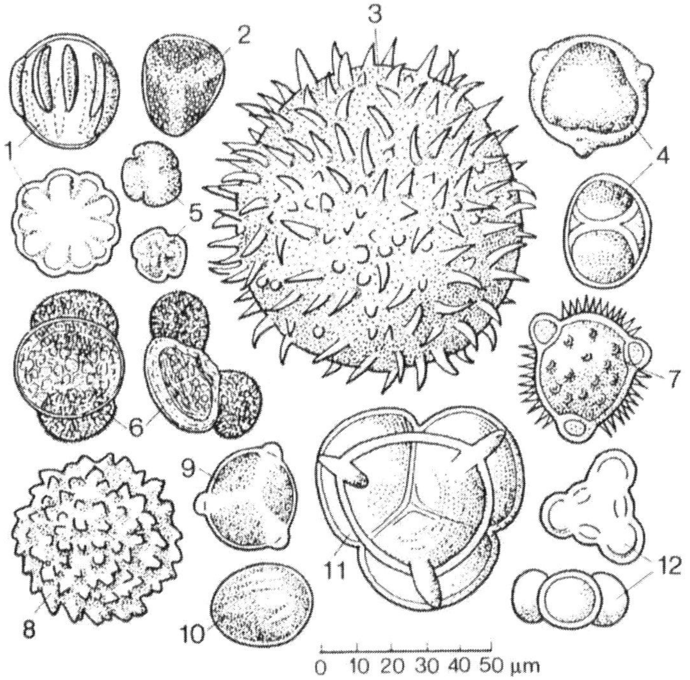

Pollen grains that are found in faeces and which are often mistaken for parasite eggs. *1* borage; *2* mullein; *3* mallow and marshmallow; *4* linden tree; *5* poppy; *6* fir and other conifers; *7* coltsfoot; *8* artichoke; *9* violet; *10* cabbage; *11* rhododendron; *12* broom. From original drawings by Rondeau du Noyer; from Brumpt et al. (1951).

I. Microscopic Examination of the Blood (Malaria Parasites, Trypanosomes, Microfilariae)

When examining the blood for parasites by light microscopy, thin and thick blood films should always be prepared.

The **thin Blood Film** (staining method can also be used for organ and biopsy impression smears):

A small drop of blood is placed at one end of a clean glass slide. A second slide (or cover glass) is placed at an angle on the first slide in contact with the drop of blood. With a steady movement the second slide is drawn toward the other end of the first slide, leaving a blood film behind. The drop of blood should be small enough to allow the film to terminate before it reaches the end of the slide.

The procedure is then as follows:
1. Fix the air-dried smear for 3 min with methanol;
2. Dry in air;
3. Stain with GIEMSA (5 drops of stock solution to 5 ml neutral or buffered distilled water per preparation); duration of staining 30 min;
4. Rinse off the staining solution with water; allow to dry.

Examination of the slides for protozoa should be done with an oil immersion lens. If filariae are suspected, the examination is carried out with a high-dry lens, after the stained smear has been covered with a layer of immersion oil and a cover glass. For staining microfilariae DELAFIELD's haematoxylin procedure is preferred or GIEMSA. A special concentration procedure is described on p. 276. Impression smears are recommended for the demonstration of tissue parasites. For this the smoothly cut surface of an organ (liver, lung) is pressed vertically on a slide and the smear air-dried, and processed like a blood film (see above).

The **"Thick film"**
A relatively small drop of blood is put on a clean, grease-free slide and spread with the corner of another slide, using a circular motion, to about the size of a thumb nail. The blood is stirred for at least 30 seconds to improve coagulation and air-dried. Without prior fixation or heating it is placed in a dish (film side down) of tap or distilled water for 5–10 min to remove the haemoglobin, one end of the slide being raised on a support. With freshly prepared preparations haemolysis takes only a few minutes; with preparations older than about 3 weeks it is advisable to add a few drops of acetic acid, however, the acid must be removed before the staining, e.g. with tap water (neutralized). After removal of the haemoglobin the film is stained with GIEMSA (without fixation).
With fresh preparations haemolysis can also be combined with staining. For staining the blood film is covered with GIEMSA solution (stock solution diluted

1:20 with water) for 20–30 min. Then the stain is carefully rinsed off with distilled or buffered water, or if necessary rain water. The thick films should be dried in air (not between filter paper).

It is important that water of neutral pH (7.0) be used – either double-distilled water or WEISE's original buffer mixture should be used in accordance with the instructions.

II. Examination of Stool Samples

If there is any suspicion of infection with intestinal parasites, at least three stool samples – obtained on different days – should be examined. In this way the probability of detecting all the parasites present is about 90%, and with five stool samples about 95% (this varies according to the species; HÖFLER et al., 1985). Each stool sample should first be examined macroscopically, looking for roundworms, hookworms, threadworms or tapeworms passed spontaneously or after medication. Tapeworm segments often still show spontaneous movement. They are therefore often mistaken for whole worms. The use of suitable concentration techniques is recommended for the microscopic examination, utilizing both fresh and fixed stool samples (see p. 308 ff.).
In a fresh preparation the characteristic movements of protozoa (amoebae, flagellates) and also their typical morphology can be clearly recognized on microscopy (e.g. *Trichomonas, Giardia;* oocysts of coccidia). If lively, motile small worms (larvae) are seen with microscopy, then infection with *Strongyloides stercoralis* is undoubtedly present. If however, the stool samples possible were kept for 1–2 days at room temperature (over 20 °C), larvae could already have hatched out of the eggs of hookworms or other nematodes (for this see the key to the identification of the filariform nematode larvae, p. 240/241).
When stool samples are stained (e.g. with iodine solution), the cytoplasmic organelles and nuclei of protozoa can be recognized; both parts are of importance for species differentiation (Plate IV).

As a general rule, a faecal smear should be so thin that print can be read through it. Eggs with colourless shells are more easily recognizable, when the faecal sample is mixed with LUGOL's solution or 2% eosin solution, instead of water.

Examination of Stools for Intestinal Protozoa

(see Plates IV and XXXII)

Coccidian oocysts which may be found in humans can be demonstrated more reliably using concentration techniques, such as those used for the demonstration of worm eggs, than by simple microscopic examination. The demonstration of the oocysts of cryptosporidia requires special methods (see p. 87).

1. *The Untreated Stool Specimen.* Place a sample of fresh faeces (preferably still warm) the size of a lentil, under a cover slip with a drop of physiological saline solution and examine at moderate magnification (× 400–500). Vegetative stages of intestinal flagellates and amoebae can be detected easily by their movements.

2. *Iodine Staining.* A small stool sample is placed between two cover slips, which are carefully pressed against each other. The two cover slips are drawn away from each other in opposite directions and the coated side is immediately placed in prepared drops of 4% iodine solution. After a few seconds the two smears will be saturated and any vegetative cells and cysts present will be stained (see Plate IV; see also concentration techniques, p. 309).

3. *The Stained Stool Preparation* (method of HEIDENHAIN). These smears are prepared either on a slide or on three to four cover slips and are fixed whilst still moist in sublimate alcohol.

The preparations (without being allowed to dry) are then treated as follows:

1. 30 min in iodine-alcohol (70% ethanol and iodine tincture or LUGOL's solution, roughly the colour of cognac).
2. At least 1 h in 70% ethanol
3. Rinse briefly in water
4. 1 h in 4% aqueous ferric ammonium sulfate solution (violet crystals should only be dissolved in distilled water)
5. Rinse briefly in water
6. 1 h in HEIDENHAIN's haematoxylin (1 g haematoxylin in 10 ml 96% ethanol and 90 ml distilled water; the solution must ripen for at least 4 weeks with the admission of air in a loosely stoppered clear glass bottle)
7. Rinse briefly in water
8. Differentiate by agitation in 2% aqueous ferric ammonium sulfate solution for 1–4 min.
9. Rinse for at least 30 min in running (tap) water.
10. Mount in balsam after passing through an ascending alcohol series and xylene (particularly important for the differentiation of amoebae species; see under *E. histolytica*, p. 53).

When performing microscopy the most favourable areas in the preparation must be sought; the quality of differentiation of the nuclear structures varies according to the thickness of the smear (see point 8).

Examination of Stools for Worm Eggs, Including Concentration Techniques (In part also suitable for Protozoan cysts)

1. *Direct Examination.* A piece of faecal material roughly the size of a lentil (fresh or fixed) is mixed on the slide with a little tap water or physiological saline. The resulting suspension should be as thin as possible. The sample is spread out and

covered with a cover slip, and examined microscopically at a magnification of × 100–200.

When looking for worm eggs, the mucus adherent to the faeces, in particular blood-stained specks of mucus, should always be investigated microscopically as well. Concentration techniques are often used because the eggs are frequently present in small numbers (see under 2–5).

2. *Concentration Technique using Saturated Saline Solution* (suitable only for the demonstration of nematode eggs, in particular hookworm eggs). A faecal sample of about the size of a hazelnut, roughly 1 g, is mixed with 20 times the amount of concentrated saline solution, which is added slowly. Coarse particles (plant residues etc.) are removed through a coarse-meshed filter or skimmed off the surface. After 20–40 min the eggs which have risen to the top are taken from the surface of the solution with a round wire loop (about 1 cm in diameter) bent at right angles. Several of the fluid films adherent to the wire loop are placed on a slide. The samples can be examined at a magnification of × 100–200. After use the wire loop should be flamed.

3. *Zinc Sulphate Concentration Technique (Flotation Method)*. 4–5 g faeces is mixed with 0.1% aqueous Triton X-100 solution, passed through a gauze filter in order to remove the coarse plant components, and centrifuged at 2000 rpm for 3 min. The resulting sediment is resuspended in 33% zinc sulphate solution (specific gravity 1.180) and is again centrifuged for 3 min at 2000 rpm. The eggs now on the surface can be removed with a loop and transferred to a slide (the addition of one drop of iodine solution makes the differentiation easier). This procedure is suitable both for worm eggs and larvae and also for amoebic cysts and oocysts of the intestinal coccidia of man and animals.

Instead of the zinc sulphate solution, a sucrose solution (56 g to 100 ml of water) can be used, to which a disinfectant has been added (e.g. 1.3 g phenol) to prevent fungal growth (from FRENKEL). This method is recommended for the demonstration of coccidian oocysts.

4. TELEMANN *Concentration Technique* (universal procedure for all worm eggs). A sample of faecal material roughly the size of a bean is suspended in a glass beaker with 7 ml semi-dilute hydrochloric acid (16%–18%). An equal amount of ether is added and the mixture stirred until it is a homogeneous emulsion. The mixture is poured through a wire gauze sieve (mesh width 1 mm) or a double layer of muslin into a centrifuge tube, with the aid of a funnel, and is centrifuged for 1 min. Four layers develop: a yellowish zone of ether at the surface, then a plug of detritus, a zone of hydrochloric acid, and at the bottom a small sediment. The latter contains pieces of muscle and cellulose as well as the worm eggs (caution is required because of the danger of explosion with ether). The sediment is transferred to a slide with a pipette and, after a cover slip has been applied, is examined microscopically.

5. *Concentration by the MIFC Technique*[1] (merthiolate-iodine-formaldehyde concentration; BLAGG et al., 1955). Two stock solutions are required:

A 250 ml distilled water, 200 ml Thimerosal (1:1000 in distilled water), 25 ml concentrated formalin, 5 ml glycerin (=480 ml)

B Fresh 5% iodine solution (5% iodine in 10% potassium iodide solution in 100 ml distilled water) which should not be above 3 weeks old

Both solutions should be stored in brown bottles.

Immediately prior to the processing of a stool sample, 4 ml of stock solution A are mixed with 1 ml solution B (or a multiple of these).

A stool sample about the size of a hazelnut is mixed with the specified amount of the mixture of solutions A and B. The sample, fixed and stained in this way, is passed through a double gauze filter and is mixed in a centrifuge tube with 7 ml ether (which should be stored in the refrigerator at 4 °C). This mixture is vigorously shaken so that no ether is to be found on the surface. The tube is left standing for 2 min and is then centrifuged for 5 min (approx. 3000 rpm). The plug of detritus between the ether and the MIF zones is gently detached from the sides of the tube with a rod and is poured out with the liquid component. At the bottom of the tube worm eggs, protozoan cysts and their vegetative forms are found in the sediment (caution is required because of the danger of explosion with ether). The sediment should be examined quantitatively under the microscope (×400).
For the dispatch of a stool sample, 0.25 g is first mixed with 0.15 ml solution B and shortly after with 2.35 ml solution A, and then stirred to form a homogeneous suspension. Stool material preserved in this way can still be examined microscopically after a period of months.
When looking for *Enterobius vermicularis* a stool sample is unsuitable – apart from macroscopic examination – because the eggs adhere to the edge of the anus and must be obtained by washing them off or detaching them with a strip of adhesive cellophane (see p. 218).

III. Microscopic Examination of Urine and Sputum

The following parasites can be found in the urinary sediment: *Trichomonas vaginalis* (see p. 42), eggs of *Schistosoma haematobium*, sometimes (rarely) *S. mansoni* and *S. intercalatum* (see p. 168), and also, rarely, microfilariae (see p. 283). *Enterobius* species eggs have occasionally been found in samples from female subjects. The eggs were probably carried into the vagina and then flushed out in the urine (see p. 219). Involvement of the kidneys and bladder by *Paragonimus* species (eggs) and *Echinococcus* species (protoscolices) when disseminated is also possible.

---

[1] Modified in accordance with method used in the Institute for Medical Parasitology at Bonn University.

Contamination from outside sources must always be considered when dealing with urinary sediments. Pollen grains often give rise to false interpretations and fly larvae are occasionally found in the urine (pseudomyiasis). These have always been introduced accidentally from outside.

To carry out the urine examination in practice a clean urine sample, preferably a morning sample, is centrifuged at about 2500 rpm for 3–5 min and the sediment is examined under the microscope. Initially a magnification of ×16–30 is used followed by a higher magnification to clarify suspect or obscure structures.

When the lungs are infected with *Paragonimus* species, eggs are found in the sputum (see p. 157). Cysts of *Pneumocystis carinii* can also be found in the lungs (p. 91). For this the sample of sputum is added to 3 times its volume of physiological saline, is well mixed and then centrifuged. The parasites can be found in the sediment. Protoscolices or individual scolex hooks can also be found in cases of pulmonary echinococcosis.

IV. General Comments on Serological Diagnosis

Intracellular parasites or those which live in close contact with an organ (e.g. *Leishmania, Trichinella*) mostly stimulate their host to marked antibody production. The quality of the antibody detection procedure largely depends – (assuming correct technical execution) – on the quality of the antigens, these should be homologous as far as possible. A deficiency in most of the serological methods used in practice to date is the absence of comparability of antigens and of internationally recognized standard sera. Apart from the use of polyclonal antibodies for the detection of parasites – the customary method hitherto – attempts are increasingly being made to detect soluble antigens in serum and stools as well. Finally, new methods are becoming available using monoclonal antibodies, which are expected to give greater specificity.

Monoclonal antibodies are produced by the hybridoma technique, an in vitro procedure based on the idea of creating a combination of the properties of basically different cells through fusion to form hybrid cells. In the practical procedure for this, mice or rats are first infected with the desired antigen so as to stimulate the formation of specific lymphocytes. These are predominantly found in the spleen and can be obtained from this organ. The lymphocytes are encouraged to fuse with a certain kind of tumour cell. As is well known, tumour cells have the property of unlimited growth and they therefore introduce this property into any hybrid cells that form. In this way a new uniform cell develops which now has the highly desirable property of multiplying indefinitely in vitro and also releases antibodies into the surrounding nutrient medium. These monoclonal antibodies, i.e. antibodies which have developed from one cell (clone), can now be used for the detection of circulating antigens from parasites which may be found in the serum. It has already been possible to use this procedure for some parasites, e.g. for *Toxoplasma gondii*. For example SETHI et al. (1984) have been able to demonstrate six individual antibodies produced from hybridomas, which attach in each case to specific regions of the *T. gondii* cells and can be visualized by fluorescence microscopy.

Despite the limitations mentioned, there have been considerable advances in the past 10 years in the serological diagnosis of parasitic diseases. These are due both to the higher degree of purity of the antigens and to the introduction of new, sometimes very specific, procedures, such as the indirect immunofluorescence test and ELISA (see Tables 2, 3).
The interpretation of a single case based on a single diagnostic method is often not possible, but the use of several methods will allow clear conclusions to be drawn.

Table 1. Prepatent period of various helminths, i.e. the period that the parasite is present in the body before it gives rise to parasitological evidence of its presence, for example eggs in the faeces or microfilariae in the blood (B) (Compiled from various data in the literature)

	Developmental stage		Prepatent period (days)	Lifespan in man (extreme values)
Trematodes				
Fasciolopsis buski	Egg		ca. 30–90	6 months
Echinostoma ilocanum	Egg			
Metagonimus yokogawai	Egg		10–14	2 years
Heterophyes heterophyes	Egg		7–8	2–4 months (?)
Schistosoma mansoni	Egg (lateral spine)		49	up to 30 years
Schistosoma intercalatum	Egg (terminal spine)		50–55	up to 25 years
Schistosoma japonicum	Egg } Small lateral knob		20–26	
Schistosoma mekongi	Egg } Small lateral knob		35	
Cestodes				
Taenia saginata	Proglottid		77–84	up to 20 years
Taenia solium	Proglottid		35–74	up to 25 years
Hymenolepis nana	Egg		14–28	2 weeks to months
Diphyllobothrium latum	Egg		18–21	15–20 years
Dipylidium caninum	Proglottid		20	
Nematodes				
Trichuris trichiura	Egg		60–90	Several years
Enterobius vermicularis	Egg		37–101	ca. 100 days
Ascaris lumbricoides	Egg		50–80	1–1.5 years
Ancylostoma duodenale	Egg		35–42	5–12 years
Necator americanus	Egg		35–42	5–8 years
Strongyloides stercoralis	Larva		17–28	20 years
Trichostrongylus orientalis	Egg		25–30	
Wuchereria bancrofti	Microfilariae	B	ca. 1 year	17 years
Brugia malayi	Microfilariae	B	50–60	8–10 years
B. timori	Microfilariae	B	90	
Loa loa	Microfilariae	B	ca. 1 year	17 years
Onchocerca volvulus	Microfilariae	Skin	12–15 months	15–18 years
Mansonella ozzardi	Microfilariae	B	?	
Dipetalonema perstans	Microfilariae	B	8–12 months	
D. streptocerca	Microfilariae	Skin	3–4 months?	

Table 2. Summary of the pathogenic intestinal parasites with details of clinical and diagnostic data

Specimen for examination	Parasite species	Developmental stage in the intestine; ⊕ with migration of larvae in bloodstream via heart, lungs (size)	Site of parasite	Typical clinical symptoms (apart from the more frequent latent infections)
1	2	3	4	5
Stools (intestinal) content)	1. *Giardia lamblia* [d]	Trophozoites and cysts	Small intestine ↓	Giardia dysentery
	2a. *Isospora belli* [d]	All developmental stages		Coccidian dysentery
	2b. *Cryptosporidium* [d] species	All developmental stages		Enterocolitis, gastroenteritis, watery stools
	3. *Sarcocystis suihominis, S. bovihominis*	Only sexual development in man		Only with *S. suihominis* enterocolitis of short duration
	4. *Balantidium coli*	Trophozoites and cysts	Large intestine ↓	Acute balantidium dysentery
	5. *Entamoeba histolytica* [d]			Acute amoebic dysentery with haematophagic amoebae; chronic course
	6. *Ancylostoma duodenale* [a]	Sexually mature worms (up to 13 mm) ⊕	Small intestine ↓	Iron deficiency anaemia, cachexia, often greatly increased blood eosinophilia [c]
	7. *Necator americanus* [a]	(up to 12 mm) ⊕		
	8. *Strongyloides stercoralis* [a, d]	(up to 2.2 mm) ⊕		Abdominal pain [c]; sometimes very severe course
	9. *Ascaris lumbricoides*	Sexually mature worms (200–300 mm) ⊕		As for 14–17; transient eosinophilic pulmonary infiltrate [c]
	10. *Trichostrongylus orientalis*	(4.5–9 mm)		General toxicity (p. 316, Nr. 14–17)

See footnotes on p. 316, 317

and also detection procedures of practical importance (see also Table 3, p. 318 ff.)

Extra-intestinal mani-festation	Detection procedures									Special comments	Chemotherapy (indications)
	a Microscopy (fresh stool sample)	*b* Culture	*c* Animal experiments	*d* Complement fixation	*e* Precipitation reaction	*f* Indirect immunofluorescence	*g* Indirect haemagglutination	*h* ELISA	*i* Skin test		
6	7									8	9
Bile ducts	Trophozoites; cysts ~12 µm	+		+?		+?				Zinc sulphate concen-tration	Metronidazole, et, similars comp.; chloroquine
None	Oocysts a ~30 µm										Sulphonamides + trimethoprim
	b ~5.5 µm									Carbol-fuchsin staining	
None	Sporocysts ~14 × 8 µm Oocysts ~22 × 12 µm										
None	Trophozoites ~50–150 µm, cysts ~35 µm	+	+?								Nitroimidazole, sulphadiazine, tetracycline
Abscesses in liver, lungs and brain; rarely, vaginal discharge	Trophozoites within erythrocytes ~20–40 µm, cysts ~12 µm	+		+		+	+	+		MIFC procedure, zinc sulphate concentra-tion, latex test	Nitroimidazole, chloroquine, diloxanitfuroate, bephenium, paromomycine, dehydroemetine
None	Eggs ~60 µm Concentration procedure, see p. 308 ff.				+?					Faecal culture (reagent glass method; see p. 238)	Bephenium salts, pyrantelembonate, mebendazole, albendazole
	~60–70 µm										
None	Larva 400–500 µm					+		+			Thiabendazole, tetramisole, mebendazole
Bile duct, liver rare	Eggs ~60 µm				Micro-precipi-tation reaction for nos. 6–10					Macroscopic demonstra-tion of worms in stool, pos-sibly also in vomit	Mebendazole, albendazole
None	Egg ~80 µm									See under nos. 6–8	Mebendazole, bephenium, and piperazine preparations

Table 2 (continued)

Specimen for examination	Parasite species	Developmental stage in the intestine; ⊕ with migration of larvae in bloodstream via heart, lungs (size)	Site of parasite	Typical clinical symptoms (apart from the more frequent latent infections)
1	2	3	4	5
Musculature	11. *Trichinella spiralis*	Intestinal stage 2–4 mm, muscle stage up to 1 mm	Small intestine and musculature	Transient dysentery [c], muscle pain, oedema
Stool (intestinal content)	12. *Trichuris trichiura*	(up to 50 mm)	Large intestine	as for 14–17 [c]
Anus	13. *Enterobius vermicularis*	(up to 12 mm)	↓	Anal itching
Stool (intestinal content)	14. *Heterophyes heterophyes* [a]	(up to 1.7 mm)	Small intestine	Varying non-specific intestinal symptoms depending on severity of infection and individual predisposition
	15. *Metagonimus yokogawai* [a]	(up to 2.5 mm)	↓	↓
	16. *Fasciolopsis buski* [a]	(up to 75 mm)	↓	↓
	17. *Echinostoma ilocanum* [a]	(up to 6 mm)	↓	↓
	18. *Taenia saginata*	Sexually mature tapeworms (up to 10 m long)	↓	↓
	19. *Taenia solium*	(3–4 m)	↓	↓
	20. *Dipylidium caninum*	(200–400 mm)	↓	↓
	21. *Diphyllobothrium latum*	(10 m or more)	↓	Pernicious anaemia Vitamin B_{12} deficiency
	22. *Hymenolepis nana* *H. diminuta*	(up to 40 mm)	↓	as for 14–20

[a] Species occurring or acquired exclusively in tropical countries (and mines)
[b] The data given in columns 7d–i (Nr. 19) apply only to cysticercosis, not to intestinal infection

Extra-intestinal manifestation	Detection procedures: a Microscopy (fresh stool sample)	b Culture	c Animal experiments	d Complement fixation	e Precipitation reaction	f Indirect immunofluorescence	g Indirect haemagglutination	h ELISA	i Skin test	Special comments	Chemotherapy (indications)
6	7									8	9
Muscle biopsy (muscle trichinella)	None		+?	+	+	+	+	+	+	Latex test, micro-precipitation test	Symptomatical corticosteroids, mebendazole, ivermectin
None	Eggs ~50 µm									Concentration procedure	Mebendazole
↓	Eggs ~55 µm									Anal smear	Pyrantelembonate, mebendazole, pyrvinium embonate
↓	Eggs ~30 µm										Praziquantel, niclosamide
↓	Eggs ~25 µm										
↓	Eggs ~140 µm										
↓	Eggs ~90 µm										
Cysticercosis[b], mostly with *T. solium*; musculature, CNS	Eggs rare, proglottids 1 cm long									Species characterized by number of uterine branches (Fig. XXI)	Praziquantel (also niclosamide in cysticercosis)
	as no. 18			+		+	+	+	+	Cysticercosis	
	Eggs ~25 µm									Eggs in small packets	
	Eggs ~70 µm (with operculum)										
Cysticercoid in intestinal villi	Eggs ~50 µm									Concentration procedure (see p. 308)	

[c] High eosinophilia, sometimes more than 50%

[d] Special risk in immunocompromised patients (e.g. AIDS patients)

Table 3. Extraintestinal blood and tissue parasites

Specimen for microscopic diagnosis	Parasite species	Name of disease, typical clinical symptoms (apart from the more frequent latent infections	Diagnostically important developmental stages in man
1	2	3	4
Blood	1. *Trypanosoma brucei gambiense* [a] *T. b. rhodesiense* [a]	African sleeping sickness	Flagellate stages (~10–30 µm)
	2. *T. cruzi* [a]	CHAGAS' disease	Flagellate stages (~20 µm)
	3. *Plasmodium vivax* [a]	Benign tertian malaria	Erythrocytic schizonts and gametocytes
	P. malariae [a]	Quartan malaria	
	P. falciparum [a]	Malignant tertian malaria	
	P. ovale [a]	Ovale malaria	
	4. *Wuchereria bancrofti* [a]	Elephantiasis	Microfilariae (~200–275 µm) of periodic occurrence (see table, p. 286/287): nocturnal
	5. *Brugia malayi* *B. timori* [a]		nocturnal
	6. *Loa loa* [a]	Transient skin swelling	diurnal
CSF	7. *Toxoplasma gondii*	Toxoplasmosis in newborn, congenitally; otherwise lymphadenitis, sometimes cerebral form in adults [d]	Free toxoplasmas in the CSF; otherwise intracellular (~3–7 µm)
	8. *Trypanosoma brucei gambiense* [a]	African sleeping sickness	Trypanosomes (~15–30 µm)
	T. b. rhodesiense [a]		

See footnotes on p. 316, 317

Detection procedures									Special comments and special procedures	Chemotherapy (indications)	Figure
a Special microscopical methods	*b* Culture	*c* Animal experiments	*d* Complement fixation	*e* Precipitation reaction	*f* Indirect immunofluorescence	*g* Indirect haemagglutination	*h* ELISA	*i* Skin test			
5									6	7	8
Thin film and "thick film"	+	+	+		+	+	+		Blood incubation test, agglutination reaction; raised IgM	Suramin, pentamidine, tryparsamide, Mel B, Mel W, nitrofurazone	I
	+	+	+		+	+	+		Xenodiagnosis	Benznidazole, nifurtimox	II
Thin film and thick film			+?		+	+			see *Babesia* p. 108	Chloroquine, quinine, primaquine, pyrimethamine, proguanil, mefloquine, halofantrin?	XI
					+	+					
			+?		+	+					
			+?		+	+					
Fresh preparation, concentration procedure and "thick film"			+		+	+	+		High eosinophilia in peripheral blood	Diethylcarbamazine citrate (DEC-C), ivermectin?	XXIX
			+		+	+		+			
			+		+	+		+			
Sediment smear		+!	+		+	+	+		SABIN-FELDMAN dye test; (see also p. 81)	Sulphonamides in combination with pyrimethamine, spiramycin	VIII
CSF sediment	+	+	+		+		+		Agglutination reaction; see no. 1, col. 6	See no. 1	I
	+	+	+		+						

Table 3 (continued)

Specimen for microscopic diagnosis	Parasite species	Name of disease Typical clinical symptoms (apart from frequent latent infections)	Diagnostically important developmental stage in man
1	2	3	4
Tissues, skin	9. *Leishmania donovani* [a]	Visceral leishmaniasis (kala-azar)	Amastigote stages (~2–5 μm)
	10. *L. tropica* [a]	Cutaneous leishmaniasis (oriental sore)	
	11. *L. mexicana* [a]	Mexican cutaneous leishmaniasis	
	12. *L. braziliensis* [a]	South American mucocutaneous leishmaniasis	
Tissues (skin, muscle, parenchyma)	13. *Trypanosoma cruzi* [a]	CHAGAS' disease (South America)	Dividing stage (~2–4 μm), amastigote
	14. *Pneumocystis carinii* [d]	Interstitial cellular pneumonia, *Pneumocystis* pneumonia	Cysts; 8-cell stage (~6–8 μm) in pulmonary alveoli
	15. *Toxoplasma gondii* [a, d]	Toxoplasmosis, lymphadenitis	Intracellular trophozoites (~3–7 μm)
	16. *Onchocerca volvulus* [a]	Onchocerciasis, subcutaneous nodules (Onchocercoma)	Microfilariae (~300 μm) in excised epidermis
	17. *Echinococcus, granulosus, E. multilocularis*	Hydatid disease echinococcosis echinococcosis	Larval stage, echinococcus cysticus echinococcus alveolaris
	18. *Taenia solium*	Cysticercosis	Larval stage, cysticercus cellulosae

See footnotes on p. 316, 317

Detection procedures									Special comments and special procedures	Chemotherapy (indications)	Plate
a Special microscopy methods	*b* Culture	*c* Animal experiments	*d* Complement fixation	*e* Precipitation reaction	*f* Indirect immunofluorescence	*g* Indirect haemagglutination	*h* ELISA	*i* Skin test			
5									6	7	8
Biopsy of liver, spleen, sternum	+	+	+		+	+	+	+	Blood smear, serum liability reaction	Antimony preparations, glucantime, diamidine, stilbamidine, pentamidine, amphotericin B, chloroquine	III
Fresh material from skin lesions	+	+			+	+		+	Immunity		
	+	?				+		+	Culture forms for nos. 9–12, flagellate (promastigote form) (NNN agar)		
	+				+	+		+		Glucantime, diamidine, chloroquine, amphotericin B	
Muscle biopsy	+	+	+		+	+		+	Xenodiagnosis (see *T. rangeli* p. 25)	See no. 2	II
Lung irrigation, sputum; lung autopsy (see col. 6)			+?		+?	+?			Lung swab preparation	Trimethoprim comp.; pentamidine isethionate	X
Lymph node puncture, biopsy?	+	+	+		+	+	+		SABIN-FELDMAN dye test	See no. 7	VIII
Skin ("snip")					+	+		+	MAZZOTTI test, Surgical removal of skin nodules	DEC-C (see nos. 4–6), ivermectin, suramin	XXX
Surgical specimens, chiefly from liver, lung, brain			+		+	+	+	+?	Caution with operation (see no. 28)	Mebendazole, albendazole	XXII
			+		+	+	+	+?			
Muscle, CNS			+		+	+	+	+	Calcified cysts on X-ray film	Praziquantel	XXI

Table 3 (continued)

Specimen for microscopic diagnosis	Parasite species	Name of disease Typical clinical symptoms (apart from frequent latent infections)	Diagnostically important developmental stages in man
1	2	3	4
Continuation from p. 320	19. *Angiostrongylus cantonensis*[a]	Eosinophilia, meningitis	Larval stages
	20. *Anisakis* sp.[c]	Anisakiasis, ileus etc,	Larval stages
	Toxocara sp.[c]	Toxocariasis, eye disease	Larval stages
Naso-pharyngeal space (in vivo)	21. *Acanthamoeba castellanii*	Meningitis, keratitis	Vegetative stages (~40 μm)
	22. *Naegleria fowleri*	Primary amoebic meningoencephalitis	Vegetative stages (~22 × 17 μm)
Urine	23. *Schistosoma haematobium*[a]	Schistosomiasis of bladder, blood in urine	Egg with terminal spine (~150 μm)
	24. *Trichomonas vaginalis*	Urethritis, prostatitis, trichomoniasis, colpitis with vaginal discharge	Lively, motile flagellate (~15–30 μm), no cysts
Vagina	*T. vaginalis*		
	25. *Entamoeba histolytica*[a, d]	Vaginal discharge, rare	Trophozoites (~ 20–30 μm)
Sputum	26. *Pneumocystis carinii*[d]	Interstitial plasma cell pneumonia, pneumocystis pneumonia	Cysts, 8-cell stage (~6–8 μm)
	27. *Paragonimus* sp.[a]	Paragonimiasis, pneumonia	Egg (~90 μm)

See footnotes on p. 316, 317

Detection procedures									Special comments and special procedures	Chemotherapy (indications)	Plate
a Special microscopical methods	*b* Culture	*c* Animal experiments	*d* Complement fixation	*e* Precipitation reaction	*f* Indirect immunofluorescence	*g* Indirect haemagglutination	*h* ELISA	*i* Skin test			
5									6	7	8
Autopsy material		+					+	+	Infection of the brain	Ivermectin?	XXVII
Intestinal tissue (operation and autopsy material)			+? +?	 +	+ +	+? +	 +	+	Microprecipitation test	Thiabendazole, albendazole, Ivermectin	
Pharyngeal smear	+		+		+					Amphotericin B, sulphadiazine	VI
Autopsy material	+		+								
Miracidium hatching technique			+	+	+	+	+	+	Cercarien-Hüllenreaction, circumoval precipitation	Praziquantel, metrifonate	XVIII
Sediment smear; dark field microscopy	+		+		+	+	+				IV
Vaginal or urethral smear	+		+		+				Urethral, prostatic secretions by culture		
Fresh preparation	+		+		+	+	+		MIFC procedure	See Table 2 no. 5	IV, V
bronchoalveolar lavage; sputum; lung autopsy			+?		+?	+?				See no. 14	X
Concentration technique recommended			+		+	+	+	+		See no. 29	XVII

Table 3 (continued)

Specimen for microscopical diagnosis	Parasite species	Name of disease Typical clinical symptoms (apart from frequent latent infections)	Diagnostically important developmental stages in man
1	2	3	4
Sputum	28. *Echinococcus granulosus*	Echinococcosis, hydatidosis of liver, lungs, brain and other organs	Larvae as echinococcus cysticus, scolex hooks, vesicular hydatids up to approx. 20 cm diameter
Stools (intestinal contents)	29. *Paragonimus westermani* [a] *P. kellicotti* [a] *P. africanus*	Lung fluke disease, pneumonia	Eggs (~90 μm) (Nr. 29–33 with operculum)
	30. *Dicrocoelium dendriticum*	Liver fluke diseases, inflammation of gall bladder, hepatic cirrhosis	Eggs (~40 μm)
	31. *Fasciola hepatica*		Eggs (~140 μm)
	32. *Clonorchis sinensis* [a]		Eggs (30 μm)
	33. *Opisthorchis felineus*		Eggs (30 μm)
	34. *Schistosoma mansoni* [a]	Schistosomiasis (intestinal) ulceration, dysentery, bloody stools, hepatosplenomegaly, (portal hypertension); adult worms in mesenteric vessels	Eggs (150 μm) (with lateral spine)
	35. *S. intercalatum* [a]		Eggs (~140 μm) (with terminal spine)
	36. *S. japonicum* [a] *S. mekongi* [a]		Eggs (~90 μm with minute lateral knob) Eggs (~40–45 μm)

See footnotes on p. 316, 317

Detection procedures									Special comments and special procedures	Chemotherapy (indications)	Plate
a Special microscopical methods	*b* Culture	*c* Animal experiments	*d* Complement fixation	*e* Precipitation reaction	*f* Indirect immunofluorescence	*g* Indirect haemagglutination	*h* ELISA	*i* Skin test			
5									6	7	8
Operation material, mostly liver, lung, brain		+?	+		+	+	+	+	Caution with surgery, cyst contents sometimes lead to anaphylactic reaction	Mebendazole, albendazole	XXII
Concentration procedure recommended			+		+	+	+	+			XVII
			+		+	+	+	+			
			+		+	+	+	+			
TELEMANN's concentration technique,										Praziquantel	XV
B bile, duodenal juice			+		+?	+	+	+			
Stools or intestinal fluid, TELEMANN's concentration technique			+	+	+	+	+	+			XVI
							+				
Miracidium hatching technique			+	+	+	+	+	+	Cercarien-Hüllen reaction, Latex flocculation test, Miracidial immobilizations-test, circumoval precipitation	Oxamniquine	XVIII
			+	+	+	+	+	+			
			+	+	+	+	+	+		Praziquantel	XVIII

Table 4. Infection routes and development of intestinal worms

Entry route	Vehicle of infection	Invasive stage	Species	1st or sole intermediate host	2nd intermediate host	Definitive hosts other than man (animal reservoir)
Oral ↓	Contact	Eggs containing larvae	*Enterobius vermicularis*	—	—	None
			Hymenolepis nana	Flea, mealworm larva	—	Rat, mouse
	Mostly raw vegetables		*Trichuris trichiura*	—	—	None
			Ascaris lumbricoides	—	—	None (pig?)
		Free larvae	*Trichostrongylus* sp.	—	—	Plant-eating domestic and wild animals
		Metacercariae	*Fasciolopsis buski*	Snail	Water plants	Pig, goat, dog
	Raw meat	Cysticercus	*Taenia saginata*	Cattle	—	None
			T. solium	Pig	—	None
		Muscle trichinella	*Trichinella spiralis*	Various mammals	—	Pig, rat etc.
	Raw fish	Plerocercoid (sparganum)	*Diphyllobothrium latum*	Copepod	Freshwater fish	Dog, cat, fox
		Metacercariae	*Metagonimus yokogawai*	Snail		Mammals, birds
			Heterophyes heterophyes			Dogs, cats
	Raw snails		*Echinostoma ilocanum*		Snail, mussels	Dogs, rats
	Ingestion of fleas etc.	Cysticercoid	*Dipylidium caninum*	Flea larvae, hair lice	—	Dogs, cats
Percutaneous ↓	Free-living larvae in moist soil	Filariform larvae	*Ancylostoma duodenale*	—	—	None
			Necator americanus	—	—	None
			Strongyloides stercoralis			Dog?, cat?
	Free-swimming larvae	Cercariae	*Schistosoma japonicum, S. mansoni, S. intercalatum, S. mekongi*	Snail	—	Domestic animals and rodents, laboratory animals

Table 5. Identification of the most important intestinal protozoa cysts (from BRUMPT and NEVEU-LEMAIRE, 1951 and ERHARDT, 1951)

Nuclei	Cyst characteristics		Size	Species
1 nucleus	*Pear-shaped* and light-refractive cysts . .		6 × 8.5 µm,	*Chilomastix mesnili*
1 nucleus	*Spherical* cysts, nucleus with large spherical karyosome		9–15 µm,	*Iodamoeba* (*Pseudolimax*)
4 nuclei	Egg-shaped cysts more or less elongated		8–9 × 10–13 µm,	*Giardia lamblia*
4 nuclei	Spherical cysts	Nucleus *with* chromatin at the nuclear membrane	10–14 µm,	*Entamoeba histolytica*
			5–10 µm,	*E. hartmanni*
4 nuclei	Spherical cysts	Nucleus *without* chromatin at the nuclear membrane	7–10 µm,	*Endolimax nana*
8 nuclei	Spherical cysts with distinct cyst wall . .		14–20 µm,	*Entamoeba coli*

Table 6. Identification of the most important helminth eggs (from BRUMPT and NEVEU-LEMAIRE, 1951 and ERHARDT, 1951)

Shell	Group	Subgroup	Eggs	Shape	Size	Species
Shell without operculum	Very thick shell, eggs contain only one eggcell	Shell covered with small knobs, eggs more or less oval			60 × 50 µm,	*Ascaris lumbricoides*
		Smooth shell; lemon-shaped eggs with bright, mucous plugs at each pole, yellow-brown			55 × 25 µm,	*Trichuris trichiura*
	Moderately thick colourless shell with double contour, oval eggs, asymmetrical, containing one larva				55 × 30 µm,	*Enterobius vermicularis*
	Thin and transparent shell	No spine on the shell	Eggs contain 2, 4 or 8, 32 cells, oval eggs	ovoid . . .	60 × 40 µm,	*Ancylostoma duodenale*
				ovoid . . .	70 × 40 µm,	*Necator americanus*
				elongated .	80 × 40 µm,	*Trichostrongylus* sp. [a]
				with larvae	54 × 32 µm,	*Strongyloides* sp. [a]
		Shell with developed spines	oval eggs	spine at 1 pole	150 × 60 µm,	*S. haematobium*
					140 × 50 µm,	*S. intercalatum*
				large lateral spine	150 × 60 µm,	*S. mansoni*
			Eggs almost spherical with very small lateral knob . .		45 × 40 µm,	*S. mekongi*
					90 × 50 µm,	*S. japonicum*
	Two "egg-shells"; oncospheres				50 × 45 µm,	*Hymenolepis* sp.
Shell with operculum	Oval eggs; not very unconspicuous operculum (particularly recognizable by their size)	Eggs furrowed with immature embryo			135 × 75 µm,	*Fasciolopsis buski*
					140 × 80 µm,	*Fasciola hepatica*
					95 × 55 µm,	*Paragonimus* sp.
					70 × 45 µm,	*Diphyllobothrium latum*
		Eggs fully embryonated . .			40 × 25 µm,	*Dicrocoelium dendriticum*
	Very protuberant operculum with knobs on opposite pole	. . . .			28 × 16 µm,	*Clonorchis sinensis*
		. . . .			28 × 13 µm,	*Opisthorchis felineus*

[a] Rarely as egg, mostly as free larvae, p. 240

References Including Textbooks and Summaries *

Abaru D E, Liwo D A, Isakina D, Okori E E (1984) Retrospective long-term study of effects of Berenil by follow-up of patients treated since 1965. Tropenmed Parasit 35: 148–150

Abdel-Wahab M F, Strickland G T, El-Sahly A, El-Kady N, Zakaria S, Ahmed L (1979) Changing pattern of schistosomiasis in Egypt 1935–79. Lancet 242–244

Adolphs H-D, Weissbach L, Thiele J (1982) Manifestation der Bilharziose am Harntrakt. Act Urol 13: 277–282

Albiez E J, Büttner D W (1984) Corium attached onchocercomata: clinical and histological findings. Zbl Bakt Hyg A 258: 388

Albiez E J, Büttner D W, Schulz-Key H (1984) Studies on nodules and adult *Onchocerca volvulus* during a nodulectomy trial in hyperendemic villages in Liberia and Upper Volta. II. Comparison of the macrofilaria population in adult nodule carriers. Tropenmed Parasit 35: 163–166

Alicata J E (1965) Biology and distribution of the rat lungworm, *Angiostrongylus cantonensis*, and its relationship to eosinophilic meningoencephalitis and other neurological disorders of man and animals. Adv Parasit 3: 223–248

Alicata J E, Jindrak K (1970) Angiostrongylosis in the Pacific and Southeast Asia. Thomas C C, Springfield

Alizadeh H, Wakelin D (1981) Mechanism of rapid expulsion of *Trichinella spiralis* from mice. In: Kim C W, Ruitenberg E J, Teppema J S (Hrsg) Trichinellosis. Reedbooks Ltd, Windsor Berks, S 81–84

Ambroise-Thomas P, Desgeorges P T (1979) L'hemagglutination indirecte dans le sero-diagnostic de l'hydatidose comparaison avec l'immuno-fluorescence indirecte et la technique ELISA. Lyon Médical 241: 755–759

Ament M E, Rubin C E (1972) Relation of giardiasis to abnormal intestinal structure and function in gastrointestinal immunodeficiency. Gastroenterology 62: 216–226

Ammann R, Akovbiantz A, Eckert J (1979) Chemotherapie der Echinococcose des Menschen mit Mebendazol (Vermox®). Schweiz Med Wschr 109: 148–151

Ancelle T, Dupony-Camet J, Lavarde C, Lapierre J (1986) The 1985 trichinosis outbreaks due to horse meat in France. IX. Int Congr Infect Parasit Dis München: Abstr 757

Anderson J, Flugsand H, Hamilton P J S, de Marshall T F (1974) Studies on onchocerciasis in the United Cameroon Republic II: Comparison of onchocerciasis in rainforest and sudan savenna. Trans Roy Trop Med Hyg 68: 209–222

Anderson R M, Schad G A (1985) Hookworm burdens and faecal egg counts: An analysis of the biological basis of variations. Trans Roy Soc Trop Med Hyg 79: 812–825

Andrade S G (1982) The influence of the strain of *Trypanosoma cruzi* in placental infections in mice. Trans Roy Soc Trop Med Hyg 76: 123–128

Andrews P (1981) Die Wirksamkeit von Praziquantel gegen Schistosomen im Tierversuch und Bemerkungen zur Wirkungsweise. Arzneim Forsch Drug Res [Biltricide Symposium] 31 (I)

Arzneimittelkommission der deutschen Ärzteschaft: Arzneiverordnungen (1984) 15. Aufl. Deutscher Ärzte-Verlag, Köln, S 77–88

Asami K (1952) Bacteria-free cultivation of *Trichomonas vaginalis*. Kitasato Arch Exp Med 25: 149–156

Asami K, Nakamura M (1955) Experimental inoculation of bacteria-free *Trichomonas vaginalis* into human vaginae and its effect on the glycogen content of vaginal epithelia. Am J Trop Med Hyg 4: 254–258

Asami K, Watanuki T, Sakai H, Imano H, Okamoto R (1965) Two cases of stomach granuloma caused by *Anisakis*-like larval nematodes in Japan. Am J Trop Med Hyg 14: 119–123

Ashby B S, Appleton P J, Dawson I (1964) Eosinophilic granuloma of gastrointestinal tract caused by herring parasite *Eustoma rotundatum*. Brit med J I: 1141–1145

Aspöck H (1983) Überwachung von Toxoplasmose während der Schwangerschaft. Gynäk Rdsch 23: 57–65

Aspöck H, Flamm H, Picher O (1973) Darmparasiten in menschlichen Exkrementen aus prähistorischen Salzbergwerken der Hallstatt-Kultur (800–350 v Chr). Zbl Bakt Hyg I Abt Orig A 223: 549–558

Awadzi K, Dadzie K Y, Schulz-Key H, Haddock D R W, Gilles H M, Aziz M A (1985) The chemotherapy of onchocerciasis. X. An assessment of four single dose treatment regimes of MK-933 (ivermectin) in human onchocerciasis. Ann Trop Med Parasitol 79: 63–78

Aziz M A (1986) Ivermectin and onchocerciasis. Trop Med Parasit 37 Suppl. Nr. 3: 68

Baermann G (1917) Über Ankylostomiasis, deren Ausbreitungsbedingungen durch die Bodeninfektion und deren Bekämpfung. Geneesk tsch Ned-Indie 57: 579

* Bähr R (Hrsg) (1982) Probleme der Echinokokkose unter Berücksichtigung parasitologischer und klinischer Aspekte. Aktuelle Probleme der Chirurgie und Orthopädie Bd 23. Hans Huber, Bern Stuttgart Wien

Bähr R, Ammann R, Bircher J, Eckert J (1984) Die Chemotherapie der menschlichen Echinokokkose. Chirurg 55: 114–116

Barker D C, Butcher J (1983) The use of DNA probes in the identification of leishmanias: discrimination between isolates of the *Leishmania mexicana* and *L. braziliensis* complexes. Trans Roy Soc Trop Med Hyg 77: 285–297

Barnham M (1977) Is *Chilomastix* harmless? Lancet II: 1077–1078

Bassil S, Farid Z, Higashi G I, Watten R H (1979) Low-dose niridazole in the treatment of *Schistosoma mansoni*. Ann Trop Med Parasit 73: 295–296

Beaver P C, Gadgil R K, Morera P (1979) *Sarcocystis* in man: a review and report of five cases. Am J Trop Med Hyg 28: 819–844

* Beaver P C, Jung R C, Cupp E W (1984) Clinical Parasitology. 9. Aufl. Lea & Febiger, Philadelphia

Becker B, Mehlhorn H, Andrews P, Thomas H (1980) Scanning and transmission electron microscope studies on the efficacy of praziquantel of *Hymenolepis nana* (Cestoda) in vitro. Z Parasitenkd 61: 121–133

Becker B, Mehlhorn H, Andrews P, Thomas H (1981) Ultrastructural investigations on the effect of praziquantel on the tegument of five species of Cestodes. Z Parasitenkd 64: 257–269

Becker B, Mehlhorn H, Andrews P, Thomas H, Eckert J (1980) Light and electron microscopic studies on the effect of praziquantel on *Schistosoma mansoni, Dicrocoelium dendriticum*, and *Fasciola hepatica* (Trematoda) in vitro. Z Parasitenkd 63: 113–128

Becker-Feldmann H, Maier W A, Seitz H M (1985) Electron microscope observations on the pathology of the midgut epithelial cells of *Anopheles stephensi* after infection with *Plasmodium yoelii nigeriensis*. Trop Med Parasit [Suppl II] 36: 5–6

Beier A (1983) Sozialmedizinische Aspekte des Bandwurmbefalls (*Taenia saginata*) bei türkischen Mitbürgern in Berlin (West). Bundesgesundheitsblatt 26: 168–170

Bell R M S, Daly J, Kanengoni E, Jones J J (1973) The effects of endemic schistosomiasis and of hycanthone on the mental ability of African school children. Trans Roy Soc Trop Med Hyg 67: 694–701

Bennett J L, Williams J F, Dave V (1988) Pharmacology of ivermectin. Parasitology Today 4: 226–228

Benwick W J, Erlandsen S L (1988) Giardiasis – is it really a zoonosis? Parasitology Today 4: 69–71

Betke K (1985) Angeborene Blutkrankheiten als Verhängnis und Vorteil. Naturw Rdsch 38: 189–197

Beverley J K A, Watson W A (1959) Ovine abortion due to toxoplasmosis. Nature 184: 2041

Bickle Q D, James E R (1978) Resistance against *Schistosoma mansoni* induced by immunization of mice with cryopreserved schistosomula. Trans Roy Soc Trop Med Hyg 72: 677–678

Bijkerk H (1969) Haringwormziekte (anisakiasis). Ned T Geneesk 113: 906–907

Blagg W, Schloegel E L, Mansour N S, Khalaf G I (1955) A new concentration technic for the demonstration of protozoa and helminth eggs in feces. Am J Trop Med Hyg 4: 23–28

* Boch J, Supperer R (1983) Veterinärmedizinische Parasitologie. 3. Aufl. Paul Parey, Berlin Hamburg

* Böckeler W, Wülker W (Hrsg) (1983) Parasitologisches Praktikum. Verlag Chemie, Weinheim, Deerfield Beach, Fla T, Basel

Böhle F, Janitschke K (1984) Evaluation of *Echinococcus* antigens from different intermediate hosts for serodiagnosis of human hydatidosis. Zbl Bakt Hyg A 258: 415

Böker C A, Schaub G A (1984) Scanning electron microscopic studies of *Trypanosoma cruzi* in the rectum of its vector *Triatoma infestans*. Z Parasitenkd 70: 459–469

Bonilla-Musoles F (1985) The destructive effect of Solco-Trichovac-Induced. In: Rüttgers H (Hrsg) Immunotherapy of vaginal infections. Gynäk Rdsch [Suppl 3] 24

Botto C, Arango M, Yarzábal L (1984) Onchocerciasis in Venezuela: prevalence of microfilaraemia in Amerindians and morphological characteristics of the microfilariae from the Upper Orinoco focus. Tropenmed Parasit 35: 167–173 ...

* Brand T von (1972) Parasitenphysiologie. Gustav Fischer, Stuttgart

Brass K (1973) Meningo-encefalitis amebiásica primaria (por Naegleriasis). Arch Venez Med Trop Méd 5: 291–305

Breckenbridge A (1981) Excretion of diethylcarbamazine (DEC) in human urine and its relationship to urinary pH. WHO ONCHO 81.159, WHO FIL 81.164

Bruce-Chwatt L J (1970) Imported malaria – a growing world problem. Trans Roy Soc Trop Med Hyg 64: 201–209

Brumpt E, Neveu-Lemaire M (1951) Praktischer Leitfaden der Parasitologie des Menschen. Übersetzt und bearbeitet von Erhardt A. 2. Aufl. Springer, Berlin Heidelberg New York

Brun R, Jenni L (1984) Wirtsspezifische Beeinflussung der Humanserum Resistenz von *Trypanosoma (T.) brucei*. In: Boch J (Hrsg) Tropenmedizin. Parasitologie 91

Brun R, Jenni L, Schönenberger M, Schell K-F (1981) In vitro cultivation of bloodstream forms of *Trypanosoma brucei*, *T. rhodesiense*, and *T. gambiense*. J Protozool 28: 470–479

Bundesgesundheitsamt (1966) Zur einheitlichen Laboratoriumsdiagnostik der Toxoplasmose. Bundesgesundheitsblatt 9: 354–357

Bundesgesundheitsamt (1975) Zur Vereinheitlichung der Laboratoriumsdiagnostik der Toxoplasmose. Bundesgesundheitsblatt 18: 170–171

Bundesgesundheitsamt (1981) Zur Vereinheitlichung der Laboratoriumsdiagnostik der Echinokokkose. Bundesgesundheitsblatt 24: 310–311

Bundesgesundheitsamt (1986) Empfehlungen zur Laboratoriumsdiagnostik der Amöbiasis, Giardiasis, Kryptosporidiose und weiterer Kokzidiosen. Labmed 10: 118–121

Bundesgesundheitsamt (1986) Empfehlungen zur Laboratoriumsdiagnostik der Pneumocystose. Labmed 10: 122–123

Burchard G D, Winkler E, Kern P (1985) *Cryptosporidia* in patients with and without AIDS. Trop Med Parasit [Suppl II] 36: 7

Burrows R B, Swerdlow M A (1956) *Enterobius vermicularis* as a probable vector of *Dientamoeba fragilis*. Am J Trop Med Hyg 5: 258–265

Butt C G, Baro C, Knorr R W (1968) *Naegleria* (sp.) identified in amoebic encephalitis. Amer J Clin Path 50: 568–574

Butterworth A E, Taylor D W, Veith M C, Vadas M A, Dessein A, Sturrock R F, Walls E (1982) Studies on the mechanisms of immunity in human schistosomiasis. Immunological Reviews 61: 5–39

Butterworth A E, Dalton P R, Dunne D W, Mugambi M, Ouma J H, Richardson B A, Arap Siongok T K, Sturrock R F (1984) Immunity after treatment of human schistosomiasis mansoni. I. Study design, pretreatment observations and the results of treatment. Trans Roy Soc Trop Med Hyg 78: 108–123

Bylund G, Bang B, Wikgren K (1977) Tests with a new compound (praziquantel) against *Diphyllobothrium latum*. J Helminth 51: 115–119

Camargo E P, Mattei D M, Yoshida N, Caulada Z (1983) Staphylococci adherence to trypanosomes exposed to immune sera as a method for the diagnosis of Chagas' disease. Trans Roy Soc Trop Med Hyg 77: 825–827

Campbell W C (1985) Ivermectin: An update. Parasitology Today 1: 10–16

Campbell, W C (1988) Trichinosis revisited – another look at modes of transmission. Parasitology Today 4: 83–86

Champ R R, Mattern C F T, Honigberg B M (1974) Study of *Dientamoeba fragilis* Jepps & Dobell. I. Electronmicroscopic observations of the binucleate stages. II. Taxonomic position and revision of the genus. J. Protozool 21: 69–82

Cançado J R, Brener Z (1979) Terapeutica. In: Brener Z, Andrade Z (Hrsg) *Trypanosoma cruzi* e Doença de Chagas. Guanabara, Rio de Janeiro, S 362–424

Capon A G, Borcham P F L (1986) Variation between different stocks of *Giardia intestinalis* IX. Int Congr Infect Parasit Dis München: Abstr 413

Carter R F (1972) Primary amoebic meningo-encephalitis. An apparaisal of present knowledge. Trans Roy Soc Trop Med Hyg 66: 193–213

Casemore D P (1977) Free-living amoebae in home dialysis unit. Lancet XI: 1078

Cerisola J A, Del Prado C E, Rohwedder R, Bozzini J P (1971) *Blastocrithidia triatomae* n. sp. found in *Triatoma infestans* from Argentina. J Protozool 18: 503–506

Cerisola J A, Russo M C, Del Prado C E, Jozami L B de, Rohwedder R W (1972) Estudio comparativo de diversos méthodos parasitológicos en la enfermedad de Chagas aguda. Simposio Intern. sobre Enfermedad de Chagas. ler Congr. Arg. de Parasitología. Buenos Aires. Soc. Arg. de Parasitología. Dic 97–100

Červa L (1966) Use of fluorescent-antibody technique to identify pathogenic hartmannellae in tissue of experimental animals. Folia Parasitol Prag 13: 328–331

Červa L (1967) Immunological studies on Hartmannellid amoebae. Folia Parasit 14: 19

Červa L (1969) Amoebic meningoencephalitis: axenic culture of *Naegleria*. Science 163: 576

Červa L, Nowăk K (1968) Amoebic meningoencephalitis; sixteen fatalities. Science 160: 92

Červa L, Serbus C, Skočil V (1973) Isolation of *Limax* amoebae from the nasal mucosa of man. Folia Parasitol 20: 97–103

Chacin-Bonilla L (1980) Successful treatment of human *Entamoeba polecki* infection with metronidazole. Am J Trop Med Hyg 29: 521–523

Chance M L, Peters W, Scchory L (1974) Biochemical taxonomy of *Leishmania*. I: Observations on DNA. Ann Trop Med Parasit 68: 307–316

Changbrumrung S, Migasena P, Supawan V, Buavatana T, Migasena S (1982) α_1-Antitrypsin, α_1-antichymotrypsin and α_2-macroglobulin in human liver fluke (opisthorchiasis). Tropenmed Parasit 33: 195–197

Chen S N (1986) Enzyme-linked immunosorbent assay (ELISA) for the detection of antibodies to *Angiostrongylus cantonesis*. Trans Roy Soc Trop Med Hyg 80: 398–405

Cheng S Z et al. (1981) Hepatic lesions in 69 children infected with *Paragonimus westermani* in Southern Anhui. WHO/Helm, 82.5

* Cheng Th C (1974) General Parasitology. Academic Press, New York London

Choromanski L, Beat D A, Nordin J H, Pan A A, Honigberg B M (1985) Further studies on the surface saccharides in *Trichomonas vaginalis* strains by fluorescein-conjugated lectins. Z Parasitenkd 71: 443–458

* Cohen S, Sadun C H (1976) Immunology of Parasite Infections. Blackwell Scientific Publications, London

Constantinescu G, Căprarŭ T (1980) The microprecipitation test in trichinosis diagnosis. Arch Roúm Path Exp Microbiol 39: 41–47

Cooper E S, Bundy D A P (1988) Trichuris is not trivial. Parasitology Today 4: 301–306
Corliss J O (1984) The kingdom Protista and its 45 Phyla. Bio Systems 17: 87–126
Coumbares A (1966) La distomatose hépatique en Algerie. Ann Parasitol 41: 71–77
Craig P S, Zeyhle E, Romig T (1986) Hydatid disease: Research and control in Turkana. II. The role of immunological techniques for the diagnosis of hydatid disease. Trans Roy Soc Trop Med Hyg 80: 183–192
Critchley E M R, Vakil S D, Hutchinson D N, Taylor P (1982) *Toxoplasma, Toxocara*, and epilepsy. Epilepsia 23: 315–321
Crompton D W T (1988) The prevalence of ascariasis. Parasitology Today 4: 162–169
Crum E D, Despommier D D, McGregor DD (1977) Immunity to *Trichinella spiralis*. I. Transfer of resistance by two classes of lymphocytes. Immunology 33: 787–795
Cuckler A C, Egerton J R, Alicata J E (1965) Therapeutic effect of Thiabendazole on *Angiostrongylus cantonensis* infections in rats. J Parasit 51: 392–396
Current W L, Haynes T B (1984) Complete development of *Cryptosporidium* in cell-culture. Science 224: 603–605 ...
Dahl R J, Johnson A M (1984) Specificity of the enzyme-linked immunosorbent assay (ELISA) for *Toxoplasma* IgG antibody. Trans Roy Soc Trop Med Hyg 78: 661–662
D'Alessandro A (1976) Biology of *Trypanosoma (Herpetosoma) rangeli* Tejera 1920 (Hoare 1972). In: Lumsden W H R, Evans D A (Hrsg) Biology of the Kinetoplastida Vol 1. Academic Press, New York London San Francisco, S 327–434
D'Alessandro A, Rausch R L, Cuello C, Aristizabal N (1979) *Echinococcus vogeli* in man, with a review of polycystic hydatid disease in Colombia and neighboring countries. Am J Trop Med Hyg 28: 303–317
Datry A, Lecso G, Rozenbaum W, Danis M, Gentilini M (1984) Cerebral toxoplasmosis in AIDS: a simple laboratory technique for diagnosis. Trans Roy Soc Trop Med Hyg 78: 679–680
Denis M, Chadee K (1988) Immunopathology of Entamoeba histolytica infections. Parasitology Today 4: 247–252
Desmonts G (1982) Les recherches d'anticorps IgM spécifiques par immuno-adsorption (Immunosorbent assay ou I.S.A.). Lyon Médical 248: 37–41
Despommier D D, McGregor D D, Crum E D, Carter P B (1977) Immunity to *Trichinella spiralis*. II. Expression of immunity against adult worms. Immunology 33: 797–805
Diamond L S, Harlow D R, Cunnick C C (1978) A new medium for the axenic cultivation of *Entamoeba histolytica* and other *Entamoeba*. Trans Roy Soc Trop Med Hyg 72: 431–432
* Dietrich M, Kern P (1983) Tropenlabor. Gustav Fischer, Stuttgart New York
Disko R, Mielcarek R (1984) The behaviour of the ascorbic acid in the snail *Biomphalaria glabrata* after infection with *Schistosoma mansoni*. Zbl Bakt Hyg A 258: 390–391
Disko R, Schinkel T (1979) Zur Diagnose des Amöben-Leberabszesses. Eine Studie an 107 Fällen. Münch Med Wschr 121: 1536–1538
Disko R, Winter W, Beier M (1984) Möglichkeiten der Vereinfachung der Echinokokken-Immunfluoreszenz. In: Boch J (Hrsg) Tropenmedizin Parasitologie. Medizin in Entwicklungsländern Bd 16. Peter Lang, Frankfurt Bern New York, S 497–499
* Dönges J (1980) Parasitologie, mit besonderer Berücksichtigung humanpathogener Formen. Georg Thieme, Stuttgart
Doumenge J P, Mott K E, Cheung C, Villenave D, Chapuis O, Perrin M F, Reaud-Thomas G (198) Atlas de la répartition mondiale des schistosomiasis. Atlas of the global distribution of schistosomiasis. CEGET-CNRS/OMS-WHO, Presses Universitaires de Bordeaux, 1987
Duarte M I S, Silva M R R, Goto H, Nicodemo E L, Amato Neto V (1983) Interstitial nephritis in human kala-azar. Trans Roy Soc Trop Med Hyg 77: 531–537
Düwel D (1978) Activity of Fenbendazole on metacestodes of different tapeworms in small and domestic animals. Current Chemotherapy 142–144

Duma R J, Ferrell H W, Nelson E C, Jones M M (1969) Primary amoebic meningoencephalitis. N Engl J Med 281: 1315–1323

Dumke K, Janitschke K (1981) Beitrag zur Morphologie und Pathogenese der eosinophilen Kolitis. Z Gastroenterologie 19: 646–654

Ebert F (1986) Vergleichende Isoenzym Untersuchungen an *Trypanosoma rangeli*-Stocks und ihre Beziehung zu anderen Trypanosomen aus Triatomen. 12. Tagung der Deutschen Gesellschaft für Parasitologie, Wien: Referat Nr. 28

Ebrahimzadeh A, Jones T C (1983) A comparative study of different *Leishmania tropica* isolates from Iran: correlation between infectivity and cytochemical properties. Am J Trop Med Hyg 32: 694–702

* Eckert J (1982) Parasiten. In: Wiesmann E (Hrsg) Medizinische Mikrobiologie. 5. Aufl. Georg Thieme, Stuttgart New York, S 366–494

Eckert J, Lämmler G (1972) Angiostrongylose bei Mensch und Tier. Z Parasitenkd 39: 303–322

Eckert J, Thompson R C A, Mehlhorn H (1983) Proliferation and metastases formation of larval *Echinococcus multilocularis*. I. Animal model, macroscopical and histological findings. Z Parasitenkd 69: 737–748

Edungbola L D, Oni G A, Aiyedun B A (1983) Babana Parasitic Diseases Project. I. The study area and a preliminary assessment of onchocercal endemicity based on the prevalence of 'leopard skin'. Trans Roy Soc Trop Med Hyg 77: 303–309

Ehrich J H H, Horstmann R, Beck E J, Dietrich M (1984) Todesursache bei 25 Patienten mit Malaria tropica. In: Boch J (Hrsg) Tropenmedizin Parasitologie. Medizin in Entwicklungsländern Bd 16. Peter Lang, Frankfurt Bern New York, S 155

Eichenlaub D, Karow J, Keller F, Kittler R, Schäfer J-H (1984) Algide Malaria, eine verkannte tödliche Erkrankung. Mitt Österr Ges Tropenmed Parasitol 6: 167–173

El-Masry N A, Trabolsi B, Bassily S, Farid Z (1983) Albendazole in the treatment of *Ancylostoma duodenale* and *Ascaris lumbricoides* infections. Trans Roy Soc Trop Med Hyg 77: 160–161

Elsdon-Dew R (1968) The epidemiology of amoebiasis. Adv Parasitol 6: 1–62

Enders B (1984) Die Chagas-Krankheit des Menschen – Versuche zur Entwicklung eines Impfstoffes. Arzneim-Forsch Drug Res (Abstracts) 34: 104

Enders B, Hermentin P (1983) Neuere Perspektiven bei der Entwicklung eines Malaria-Impfstoffes. Forum Mikrobiologie 6: 10–19

Enzensberger W, Helm E B, Fischer P-A, Stille W (1985) Toxoplasmosis of the CNS: an important neurological complication of AIDS. Trop Med Parasit [Suppl II] 36: 19

Erhardt A (1951) s. Brumpt E, Neveu-Lemaire M

Evans D A, Ellis D S (1983) Recent observations on the behaviour of certain trypanosomes within their insect hosts. Adv Parasitol 22: 2–42

Falkner von Sonnenburg F, Krampitz H E, Löscher T, Prüfer L, Weiland G (1979) Zur Diagnose eingeschleppter viszeraler Leishmaniosen (Kala-Azar). Münch med Wschr 121: 1353–1356

Farthing M J G, Varon S R, Keusch G T (1983) Mammalian bile promotes growth of *Giardia lamblia* in axenic culture. Trans Roy Soc Trop Med Hyg 77: 467–469

Fayer R (1970) *Sarcocystis:* development in cultured avian and mammalian cells. Science 168: 1104–1105

Fayer R (1972) Gametogony of *Sarcocystis sp.* in cell culture. Science 175: 65–67

Feldheim W, Knobloch J (1982) Serodiagnosis of *Opisthorchis viverrini* infestation by an enzyme immuno-assay. Tropenmed Parasit 33: 8–10

Feldmeier H, Bienzle U, Dietrich M (1979) Combination of viability test and a quantification method for *Schistosoma haematobium* eggs (Filtration – trypan blue-staining-technique). Tropenmed Parasit 30: 417–422

Feldmeier H, Bienzle U, Schuh D (1981) Combination of techniques for concentration and identification of microfilariae from peripheral blood. Trans Roy Trop Med Hyg 75: 251–253

Feldmeier H, Doehring E, Daffalla A A, Omer A H S, Dietrich M (1982) Efficacy of metrifonate in urinary schistosomiasis in light and heavy infections. Tropenmed Parasit 33: 102–106

Fernex M, Leimer R (1985) Treatment of multi-resistant *Plasmodium falciparum* malaria with the combination of mefloquine with sulfadoxine/pyrimethamine. Trop Med Parasit [Suppl III] 36: 2

Ferreira L F, de Araújo A J G, Confalonieri U E C (1980) The finding of eggs and larvae of parasitic helminths in archaeological material from Unai, Minas Gerais, Brazil. Trans Roy Soc Trop Med Hyg 74: 798–800

Frandsen F (1979a) Further studies on the compatibility between *S. intercalatum* from Cameroon and Zaire and species of *Bulinus*. Z Parasitenkd 58: 161–167

Frandsen F (1979b) Discussion of the relationships between *Schistosoma* and their intermediate hosts, assessment of the degree of host-parasite compatibility and evaluation of schistosome taxonomy. Z Parasitenkd 58: 275–296

Frandsen F, Monrad J, Ørnbjerg Christensen N (1978) Sheep as a potential reservoir host for *Schistosoma intercalatum*. J Parasitol 64: 1136

* Frank W (1976) Parasitologie. Eugen Ulmer, Stuttgart

Frank W (1982) Biologie und Epidemiologie des *Echinococcus granulosus* und des *Echinococcus multilocularis*. In: Bähr R (Hrsg) Aktuelle Probleme in der Chirurgie und Orthopädie Bd 23. Hans Huber, Bern Stuttgart Wien, S 12–25

Frank W (Hrsg) (1982) Immune Reactions to Parasites. Zbl Bakt [Suppl 12] Fortschritte der Zoologie Bd 27. Gustav Fischer, Stuttgart New York ...

* Frank W (1984) Non-hemoparasitic Protozoans. In: Hoff G L, Frye F L, Jacobson E R (Hrsg) Diseases of Amphibians and Reptiles. Plenum Publishing Corporation, S 259–384

* Franz J M, Krieg A (1982) Biologische Schädlingsbekämpfung. Pareys Studientexte 12. Paul Parey, Berlin Hamburg

Franz M (1982) The fine structure of adult *Onchocerca volvulus*. I. The cuticle, the hypodermis and the muscle cell of the male worm. Tropenmed Parasit 33: 69–75

Franz M, Büttner D W (1983) The fine structure of adult *Onchocerca volvulus*. IV. The hypodermal chords of the female worm. Tropenmed Parasit 34: 122–128

Frayha G J, Smyth J D (1983) Lipid metabolism in parasitic helminths. Adv Parasitol 22: 309–387

* Frenkel J K (1971) Toxoplasmosis. Ergebnisse der Pathologie 54: 28–75

Frenkel J K (1976) *Pneumocystis jiroveci* n. spec. from man: morphology, physiology and immunology in relation to pathology. Natl Cancer Inst Monogr 43: 13–30

Frenkel J K (1981) False-negative serologic tests for *Toxoplasma* in birds. J Parasitol 67: 952–953

Frenkel J K (1981) Congenital toxoplasmosis: Prevention or palliation. Am J Obstet Gynecol 141: 359–361

Frenkel J K (1988) Pathophysiology of toxoplasmosis. Parasitology Today 4: 273–278

Frenkel J K, Ruiz A (1981) Endemicity of Toxoplasmosis in Costa Rica. Transmission between cats, soil, intermediate hosts and humans. Am J Epidemiol 113: 254–269

Fröscher W, Gulotta F, Saathoff M (1982) Chronische Trichinose und neuromuskuläre Erkrankungen. Dtsch med Wschr 107: 1432–1437

Gameel A A (1982a) Tissue ascorbic acid concentrations in rats experimentally infected with *Fasciola hepatica*. Z Parasitenkd 68: 181–184

Gameel A A (1982b) *Fasciola hepatica:* plasma ascorbic acid, plasma iron and iron-binding capacity in experimentally infected sheep. Z Parasitenkd 68: 185–189

Gardener P J, Chance M L, Peters W (1974) Biochemical taxonomy of *Leishmania*. II. Electrophoretic variation of malate dehydrogenase. Ann Trop Med Parasit 68: 317–325

* Garnham P C C (1971) Progress in Parasitology. The Athlone Press, London
Garnham P C C (1984) The present state of malaria-research: An historical survey. Experientia 40: 1305–1310
* Geigy R, Herbig A (1955) Erreger und Überträger tropischer Krankheiten: Verlag f Recht und Gesellschaft AG Basel
Gentilini M, Vernes A, Gentilini J L, Richard-Lenoble D, Bourée P, Wattez A (1976) Étude enzymatique et sérologique de la trichinose humaine à propos d'une récente épidémie de la Banlieue sud de Paris. Bull Soc Path Ex 69: 525–531
Gigase P L, Van Marck E A E (1983) From Parasitic Infection to Parasitic Disease. In: Contributions to microbiology and immunology, Vol 7. S. Karger, Basel München Paris London New York Sydney
Gillespie S H (1988) The epidemiology of *Toxocara canis*. Parasitology Today 4: 180–182
Gilman R H, Chong Y H, Davis C, Greenberg B, Virik H K, Dixon H B (1983) The adverse consequences of heavy *Trichuris* infection. Trans Roy Soc Trop Med Hyg 77: 432–438
Glickman L T, Cypess R, Hiles D, Gessner T (1979) *Toxocara*-specific antibody in the serum and aqueous humor of a patient with presumed ocular and visceral toxocariasis. Am J Trop Med Hyg 28: 29–35
Glickman L T, Schantz P M, Cypess R H (1979) Epidemiological characteristics and clinical findings in patients with serologically proven toxocariasis. Trans Roy Soc Trop Med Hyg 73: 254–258
Godfrey D G (1979) The zymodemes of Trypanosomes. In: Taylor A E R, Muller R (Hrsg) Problems in the identification of parasites and their vectors, Bd 17, 31–53, Symposia of the British Society for Parasitology London School of Hygiene and Tropical Medicine, London
Godoy G A, Orihel T C, Volcan G S (1980) *Microfilaria bolivarensis:* a new species of filaria from man in Venezuela. Am J Trop Med Hyg 29: 545–547
Göbel E (1983) Kryptosporidiose bei Mensch und Tier; Nachweis und Intestinalentwicklung des Parasiten. Mitt Österr Ges Tropenmed Parasitol 5: 49–53
Gönnert R (1972) Lampit® (Nifurtimox). Arzneim-Forsch Drug Res 22: 1563–1642
Gönnert R, Andrews P (1977) Praziquantel, a new broad-spectrum antischistosomal agent. Z Parasitenkd 52: 129–150
Goldsmid J M (1968) The differentiation of *Ternidens deminutus* and hookworm ova in human infections. Trans Roy Soc Trop Med Hyg 62: 109–116
Gottstein B (1984a) Detection of circulating antigens in patients with cystic echinococcosis. Zbl Bakt Hyg A 258: 417
Gottstein B (1984b) Eine spezifische Antigen-Fraktion aus *Echinococcus multilocularis* und ihre Verwendung im ELISA. In: Boch J (Hrsg) Tropenmedizin Parasitologie. Medizin in Entwicklungsländern Bd 16. Peter Lang, Frankfurt Bern New York, S 379
Gottstein B, Eckert J, Woodtli W (1984) Determination of parasite-specific immunoglobulins using the ELISA in patients with echinococcosis treated with mebendazole. Z Parasitenkd 70: 385–389
Gottstein B, Schantz P M, Tsang V C W (1986) Antigenanalyse von *Taenia solium* und spezifische Immundiagnose der Zystizerkose des Menschen. 12. Tagung der Deutschen Gesellschaft für Parasitologie. Referat Nr. 32, Wien
* Granz W, Ziegler K (1976) Tropenkrankheiten, Grundlagen und Klinik. Joh Amb Barth, Leipzig
Granzer M (1984) Chemical stimuli of the attachment of *Schistosoma mansoni* cercariae. Zbl Bakt Hyg A 258: 392
Greenblatt C L (1980) The present and future of vaccination for cutaneous leishmaniasis. In: Mizrahi A, Hertman I, Klingberg M A, Cohn A (Hrsg) New developments with human and veterinary vaccines. Alan R Liss Inc, New York, S 259–285
Greenblatt C L (1988) Cutaneous leishmaniasis: The prospects for a killed vaccine. Parasitology Today 4: 53–54

Greene B M, Taylor H R, Cupp E W, Murphy R P, White A T, Aziz M A, Schulz-Key H, D'Anna S A, Newland H S, Goldschmidt L P, Auer C, Hanson A P, Freeman S V, Reber E W, Williams P N (1985) Comparison of Ivermectin and diethylcarbamazine in the treatment of onchocerciasis. N Engl J Med 313: 133–138

Grocott R G (1955) A stain for fungi in tissue sections and smears, using Gomori's methenamine silver nitrate technique. Am J Clin Path 25: 975

Groll E (1981) Cisticerosis humana y Praziquantel: Una apreciación panorámica de las primeras experiencias clínicas. Bol Chile Parasit 36: 29–37

Grove D I (1982) Treatment of strongyloidiasis with Thiabendazole: an analysis of toxicity and effectiveness. Trans Roy Soc Trop Med Hyg 76: 114–118

Grüntzig J (1984) Sind die Bindehautlymphgefäße für die Pathogenese der Augenonchozerkose von Bedeutung? In: Boch J (Hrsg) Tropenmedizin Parasitologie. Medizin in Entwicklungsländern Bd 16. Peter Lang, Frankfurt Bern New York, S 223–226

* Grumbach A, Bonin O (Hrsg) (1969) Die Infektionskrankheiten des Menschen und ihre Erreger Bd II 2. Aufl. Georg Thieme, Stuttgart

Gsell O (1978) Importierte Infektionskrankheiten und deren epidemiologische Auswirkungen. Zbl Bakt Hyg I Abt Orig B 166: 471–516

* Gsell O (Hrsg) (1980) Importierte Infektionskrankheiten. Epidemiologie und Therapie. 15. Symposium der Deutschen Gesellschaft für Fortschritte auf dem Gebiet der Inneren Medizin, Freiburg i. Br. 1980. Georg Thieme, Stuttgart

* Gsell O, Mohr W (Hrsg) (1972) Infektionskrankheiten BD IV: Rickettsiosen und Protozoenkrankheiten. Springer, Berlin Heidelberg New York

Haas W (1985) Bilharziose: die biologische und biotechnische Bekämpfung einer Tropenkrankheit. Verh Dtsch Zool Ges 78: 45–60

Haas W, Feiler W, Granzer M, Van de Roemer A (1984) Cercarial host-finding and invasion strategies. Zbl Bakt Hyg A 258: 393

Haas W, Granzer M, Garcia E G (1986) Wirtsfindung und Wirtserkennung der Zerkarie von *Schistosoma japonicum*. 12. Tagung der Deutschen Gesellschaft für Parasitologie: Referat Nr. 83, Wien

Habs H, Piekarski G (1976) Abschied von ‚eingeschleppten Krankheiten'. Dtsch med Wschr 101: 1891–1892

Hamperl H (1988) Lehrbuch der allgemeinen Pathologie und pathologischen Anatomie. Springer-Verlag, Berlin Heidelberg New York London Paris Tokyo Hong Kong

Harada Y, Mori O (1955) A new method for culturing hook-worm. Yonago Acta Med 1: 177–179 (Trop Dis Bull 53: 343, 1956)

Harinasuta T, Bunnag D, Wernsdorfer W H (1983) A phase II clinical trial of Mefloquine in patients with chloroquine-resistant *falciparum* malaria in Thailand. WHO Bull 61 (2): 299–305

Harnett W (1988) The anthelmintic action of praziquantel. Parasitology Today 4: 144–146

Harris A R C, Russell R J, Charters A D (1984) A review of schistosomiasis in immigrants in Western Australia, demonstrating the unusual longevity of *Schistosoma mansoni*. Trans Roy Soc Trop Med Hyg 78: 385–388

Harris J R W (1985) Double-blind comparative study of *Trichomonas vaginalis* infection: Solco-Trichovac versus placebo. Gynäk Rdsch [Suppl 3] 24

Hartmann M G (1972) Viscerale Leishmaniase: Kala Azar. In: Gsell O, Mohr W (Hrsg) Infektionskrankheiten Bd IV. Springer, Berlin Heidelberg New York

Hay J, Graham D I, Aitken P P (1984) Congenital toxoplasmosis and mental subnormity. J Roy Soc Med 77: 344

Heidrich H-G (1986): *Plasmodium falciparum* antigen as target molecules for a protective immunization against malaria: An up-todate review. Z Parasitenkd 72: 1–11

He Y, Gong Z, Ma J (1980) Scaning and transmission electron microscopy of *Schistosoma japonicum* egg shell. Chinese Medical Journal 93: 861–864; (Trop Dis Bull 78: 473 v, 1981)

Henriksen S A, Pohlenz J (1981) Staining of cryptosporidia by a modified Ziehl-Neelsen technique. Acta vet scan 22: 594–596

* Hiepe T, Buchwalder R, Nickel S (1985) Veterinärmedizinische Helminthologie. In: Hiepe T (Hrsg) Lehrbuch der Parasitologie, Bd 3. Gustav Fischer, Stuttgart

* Hiepe T, Buchwalder R, Ribbeck R (1981) Allgemeine Parasitologie. In: Hiepe T (Hrsg) Lehrbuch der Parasitologie, Bd 1. Gustav Fischer, Stuttgart New York

* Hiepe T, Jungmann R (1983) Veterinärmedizinische Protozoologie. In: Hiepe T (Hrsg) Lehrbuch der Parasitologie, Bd 2. Gustav Fischer, Stuttgart New York

Hinz E, Diesfeld H J, Gehrig H, Kirsten C (1981) Serologische Kreuzreaktionen zwischen experimentell mit *Echinococcus multilocularis* oder *Dipetalonema viteae* infizierten Nagetieren. Tropenmed Parasit 32: 247–249

Höfler W (1980a) Sulfadoxin-Pyrimethamin-resistente *falciparum*-Malaria aus Kambodscha. Dtsch med Wschr 105: 350–351

Höfler W (1980b) Hinweise zur parasitologischen Diagnose der Amöbendysenterie. Dtsch Ärztebl – Ärztliche Mittlg 77: 367–368

Höfler W, Lindner H U, Lorenz K, Hermann I (1985) Study on the efficiency of the examination for intestinal protozoa. Trop Med Parasit [Suppl II] 36: 15

Hörchner F (1983) Rinderfinnen – ein Problem? Berl Münch Tierärztl Wschr 96: 347–350

Hörchner F, Albert H (1979) Zur Bekämpfung und Diagnostik der Rinderfinnen. I. Therapie und Reinfektion. Berl Münch Tierärztl Wschr 92: 107–111

Hohorst W, Graefe G (1961) Ameisen-obligatorische Zwischenwirte des Lanzettegels (*Dicrocoelium dendriticum*). Naturwissenschaften 48: 229

Hollander N den, Riley D, Befus D (1988) Immunology of giardiasis. Parasitology Today 4: 124–131

* Honigberg B M (1978) Trichomonads of importance in human medicine. In: Kreier J P (Hrsg) Parasitic Protozoa *II:* Academic Press, New York London San Fransisco

Horii Y, Usui M (1985) Experimental transmission of *Trichuris* ova from monkeys to man. Trans Roy Soc Trop Med Hyg 79: 423

Hornbostel H (1959) Bandwurmprobleme in neuer Sicht. Ferdinand Enke, Stuttgart

Horton R J (1988) Introduction of halofantrine for malaria treatment. Parasitology Today 4: 238–239

Hughes W T (1982) Immunology of *Pneumocystis carinii.* In: Nahmias A J, O'Reilly R J (Hrsg) Immunology of human infections. Part II: Viruses and parasites; immunodiagnosis and prevention of infectious diseases. Plenum Publishing Corporation, New York, S 373–383

Hutchison W M (1965) Experimental transmission of *Toxoplasma gondii.* Nature 206: 961–962

Hutchison W M, Bradley M, Cheyne W M, Wells B W P, Hay J (1980) Behavioural abnormalities in *Toxoplasma*-infected mice. Ann Trop Med Parasit 74: 337–345

Ishikura H, Kikuchi Y, Ishikura H (1983) Cit: Oshima J (1987) Parasitology Today 3: 42

Ismail S A, Stek M jr, Leef J L (1983) Circumoval precipitin (COP) test in schistosomiasis with frozen *Schistosoma mansoni* eggs. Trans Roy Soc Trop Med Hyg 77: 809–811

Jackson T F H G, Anderson C B, Simjee A E (1984) Serological differentiation between past and present infection in hepatic amoebiasis. Trans Roy Soc Trop Med Hyg 78: 342–345

Jacobs T, Ziefer A, Seitz H M (1984) Staining techniques for *Pneumocystis carinii.* Zbl Bakt Hyg A 258: 379

Janitschke K (1982) Primäre Amöben-Meningoenzephalitis. Bundesgesundheitsblatt 25: 315–317

Janitschke K (1985) Ergebnisse und aktuelle Probleme bei der Standardisierung der Serodiagnostik von Parasitosen. Ärztl Lab 31: 11–14

Janitschke K (1986) siehe Bundesgesundheitsamt

Janitschke K, El-Kalouby A H, Braun-Munzinger R A, El-Baz H, Mahmoud M (1981) Evaluation of the ELISA-test as an epidemiological tool in schistosomiasis. J Trop Med Hyg 84: 147–154

Janitschke K, Karavias T, Werner H, Rozycki C (1981) Vergleich der Sensitivität und Spezifität verschiedener serologischer Methoden zum Nachweis einer Echinokokkose. Lab med 5: 274–277

Janitschke K, Werner H (1983) Übersicht der Möglichkeiten zum Nachweis von Antikörpern gegen Parasiten des Menschen in der Bundesrepublik Deutschland, Österreich und der Schweiz. Lab med 7: 294–296

Jelnes J E, Highton R B (1984) *Bulinus crystallinus* (Morelt, 1868) acting as intermediate host for *Schistosoma intercalatum* (Fisher, 1934) in Gabon. Trans Roy Soc Trop Med Hyg 78: 412–413

Jenni L, Brun R (1982) A new in vitro test for human serum resistance of *Trypanosoma* (*T.*) *brucei*. Acta Tropica 39: 281–284

Jenni L, Brun R (1984) Humanserum Resistenz von *Trypanosoma* (*T.*) *brucei* und ihre epidemiologische Bedeutung. In: Boch J (Hrsg) Tropenmedizin Parasitologie, Medizin in Entwicklungsländern Bd 16. Peter Lang, Frankfurt Bern New York, S 93–95

Jensen C (1984) The coprophagic infection of *Triatoma infestans* (Reduviidae) with *Blastocrithidia triatomae* (Trypanosomatidae). Zbl Bakt Hyg A 258: 379–380

Jirovec O (1954) Über die durch *Pneumocystis carinii* verursachte interstitielle Pneumonie der Säuglinge. Mschr Kinderheilk 102: 476–485

Jirovec O (1959) Über die durch *Pneumocystis carinii* verursachte interstitielle Pneumonie der Säuglinge. J Hyg Epidem (Prag) 3: 28–59

Jirovec O, Vanék J (1954) Zur Morphologie der *Pneumocystis carinii* und zur Pathogenese der *Pneumocystis*-Pneumonie. Zbl Allg Path Path Anat 92: 424–437

Johnson A M (1984) Strain-dependent, route of challenge-dependent, murine susceptibility to toxoplasmosis. Z Parasitenkd 70: 303–309

Kadlec V, Škvárová J, Červa L, Nebáznivá D (1980) Virulent *Naegleria fowleri* in indoor swimming pool. Folia Parasitol Prag 27: 11–17

Kagan I G, Norman L G (1976) Serology of pneumocystosis. Symposium on *Pneumocystis carinii* infections. Natl Cancer Inst Monogr 43: 121–125

Kasprzak W, Majewska A C (1983) Isolation and axenic growth of fresh *Giardia intestinalis* strains in TPS-1 medium. Trans Roy Soc Trop Med Hyg 77: 223–224

Kassur B, Januszkiewicz J, Poznańska H (1978) Clinic of trichinellosis. In: Kim C W, Pawłowski Z S (Hrsg) Trichinellosis. Reedbooks Ltd, Windsor Berks, S 27–44

* Katz M, Despommier D D, Gwadz R W (1982) Parasitic Diseases. Springer, Berlin Heidelberg New York

Keister D B (1983) Axenic culture of *Giardia lamblia* in TYI-S-33 medium supplemented with bile. Trans Roy Soc Trop Med Hyg 77: 487–488

Kern P, Burchard G D, Kressel K, Heinz M (1985) Schistosomal egg deposition in rectal mucosa before and after specific chemotherapy. Trop Med Parasit [Suppl II] 36: 12–13

Kessel J-F, Johnstone H G (1949) The occurrence of *Entamoeba polecki,* Prowazek 1912, in *Macaca mulatta* and in man. Amer J Trop Med 29: 311–317

Khamtsov V G (1973) Disturbance of ascorbic acid balance in patients with balantidiasis. Medskaya Parazit 42: 443–447 (russisch). (Trop Dis Bull 70: 1077 1973) ...

Kikuchi S, Hayashi S, Nakajima M (1967) Studies on anisakiasis in dolphins. Jap J Parasit 16: 156–166 (japanisch, englische Zusammenfass.). (Trop Dis Bull 65: 297 1968)

Kilkinov G I (1967) Unusual cases of *Hymenolepis nana* localization in experimental *Hymenolepis* infections. Trop Dis Bull 64: 994

* Kim C W, Pawłowski Z S (Hrsg) (1978) Trichinellosis: Proceedings of the fourth international conference on trichinellosis, Poznań Poland 1976. University Press of New England, Hanover NH, USA

* Kim C W, Ruitenberg E J, Teppema J S (Hrsg) (1981) Trichinellosis: Proceedings of the fifth international conference on trichinellosis, Noordwijk aan Zee The Netherlands 1980. Reedbooks Ltd, Windsor Berks

Kimmig P, Braun U (1980) Die Isolierung von Mikrofilarien von *Litomosoides carinii* im Ficoll-Dichtegradienten. Z Parasitenkd 63: 171–175

Kimmig P, Piekarski G, Heydorn A O (1979) Zur Sarkosporidiose (*Sarcocystis suihominis*) des Menschen (II). Immun Infekt 7: 170–177

Kingston D, Warhurst D C (1969) Isolation of amoebae from the air. J med Microbiol 2: 27–36

Kinyoun (1985) Cryptosporidiosis. J Am Vet Med Assoc 187: 1334–1335

Klein J S (1978) Treatment of severe trichinellosis. In: Kim C W, Pawłowski Z S (Hrsg) Trichinellosis. Reedbooks Ltd, Windsor Berks, S 395–406

Klenk A, Geyer E, Zahner H (1984) Serodiagnosis of human onchocerciasis: Evaluation of sensitivity and specificity of a purified *Litomosoides carinii* adult worm antigen. Tropenmed Parasit 35: 81–84

* Kloft W J (1978) Ökologie der Tiere. UTB 729. Ulmer, Stuttgart

Knobloch J, Delgado E, Alvarez A, Reymann N, Bialek R (1985) Human fascioliasis in Cajamarca, Peru. I. Diagnostic methods and treatment with praziquantel. Trop Med Parasit 36: 88–90

Knobloch J, Lederer I (1983) Immunodiagnosis of human paragonimiasis by an enzyme immunoassay. Tropenmed Parasit 34: 21–23

Knothe H (1982) Prophylaxe bei Fernreisen. Schriftenreihe der Bundesapothekerkammer, Gelbe Reihe Bd X, S 7–18

Kocięcka W, Knapen F van, Ruitenberg E J, Geleijnse M E M, Terlingen J B A (1981) *Trichinella pseudospiralis* and *T. spiralis* infections in monkeys. II. Clinical Aspects. In: Kim C W, Ruitenberg E J, Teppema J S (Hrsg) Trichinellosis: Proceedings of the fourth international conference on trichinellosis, Noordwijk aan Zee The Netherlands 1980. Reedbooks Ltd, Windsor Berks, S 205–208

Köberle F (1968) Chagas' disease and Chagas' syndromes: the pathology of American trypanosomiasis. Adv Parasitol 6: 63–116

Köberle F (1972) Chagas-Krankheit und Chagas-Leiden. In: Gsell O, Mohr W (Hrsg) Infektionskrankheiten Bd IV. Springer, Berlin Heidelberg New York, S 289–324

König J W (1984) Klinische Erfahrungen bei der Behandlung der Trypanosomiasis in Liberia. In: Boch J (Hrsg) Tropenmedizin Parasitologie. Medizin in Entwicklungsländern Bd 16. Peter Lang, Frankfurt Bern New York, S 73–76

Königk E, Putfarken B (1984) In vitro-Untersuchungen über den Wirkungsmechanismus von Chloroquine bei Chloroquine-empfindlichen und Chloroquine-resistenten *Plasmodium falciparum*. In: Boch J (Hrsg) Tropenmedizin Parasitologie. Medizin in Entwicklungsländern Bd 16. Peter Lang, Frankfurt Bern New York, S 145–146

Köster B, Seitz H M (1985) Monoclonal antibodies against cercariae of *Schistosoma mansoni*. Trop Med Parasit [Suppl II] 36: 17

Korte W (1973) Epidemiologie, Pathologie und Klinik der Infektion mit *Trichomonas vaginalis*. Med Klinik 68: 99–106

Kozicki M, Steffen R (1984) Expositionsprophylaxe der Reisediarrhoe – eine Illusion? In: Boch J (Hrsg) Tropenmedizin Parasitologie. Medizin in Entwicklungsländern Bd 16. Peter Lang, Frankfurt Bern New York, S 415

Krampitz H E (1979) Neuere Erkenntnisse über die geographische Verbreitung der Leishmaniasen in der Alten Welt. Geogr Z [Suppl] 51: 42–58

Krampitz H E (1980) Trypanosomiasis- und Leishmaniasisimport. In: Gsell O (Hrsg) Importierte Infektionskrankheiten. Georg Thieme, Stuttgart, S 42–48

Krampitz H E (1981) Elba-Trias: Harara, Lichtdermatosen, Leishmaniasis. Der ökologische Hintergrund. Der Hautarzt 32: 221–227

Krampitz H E, Buschmann H, Münchhoff P (1986) Gibt es latente Babesieninfektionen beim Menschen in Süddeutschland? Mitt Österr Ges Tropenmed Parasitol 8: 1–11

Krieg A (1984) Mit Bakterien gegen Schadinsekten. Neue Wege der Schädlingsbekämpfung. Naturw Rdsch 37: 11–13

Kubelka C F, Krammer P, Ruppel A, Gemsa D (1984) Macrophage cytotoxicity against schistosomula of *Schistosoma mansoni* following stimulation with macrophage-activating factors produced by T-cell clones. In: Boch J (Hrsg) Tropenmedizin Parasitologie. Medizin in Entwicklungsländern Bd 16. Peter Lang, Frankfurt Bern New York, S 185–186

Kuipers F C (1964) Pathogenese van de haringwormflegmone bij de mens. Ned T Geneesk 108: 304–305

Kvalsvig J D (1988) The effects of parasitic infection on cognitive performance. Parasitology Today 4: 206–208

Laarman J J, Van Der Slik-Van Der Veen J V (1961) Coccidiose bij de mens in Nederland. Ned T Geneesk 105: 1731–1735

Lacey L A, Escaffre H, Philippon B, Sékétéli A, Guillet P (1982) Large river treatment with *Bacillus thuringiensis* (H-14) for the control of *Simulium damnosum* s.l. in the onchocerciasis control programme. Tropenmed Parasit 33: 97–101

Lämmler G, Zahner H, Texdorf I (1968) Infektionsversuche mit Darmnematoden, Cestoden und Trematoden bei *Mastomys natalensis* (Smith 1834). Z Parasitenkd 31: 166–202

La Fuente C, Saucedo E, Urjel R (1984) The use of microhaematocrit tubes for the rapid diagnosis of Chagas' disease and malaria. Trans Roy Soc Trop Med Hyg 78: 278–279

Laird M (1977) Tsé-tsé: L'avenir des méthodes biologiques dans la lutte intégrée contre la Tsé-tsé. International Scientific Council for Trypanosomiasis Research and Control 15. Meeting Ref Nr 110. Banjul The Gambia

Lamina J (1970) Immunbiologischer Nachweis einer ‚Larva migrans visceralis-Infektion'. Tierexperimentelle Ergebnisse. II. Mitteilung: Der Mikropräzipitationstest an der lebenden Larve. Zbl Bakt I Abt A Orig 215: 386–397

Lamina J (1980) Larva-migrans-visceralis-Infektionen durch *Toxocara*-Arten. Dtsch med Wschr 105: 796–799

Larivière M et al. (1986) A double blind study of Ivermectin and placebo in onchocerciasis IX. Int Congr Infect Parasit Dis München 1986, Abstr 548

Lee K T, Little M D, Beaver P C (1975) Intracellular (muscle-fiber) habitat of *Ancylostoma caninum* in some mammalian hosts. J Parasitol 61: 589–598

Lejkina E S (1965) Research on ascariasis immunity and immunodiagnosis. Bull Wld Hlth Org 32: 699–708

Le Riche P D, Sewell M M H (1977) Differentiation of *Taenia saginata* and *Taenia solium* by enzyme electrophoresis. Trans Roy Soc Trop Med Hyg 71: 327–328

* Leuckart R (1863, 1876) Die menschlichen Parasiten und die von ihnen herrührenden Krankheiten. Ein Hand- und Lehrbuch für Naturforscher und Ärzte Bd I und II. C. F. Winter'sche Verlagshandlung, Leipzig Heidelberg

* Levine N D, Corliss J O, Cox F E G, Deroux G, Grain J, Honigberg B M, Leedale G F, Loeblich A R, Lom J, Lynn D, Merinfeld E G, Page F C, Poljanski G, Sprague V, Vavra J, Wallace F G (1980) A newly revised classification of the protozoa. J Protozool 27: 37–58

Lichtenberg F von, Sher A, McIntyre S (1977) A lung model of schistosome immunity in mice. Amer J Path 87: 105–120

Lim B L, Ow-Yang C K, Lie K J (1965) Natural infection of *Angiostrongylus cantonensis* in Malaysia rodents and intermediate hosts, and preliminary observations on acquired resistance. Am J Trop Med Hyg 14: 610–617

Lindsay S W, Gibson M E (1988) Bednets revisited – old idea, new angle. Parasitology Today 4: 270–272

Löscher T, Heidelmeyer C F, Lang W (1978) Diagnostische Bedeutung von IgE und Eosinophilie bei Amöben-, Plasmodien- und Helminthen-Infektionen des Menschen. Zbl Bakt I Abt Ref 257: 2

Löscher T, Nothdurft H-D, Prüfer L, Falkner von Sonnenburg F, Lang W (1981) Praziquantel in clonorchiasis and opisthorchiasis. Tropenmed Parasit 32: 234–236

López Antuñano F J, Wernsdorfer W H (1979) In vitro response of chloroquine-resistant *Plasmodium falciparum* to mefloquine. WHO Bull 57: 663–665

Loría-Cortés R, Lobo-Sanahuja J F (1980) Clinical abdominal angiostrongylosis. A study of 116 children with intestinal eosinophilic granuloma caused by *Angiostrongylus costaricensis*. Am J Trop Med Hyg 29: 538–544

Lucius R, Frank W (1978) Beitrag zur Biologie von *Dicrocoelium hospes* Looss, 1907 (Trematodes, Dicrocoeliidae). Acta Tropica 35: 161–181

Luder P J, Witassek F, Weigand K, Eckert J, Bircher J (1985) Treatment of cystic echinococcosis (*Echinococcus granulosus*) with mebendazole: Assessment of bound and free drug levels in cysts fluid and of parasite vitality in operative specimens. Eur J Clin Pharmacol 28: 279–285

Lumbreras H, Terashima A, Alvarez H, Tello R, Guerra H (1982): Single does treatment with Praziquantel (Cesol®, EmBay 8440) of human cestodiasis caused by *Diphyllobothrium pacificum*. Tropenmed Parasit 33: 5–7

* Lumsden W H R, Evans D A (Hrsg) (1976) Biology of the Kinetoplastida Vol. 1 u. 2. Academic Press, New York London San Francisco

Lund O-E, Stefani F H, Dechant W (1978) Amoebic keratitis: a clinicopathological case report. Brit J Ophtal 62: 373–375

Lunde M N, Jacobs L (1983) Antigenic differences between endozoites and cystozoites of *Toxoplasma gondii*. J Parasitol 69: 806–808

Lunde M N, Ottesen E A, Cheever A W (1979) Serological differences between acute and chronic schistosomiasis mansoni detected by enzyme-linked immunosorbent assay (ELISA). Am J Trop Med Hyg 28: 87–91

Mackerras M J, Sandars D F (1955) The life history of the rat lungworm, *Angiostrongylus cantonensis* (Chen) (Nematoda: Metastrongylidae). Aust J Zool 3: 1–25

Macpherson C N L, Romig T, Zeyhle E (1984) *Echinococcus granulosus* im Turkana-Gebiet 11. Tagung der Dtsch Ges Parasitologie Bad Harzburg; Referat Nr 5

Madsen H (1974) The principles of the epidemiology of Trichinelliasis with a new view on the life cycle. In: Kim C W (Hrsg) Trichinellosis. Intext Educational Publishers, New York, S 615–638

Malek E A (1981) Presence of *Angiostrongylus costaricensis* Morera and Céspedes 1971 in Colombia. Am J Trop Med Hyg 30: 81–83

Mannweiler E (1980) Möglichkeiten und Grenzen der Serologie parasitärer Infektionen. In: Gsell O (Hrsg) Importierte Infektionskrankheiten. Georg Thieme, Stuttgart, S 77–82

Mannweiler E (1982) Die Immundiagnostik der Echinokokkose. In: Bähr R (Hrsg) Probleme der Echinokokkose unter Berücksichtigung parasitologischer und klinischer Aspekte. Hans Huber, Bern Stuttgart Wien, S 56–60

Mannweiler E, Knobloch J (1984) Antikörperverlauf bei 216 Patienten mit Amöbenleberabszeß. In: Boch J (Hrsg) Tropenmedizin Parasitologie. Medizin in Entwicklungsländern. Bd 16 Peter Lang, Frankfurt Bern New York, S 387

Mansour N S, Soliman G N, El-Assal F M (1984) Studies on experimental mixed infections of *Schistosoma mansoni* and *S. haematobium* in hamsters. Z Parasitenkd 70: 345–357

Marcus L C, Valigorsky J M, Fanning W L, Joseph T, Glick B (1982) A case report of transfusion induced babesiosis. J Am Med Ass 284: 465–467

Marinkelle C J, Rodriguez P E (1981) Progresos en leishmaniasis. Tribuna Médica 746: 1–6

* Martinez A J (1985) Free-living Amebas. CRC Press, Inc. Boca Raton, Florida

Martinez A J, Janitschke K (1979) Amöbenenzephalitis durch *Naegleria* und *Acanthamoeba*. Vergleich und Gegenüberstellung der Organismen und der Ei krankungen. Immun Infekt 7: 57–64

Martinez A J, Janitschke K (1985) *Acanthamoeba*, an opportunistic micro-organism. A review. Infections 13: 251–256

Matsumoto Y, Yoshida Y (1984) Sporogony in *Pneumocystis carinii:* Synaptonemal complexes and meiotic nuclear divisions observed in precysts. J Protozool 31: 420

Matsumoto Y, Yamada M, Yoshida Y (1987) Light microscopical appearance and ultrastructure of *Blastocystis hominis,* an intestinal parasite of man. Zbl Bakt Abt A, 264: 379–385

Mattern C F T (1977) Viruses of *Entamoeba histolytica.* VII. Novel beaded virus. J Vir 23: 685–691

Mayers J D, Kuharic H A, Holmes K K (1977) *Giardia lamblia* infection in homosexual men. Brit J Venereal Diseases 53: 54–55

McGregor I A, Wilson M E, Billewicz W Z (1983) Malaria infection of the placenta in The Gambia, West Africa; its incidence and relationship to stillbirth, birthweight and placental weight. Trans Roy Soc Trop Med Hyg 77: 232–244

Mehlhorn H (1988) Parasitology in Focus. Springer-Verlag, Berlin Heidelberg New York London Paris Tokyo Hong Kong

* Mehlhorn H, Düwel D, Raether W (1986) Diagnose und Therapie der Parasiten von Haus-, Nutz- und Heimtieren. Gustav Fischer, Stuttgart

Mehlhorn H, Schein E (1984) The piroplasms: Life cycle and sexual stages. Adv Parasitol 23: 38–103

Mehlhorn H, Eckert J, Thompson R C A (1983) Proliferation and metastases formation of larval *Echinococcus multilocularis.* II. Ultrastructural investigations. Z Parasitenkd 69: 749–763

Mehlhorn H, Frenkel J K, Andrews P, Thomas H (1982) Light and electron microscopic studies on *Schistosoma mansoni* granulomas of mouse livers following treatment with Praziquantel. Tropenmed Parasit 33: 229–239

Mehlhorn H, Haberkorn A, Peters W (1977) Electron microscopic studies in developmental stages of *Trypanosoma cruzi* and their surface coat within the heart and skeletal muscle of experimentally infected mice. Protistologica 13: 287–298

Mehlhorn H, Heydorn A O (1978) The sarcosporidia, fine structure and life cycle. Adv Parasitol 16: 43–92

Mehlhorn H, Heydorn A O (1979) Electron microscopical study on gamogony of *Sarcocystis suihominis* in human tissue cultures. Z Parasitenkd 58: 97–113

Mehlhorn H, Heydorn A O, Frenkel J K, Göbel E (1985) Announcement of the establishment of neohepantotypes for some important *Sarcocystis* species. Z Parasitenkd 71: 689–692

* Mehlhorn H, Peters W (1983) Diagnose der Parasiten des Menschen, einschließlich der Therapie einheimischer und tropischer Parasitosen. Gustav Fischer, Stuttgart

* Mehlhorn H, Piekarski G (1985) Grundriß der Parasitenkunde. 2. Aufl. Gustav Fischer, Stuttgart

Mehlhorn H, Raether W, Schein E, Weber M, Uphoff M (1986) Licht- und elektronenmikroskopische Untersuchungen zum Entwicklungszyklus und Einfluß von Pentamidin auf die Morphologie der intraerythrozytären Stadien von *Babesia microti.* Dtsch tierärztl Wschr 93: 400–405

Mehlitz D, Zillmann U, Sachs R (1985) The domestic pig as a carrier of *Trypanosoma brucei gambiense* in West Africa. Trop Med Parasit [Suppl II] 36: 18

Merkelbach J W (1964) Een visser met haringwormziekte (anisakiasis) van het rectum. Ned T Geneesk 108: 2131–2132

Meuwissen J H E T (1976) Infections with *Pneumocystis carinii.* Symposium on *Pneumocystis carinii* infections. Natl Cancer Inst Monogr 43: 133–136

Meuwissen J H E T (1986) The development of vaccines blocking malaria transmission: A review. IX. Int Congr Infect Parasit Dis München: Abstr. 989

Meyer E A, Radulescu S (1979) *Giardia* and Giardiasis. Adv Parasitol 17: 1–47

Michel R, Röhl R, Schneider H (1982) Isolierung von freilebenden Amöben durch Gewinnung von Nasenschleimhautabstrichen bei gesunden Probanden. Zbl Bakt I Abt Orig B 176: 155–159

Michel R, De Jonckheere J F (1983) Isolation and identification of pathogenic *Naegleria australiensis* (De Jonckheere, 1981) from pond water in India. Trans Roy Soc Trop Med Hyg 77: 878

Michel R, Just H-M (1984) Acanthamoeben, Naeglerien und andere freilebende Amöben in Kühl- und Spülwasser von Zahnbehandlungseinheiten. Zbl Bakt I Abt Orig B 179: 56–72

Miles M A (1983) The epidemiology of South American trypanosomiasis – biochemical and immunological approaches and their relevance to control. Trans Roy Soc Trop Med Hyg 77: 5–23

Miller L H (1977) A critique of merozoite and sporozoite vaccines in malaria. In: Miller L H, Pino J A, McKelvey J J jr. (Hrsg) Immunity to blood parasites of animals and man. Plenum Publishing Corporation, New York, S 113–120

Miller L H, Mason S J, Clyde D F, McGinniss M H (1976) Resistance factor to *Plasmodium vivax:* duffy genotype FyFy. N Engl J Med 295: 302–304 ...

Miller L H, Mason S J, Dvorak J A, McGinniss M H, Rothman I K (1975) Erythrocyte receptors for (*Plasmodium knowlesi*) malaria: duffy blood group determinants. Science 189: 561–563

Miller T A (1979) Hookworm Infection in Man. Adv Parasitol 17: 315–384

* Minning W (1969) Die Wurmkrankheiten. In: Grumbach A, Bonin O (Hrsg) Die Infektionskrankheiten des Menschen und ihre Erreger Bd II 2. Aufl. Georg Thieme, Stuttgart, S 1853–1902

Mirelman D, Bracha R, Chayen A, Diamond L S (1986) Bacterial flora can affect virulence and zymodeme expression of non-pathogenic isolates of *Entamoeba histolytica*. IX. Int Congr Infect Parasit Dis München: Abstr. 409

Mittermayer T, Spaldonová R (1981) The use of mebendazole in the treatment of trichinellosis in man. Folia Parasitol Prag 28: 235–242

Miyazaki I, Habe S (1976) A newly recognized mode of human infection with the lung fluke, *Paragonimus westermani* (Kerbert 1878). J Parasit 62: 646–648

Mohr W (1972) Malaria. Schwarzwasserfieber. Toxoplasmose. In: Gsell O, Mohr W (Hrsg) Infektionskrankheiten Bd IV. Springer, Berlin Heidelberg New York, S 461–575. S 585–594. S 611–646

Mohr W, Racz P (1984) Laparoskopische und histologische Befunde der Leber bei Bilharziose. In: Boch J (Hrsg) Tropenmedizin Parasitologie. Medizin in Entwicklungsländern. Bd 16 Peter Lang, Frankfurt Bern New York, S 177

Morera P (1970) Investigación del huésped definitivo de *Angiostrongylus costaricensis* (Morera y Céspedes, 1971). Bol Chile Parasit 25: 133–134

Morisita T et al. (1965) A trial of the skin test in human anisakiasis. Jap J Parasit 14: 230–232 (japanisch, englische Zusammenfass.). (Trop Dis Bull 63: 305 1966)

Morris D L, Dykes P W, Dickson B, Marriner S (1983) Albendazole in human hydatid disease. In: Proceedings. 13th International Congress on Chemotherapy, Wien 1983, SE – 7.30/1–4

Müller B (1982) Überlegungen zur Epidemiologie und Prophylaxe der alveolären Echinokokkose. In: Bähr R (Hrsg) Aktuelle Probleme der Chirurgie und Orthopädie Bd 23. Hans Huber, Bern Stuttgart Wien, S 41–43

Mungelluzzi C, Bianchini C (1969) Osservazioni sulla coltivazione di *Hartmannella castellanii* (Douglas) senza flora associata. Arch ital Sci med trop 50: 151–157

Murphy M J, Pifer L L, Hughes W T (1977) *Pneumocystis carinii* in vitro. A study by scanning electron microscopy. Am J Pathol 86: 387–394

Naik S R, Kumar L, Naik S, Sehgal S, Rau N R, Vinayak V K (1979) Immunological studies in giardiasis. Ann Trop Med Parasit 73: 291–292

Naot Y, Desmonts G, Remington J S (1981) IgM enzyme-linked immunosorbent assay test for the diagnosis of congenital *Toxoplasma* infection. J Pediatr 98: 32–36

* Nauck E G (1975) Lehrbuch der Tropenkrankheiten, begründet von Nauck E G (Hrsg: Mohr W, Schumacher H-H, Weyer F) 4. Aufl. Georg Thieme, Stuttgart

Nelson G S (1970) Onchocerciasis. Adv Parasitol 8: 173–226

Nicholson N W (1978) Case report of *Balantidium coli* infection. East African Medical Journal 55: 133

Nibha Jaroonvesama (1988) Differential diagnosis of eosinophilic meningitis. Parasitology Today 4: 262–266

Nöller W (1923) Die Züchtung der tierischen Parasiten und Krankheitserreger auf künstlichen Nährböden. Kraus-Uhlenbuths Handb. mikrobiol. Technik Bd 1: 647

Nussenzweig R (1982) Vaccination against malaria. Use of monoclonal antibodies for the characterization of the protective antigens. In: Perspectives of immunization in parasitic diseases. Pontificiae Academiae Scientiarum Scripta Varia No. 47, S 33–38

Ockert G (1972) Zur Epidemiologie von *Dientamoeba fragilis*. II. Versuch der Übertragung der Art mit *Enterobius*-Eiern. J Hyg Epidemiol Microbiol Immunol 16: 222–225

Ockert G (1974) *Hartmannella* – *Naegleria* – Amöben in Nasen-Rachen-Abstrichen von Schulkindern und Jugendlichen. In: Proceedings. III. Intern Congr Parasitol München 1974, Facta Publication Vol. 1, S 190. Egermann H, Wien

Ockert G (1974) Zur Epidemiologie von *Dientamoeba fragilis*. In: Proceedings III. Intern Congr Parasitol München. Facta Publication Vol. 1, 163. Egermann H, Wien

Oshima T (1987) Anisakiasis – is the sushi bar guilty? Parasitology Today 3: 44–48

Oyanagi T (1967) Experimental studies on the visceral lesions of gastro-intestinal walls due to *Anisakis* larvae. Jap J Parasit 16: 470–493 (japanisch, englische Zusammenfass.). (Trop Dis Bull 66: 148, 1969)

Page F L (1967) A illustrated key to freshwater and soil amoebae: Freshwater Biological Association. Scientific Publication Nr. 34, Titus Wilson, Kendal

Palomino J C, Guerra H, Lumbreras H (1983) A selective liquid medium for primary isolation of South American leishmanias. Tropenmed Parasit 34: 229–232

De Paola D, Winslow D J (1967) Geographic pathology of *Schistosoma mansoni*. In: Mostofi F K (Hrsg) Bilharziasis. Springer, Berlin Heidelberg New York

Pavić R, Stojković L (1982) Vakzination mit Solco Trichovac. Immunologische Aspekte eines neuen Prinzips zur Therapie und Reinfektionsprophylaxe der Trichomoniasis bei der Frau. Gynäk Rdsch [Suppl 2] 22: 27–38

Pavlica F (1966) The first observation of congenital pneumocystic pneumonia in a fully developed still-born child. Ann Paediat 198: 177

Pawłowski Z S (1981) Control of Trichinellosis. In: Kim C W, Ruitenberg E J, Teppema J S (Hrsg) Trichinellosis. Reedbooks Ltd, Windsor Berks, S 7–20

Pawłowski Z S (1982) Ascariasis: host-pathogen biology. Rev Infect Diseases 4: 806–814

Pawłowski Z S, Ruitenberg E J (1978) Is *Trichinella pseudospiralis* likely to be a human pathogen? Lancet 24: 1357

* Pawłowski Z S, Schultz M G (1972) Taeniasis and cysticercosis (*Taenia saginata*). Adv Parasitol 10: 269–343

Pellegrino J, Lima-Costa F F, Carlos M A, Mello R T (1977) Experimental chemotherapy of schistosomiasis mansoni. XIII. Activity of Praziquantel, an isoquinoline-pyrazino derivative, on mice, hamsters, and *Cebus* monkeys. Z Parasitenkd 52: 151–168

Pellerdy L P (1974) Coccidia and Coccidiosis. 2. Aufl. Paul Parey, Berlin Hamburg

Peppersack T (1981) Zur Pathogenität der Larven von *Toxocara canis* Werner 1782 (Anisakidae) bei der Maus. Inaug Diss, Vet Med Fakt, Tierärztl. Hochschule Hannover

Pflüger W, Roushdy M Z, El Emam M (1984) The prepatent period and cercarial production of *Schistosoma haematobium* in *Bulinus truncatus* Egyptian field strain at different constant temperatures. Z Parasitenkd 70: 95–103

Phillips B P (1974) *Naegleria:* another pathogenic ameba. Studies in germfree guinea pigs. Amer J Trop Med Hyg 23: 850–855

Picq J J, Roux J (1973) Faits nouveaux dans l'onchocercose. La microfilarurie, sa répartition géographique, ses rapports avec les densités microfilariennes cutanées, l'albuminurie et la chimiothérapie. Premiers résultats. Méd Trop 33: 451–461

* Piekarski G (1965) Symbiose und Parasitismus. In: Büchner F, Letterer E, Roulet F (Hrsg) Handbuch der allgemeinen Pathologie Bd 11 Teil II. Springer, Berlin Heidelberg New York, S 1–53

Piekarski G (1969) Leishmaniasen. Balantidienruhr. Flagellaten des Darmkanals und der Genitalien. In: Grumbach A, Bonin O (Hrsg) Die Infektionskrankheiten des Menschen und ihre Erreger Bd II 2. Aufl. Georg Thieme, Stuttgart, S 1815–1830 u. 1838–1845

Piekarski G, Heydorn A O, Aryeetey M E, Hartlapp J-H, Kimmig P (1978) Klinische, parasitologische und serologische Untersuchungen zur Sarkosporidiose (*Sarcocystis suihominis*) des Menschen. Immun Infekt 6: 153–159

Piekarski G, Maier W (1984) Medizinische Parasitologie. In: Brandis H, Otte H J (Hrsg) Lehrbuch der Medizinischen Mikrobiologie 5. Aufl. Gustav Fischer, Stuttgart New York, S 515–551

* Piekarski G, Mohr W (1975) Parasitäre Infektionen. In: Otten H, Plempel M, Siegenthaler W (Hrsg) Antibiotika-Fibel. 4. Aufl. Georg Thieme, Stuttgart, S 952–1007

Piekarski G, Piekarski C (1983) Klinik der Darmparasitosen des Menschen. In: Bock H E, Gerok W, Hartmann F (Hrsg) Klinik der Gegenwart Bd VI, E115–E150. Urban u. Schwarzenberg, München Wien Baltimore, S 1–63

* Piekarski G, Seitz H M (1984) Parasitäre Erkrankungen des Darmes. In: Demling L (Hrsg) Klinische Gastroenterologie Bd I 2. Aufl. Georg Thieme, Stuttgart New York, S 545–564

* Piekarski G, Seitz H M (1987) Qualitätssicherung bei lichtmikroskopischen und serologischen Untersuchungen auf Parasiten. In: v. Boroviczény K-G, Merten R, Merten U P (Hrsg) Qualitätssicherung im medizinischen Laboratorium. Springer, Berlin Heidelberg New York 823–854 ...

Pifer L L, Hughes W T, Murphy M J (1977) Propagation of *Pneumocystis carinii* in vitro. Pediatric Res 11: 305–316

Pinder M (1988) Loa loa – a neglected filaria. Parasitology Today 4: 279–284

Piringer-Kuchinka A (1952) Eigenartige mikroskopische Befunde an excedierten Lymphknoten. Verh dtsch Ges Path 36: 352–362

Polak M F (1966) Haringwormziekte in 1965. Ned T Geneesk 110: 1029–1030

Poznańska H, Kassur B, Januskiewicz J (1981) The origin of serum enzymes in trichinellosis. In: Kim C W, Ruitenberg E J, Teppema J S (Hrsg) Trichinellosis. Reedbooks Ltd, Windsor Berks, S 253–255

Raaflaub J (1980) Multiple-dose kinetics of the trypanosomicide benznidazole in man. Arzneim-Forsch Drug Res 30: 2192–2194

Rácz P, Voelker J (1984) Experimental infections of rhesus monkeys with *Paragonimus africanus* and *P. uterobilateralis:* histopathological findings at the end of infection in comparison with the acute phase of the disease. Zbl Bakt Hyg A 258: 398–399

Rácz P, Tenner-Rácz K, Luther B, Büttner D W, Albiez E J (1984) Licht- und elektronenmikroskopische Untersuchungen der immunpathologischen Veränderungen der Lymphknoten bei der Mazzotti-Reaktion. In: Boch J (Hrsg) Tropenmedizin Parasitologie. Medizin in Entwicklungsländern. Bd 16 Pater Lang, Frankfurt Bern New York, S 219–221

Ramsdale C D, Herath P R J, Davidson G (1980) Recent developments of insecticide resistance in some Turkish anophelines. J Trop Med Hyg 83: 11–19

Rathor H R, Toqir G (1980) Malathion resistance in *Anopheles stephensi* Liston in Lahore, Pakistan. Mosquito News 40: 526–531

Reed S L, Sargeaunt P G, Braude A I (1983) Resistance to lysis by human serum of pathogenic *Entamoeba histolytica*. Trans Roy Soc Trop Med Hyg 77: 248–253

* Reichenow E (1953) Lehrbuch der Protozoenkunde. 6. Aufl. Gustav Fischer, Jena

* Reichenow E, Vogel H, Weyer F (1969) Leitfaden zur Untersuchung der tierischen Parasiten des Menschen und der Haustiere. 4. Aufl. J. A. Barth, Leipzig

Remington J S (1969) The present status of IgM fluorescent antibody technique in the diagnosis of congenital toxoplasmosis. J Pediatr 75: 1116–1124

Remington J S (1982a) A double-sandwich IgM-ELISA for diagnosis of acute acquired and congenital *Toxoplasma* infection. Lyon Médical 248: 31–35

Remington J S (1982b) Toxoplasmosis in homosexuals. Lyon Médical 248: 133–134

Renz A, Wenk P (1983) The distribution of the microfilariae of *Onchocerca volvulus* in the different body regions in relation to the attacking behaviour of *Simulium damnosum* s.l. in the Sudan savanna of northern Cameroon. Trans Roy Soc Trop Med Hyg 77: 748–752

Rickman L R, Robson J (1972) Some supplementary observations on the blood incubation infectivity test. Bull Wld Hlth Org 46: 403–404

Rickman L R, Ernest A, Dukes P, Maudlin I (1984) The acquisition of human serum resistance during cyclical passage of a *Trypanosoma brucei brucei* clone through *Glossina morsitans morsitans* maintained on human serum. Trans Roy Soc Trop Med Hyg 78: 284

Rim H-J, Lyu K-S, Lee J-S, Joo K-H (1981) Clinical evaluation of the therapeutic efficacy of praziquantel (Embay 8440) against *Clonorchis sinensis* infection in man. Ann Trop Med Parasit 75: 27–33

Rim H-J (1986) The current pathobiology and chemotherapy of clonorchiasis. Korean J. Parasitology 24: Suppl Monogr Series No 3, 1–141

Rivas-Alcala R, Mackenzie C D, Gomez-Rojo E, Greene B M, Taylor H R (1984) The effects of diethylcarbamazine, mebendazole and levamisole on *Onchocerca volvulus* in vivo and in vitro. Tropenmed Parasit 35: 71–77

* Rodenwaldt E, Jusatz H (1951/60) Weltseuchenatlas I–III. Falk, Hamburg

Rohde K (1978) The bird schistosome *Gigantobilharzia* sp. in the silver gull, *Larus novae hollandiae*, a potential agent of schistosome-dermatitis in Australia. Search 9: 40–42

Rohwedder R W (1969) Infección chagásica en dadores de sangre y las posibilidades de transmitirla por medio de la transfusión. Bol Chile Parasit 24: 88–93

Rommel M, Heydorn A O (1972) Beiträge zum Lebenszyklus der Sarkosporidien. III. *Isospora hominis* (Railliet und Lucet, 1891) Wenyon, 1923, eine Dauerform der Sarkosporidien des Rindes und des Schweines. Berl Münch Tierärztl Wschr 85: 143–145

Rommel M, Heydorn A O, Gruber F (1972) Beiträge zum Lebenszyklus der Sarkosporidien. I. Die Sporocyste von *Sarcocystis tenella* in den Fäzes der Katze. Berl Münch Wschr 85: 101–105

Rüttgers H (1982a) Epidemiologie und Klinik der Trichomoniasis. Gynäk Rdsch [Suppl 2] 22: 3–9

Rüttgers H (1982b) Klinische Erfahrungen mit Solco-Trichovac bei der Behandlung der Trichomonaden-Infektion der Frau. Gynäk Rdsch [Suppl 2] 22: 63–69

Rüttgers H (1985) Bacterial non-specific vaginitis ('Bacterial' vaginosis). Gynäk Rdsch [Suppl 3] 24: 2–4

Ruitenberg E J, Elgersma A, Kruizinga W, Leenstra F (1978b) Host protection to the intestinal phase of *Trichinella spiralis*. In: Kim C W, Pawłowski Z S (Hrsg) Trichinellosis. University Press of New England, Hanover NH USA, S 169–181

Ruitenberg E J, Knapen F van, Vermeulen C J (1978a) Enzyme-linked immunosorbent assay (ELISA) in *Trichinella spiralis* infections in pigs. In: Kim C W, Pawłowski Z S (Hrsg) Trichinellosis. University Press of New England, Hanover NH USA, S 487–499

Ruitenberg E J, Ljungström J, Steerenberg P A, Buys J (1975) Application of immunofluorescence and immunoenzyme methods in the serodiagnosis of *Trichinella spiralis* infection. Ann N Y Acad Sci 254: 296–303

Ruiz A, Frenkel J K (1980) *Toxoplasma gondii* in Costa Rican cats. Amer J Trop Med Hyg 29: 1150–1160

Saathoff M (1985) Verwendung von modifiziertem Asami-Medium als monoxensisches Kulturmedium für *Entamoeba histolytica*. Zbl Bakt Hyg A 259: 142–145

Saathoff M, Seitz H M (1985) Untersuchungen zum *Toxoplasma*-spezifischen IgM-Antikörper-Vergleich von ISAgA-(Immunosorbent Agglutination Assay) und Immunfluoreszenz-Ergebnissen. Z Geburtsh Perinat 189: 73–78

Sachs R (1984a) Improvements in the miniature anion exchange centrifugation technique for detecting trypanosomes in domestic pigs. Trans Roy Soc Trop Med Hyg 78: 561

Sachs R (1984b) The superiority of the miniature anion-exchange centrifugation technique for detecting low grade trypanosome parasitaemias. Trans Roy Soc Trop Med Hyg 78: 694–696

Sachs R, Kern P (1982) Epidemiological investigations of human lung fluke infection in Gabon, Central Africa. VII Intern Congr Inf Paras Diseases Stockholm 1982

Sachs R, Mehlitz D, Zillmann U (1984) Sleeping sickness in West-Africa: parasitological-epidemiological field studies in Liberia. Zbl Bakt Hyg A 258: 384

Sachs R, Voelker J (1982) Human paragonimiasis caused by *Paragonimus uterobilateralis* in Liberia and Guinea, West Africa. Tropenmed Parasit 33: 15–16

Sadun E H (1955) Studies on *Opisthorchis viverrini* in Thailand. Am J Hyg 62: 81–115

Saenz R E, Paz H, Johnson C M, Rogers M D, Nelson D J, Pattishall K P (1986) Efficacy and safety of Allopurinol-ribose in the treatment of American cutaneous leishmaniasis (ACL). IX. Intern Congr Infect Parasit Dis München; Abstr. Nr. 446

Sagua H, Araya J, González J, Fuentes A (1982) Seropositividad chagásica en banco de sangre de zona endémica. Algunos aspectos epidemiológicos de los hemodonantes. Bol Chile Parasit 37: 24–26

Sargeaunt P G, Baveja U K, Nanda R, Anand B S (1984) Influence of geographical factors in the distribution of pathogenic zymodemes of *Entamoeba histolytica:* identification of zymodeme XIV in India. Trans Roy Soc Trop Med Hyg 78: 96–101

Sastry S C, Kumar K J, Lakshminarayana V (1978) The treatment of dracontiasis with thiabendazole. J Trop Med Hyg 81: 32–35

Sato M, Koyama A, Iwai K, Kawabata Y, Kojima S (1985) Human pulmonary dirofilariasis with special reference to the ELISA for the diagnosis and follow-up study. Z Parasitenkd 71: 561–563

Schantz P M, Bossche H van den, Eckert J (1982) Chemotherapy for larval echinococcosis animals and humans: report of a workshop. Z Parasitenkd 67: 5–26

Schaub G A (1986) Können Vektoren der Chagas-Krankheit mit *Blastocritidia triatomae* (Trypanosomatidae) bekämpft werden? 12. Tagung der Deutschen Gesellschaft für Parasitologie; Referat Nr. 1, Wien 1986

Schenone H, Rojas A, Alfaro E, Concha L, Aranda R (1981) Estudio longitudinal de la persistencia de la acción terapéutica del nifurtimox y del benznidazol en pacientes con infección chagásica crónica. Bol Chile Parasit 36: 59–62

Schenone H jun, González H, Schenone H sen, Rojas A (1982) Infección experimental de ratas con *Trypanosoma cruzi* por vía oral. Bol Chile Parasit 37: 2–9

Schmid K, Raub C (1984) Comparative studies on the development of *Schistosoma japonicum* in laboratory rodents. Zbl Bakt Hyg A 258: 402

Scholtyseck E (1973) Die Deutung von Endodyogenie und Schizogonie bei Coccidien und anderen Sporozoen. Z Parasitenkd 42: 87–104

* Scholtyseck E (1979) Fine structure of parasitic protozoa. Springer, Heidelberg

Scholtyseck E, Teras J, Kasakova F, Sethi K K (1985) Electron microscope observations on the interaction of *Mycoplasma fermentans* with *Trichomonas vaginalis*. Z Parasitenkd 71: 435–442

Schottelius J (1982) The identification by lectins of two strains groups of *Trypanosoma cruzi*. Z Parasitenkd 68: 147–154

Schottelius J (1984) Differentiation between *Trypanosoma cruzi* and *Trypanosoma rangeli* on the basis of their sialic acid content. Tropenmed Parasit 35: 160–162

Schottelius J (1986) Thiobarbitursäure/Methylumbelliferyltest zur Unterscheidung von *Trypanosoma cruzi* and *Trypanosoma rangeli*. 12. Tagung der Deutschen Gesellschaft für Parasitologie; Referat Nr. 14, Wien 1986

Schüffner W, Swellengrebel N H (1949) Retrofection in oxyuriasis. A newly discovered mode of infection with *Enterobius vermicularis*. J Parasit 35: 138–146

Schulz-Key H, Awadzi K, Dadzie Y, Aziz M A (1984) Efficacy of Ivermectine on the micro- and macrofilariae of *Onchocerca volvulus*. Zbl Bakt Hyg A 258: 430

Schulz-Key H, Karam M (1984) Quantitative assessment of microfilariae and adults of *Onchocerca volvulus* in ethanol-fixed biopsies and nodules. Trans Roy Soc Trop Med Hyg 78: 157–159

Schulz-Key H, Kläger S, Awadzi K, Diallo S, Greene B M, Larivière M, Aziz M A (1985) Treatment of human onchocerciasis: The efficacy of Ivermectin on the parasite. Trop Med Parasit [Suppl II] 36: 20

Schulz-Key H, Greene B M, Awadzi K, Larivière M, Kläger S, Dadzie Y, Aziz M A (1986) Efficacy of Ivermectin on the reproductivity of female *Onchocerca volvulus*. Trop Med Parasit [Suppl 34] 37: 89

Seitz H M, Kersting G (1985) Toxoplasma infection in AIDS patients. Trop Med Parasit [Suppl II] 36: 15

Sepúlveda B (1980) Inducción de inmunidad antiamibiásica en primates subhumanos con antígeno lisosomal de *Entamoeba histolytica*. Archivos de investigación médica [Suppl 1] 11: 245–276

Sethi K K (1982) Monoclonal antibodies against *Toxoplasma gondii*. Lyon Médical 248: 55–58

Sethi K K, Omata Y, Brandis H (1984) Topographical localization of distinct antigenic domains of *Toxoplasma gondii* with the aid of monoclonal antibodies. Z Parasitenkd 70: 699–707

Shao X Y (1981) Paragonimiasis with hepatic lesions as main clinical manifestation. Chinese Journal of Internal Medicine 20: 396 (chinesisch) WHO/Helm/82.5

Sharma S P, Dubey J P (1981) Quantitative survival of *Toxoplasma gondii* tachyzoites and bradyzoites in pepsin and in trypsin solutions. Amer J Vet Res 42: 128–130

Siegel J P, Remington J S (1983) Comparison of methods for quantitating antigen-specific immunoglobulin M antibody with a reverse enzyme-linked immunosorbent assay. J Clin Microbiol 18: 63–70

Sivayathorn A, Kiatakrapol (1986) Albendazol: a new treatment for creeping eruption. IX. Int Congr Infect Parasit Dis München: Abstr. 1323

Smith J W, Wootten R (1978) *Anisakis* and anisakiasis. Adv Parasitol 16: 93–163

Smithers S R, Terry R J (1976) The immunology of schistosomiasis. Adv Parasitol 14: 399–422

Smyth J D (1969) The biology of the hydatid organisms. Adv Parasit 7: 327–347

Spencer H C (1985) Drug-resistant malaria – changing patterns mean difficult decisions. Trans Roy Soc Trop Med Hyg 79: 748–758

Speiser F, Gottstein B (1984) Studies on standardization of the ELISA for the serodiagnosis of human toxocariasis. Zbl Bakt Hyg A 258: 421

Spina-França A, Nobrega J P S, Livramento J A, Machado L R (1982) Administration of Praziquantel in neurocysticercosis. Tropenmed Parasit 33: 1–4

Stahel E (1984) Häufigkeit und Klinik der *Isospora belli*-Infektion. In: Boch J (Hrsg) Tropenmedizin Parasitologie. Medizin in Entwicklungsländern Bd 16. Peter Lang, Frankfurt Bern New York, S 447–451

Stallbaumer M F, Morris D L, Clarkson M J (1983) Albendazole treatment of pulmonary hydatid disease in a sheep model. Proceedings. 13th International Congress on Chemotherapy, Wien 1983 SE–7.30

Steffen R (1980) Reisekrankheitenstatistik. In: Gsell O (Hrsg) Importierte Infektionskrankheiten, Epidemiologie und Therapie. Georg Thieme, Stuttgart, S 58–68

* Steffen R (1984) Reisemedizin. Springer, Berlin Heidelberg New York

Stephenson L S (1980) The contribution of *Ascaris lumbricoides* to malnutrition in children. Parasitology 81: 221–233

Stevens A R, Willaert E (1980) Drug sensitivity and resistance of four *Acanthamoeba* species. Trans Roy Soc Trop Med Hyg 74: 806–808

Stoll N R (1947) This wormy world. With addendum. J Parasitol 33: 1–18

Street A, Taylor-Robinson D, Ackers J P, McMillan A (1983) ELISA for the detection of antibody to *Trichomonas vaginalis*. Wiadomości Parazytologiczne T. XXIX, Nr. 1–2, 99–102

Street D A, Taylor-Robinson D, Ackers J P, Hanna N F, McMillan A (1982) Evaluation of an enzyme linked immunosorbent assay for the detection of antibody to *Trichomonas vaginalis* in sera and vaginal secretions. Brit J Venereal Diseases 58: 330–333

Sturrock R F, Kimani R, Cottrell B J, Butterworth A E, Seitz H M, Siongok T K, Houba V (1983) Observations on possible immunity to reinfection among Kenyan schoolchildren after treatment for *Schistosoma mansoni*. Trans Roy Soc Trop Med Hyg 77: 363–371

Szabados A, Schierz G, Deinhardt F (1986) Nachweis der Beweglichkeit der intracystischen Körperchen in der reifen Mutterzelle von *Pneumocystis carinii*. 12. Tagung der Deutschen Gesellschaft für Parasitologie: Referat Nr. 30, Wien

* Tadros W, Laarman J J (1982) Current concepts on the biology, evolution and taxonomy of tissue cyst-forming eimeriid coccidia. Adv Parasitol 20: 293–468

Taylor S M, Pearson G R (1981) The effect of a single injection of Ivermectin on encysted muscle larvae of *Trichinella spiralis* in experimentally infected pigs. In: Kim C W, Ruitenberg E J, Teppema J S (Hrsg) Trichinellosis. Reedbooks Ltd, Windsor Berks, S 353–357

Teppema J S, Blomjous F J E M, Elgersma A, Ruitenberg E J (1981) *Trichinella pseudospiralis* and *T. spiralis* infections in monkeys III Pathological aspects. In: Kim C W, Ruitenberg E J, Teppema J S (Hrsg) Trichinellosis. Reedbooks Ltd, Windsor Berks, S 209–214

Teras J (1961) On the question of the types of *Trichomonas vaginalis*. In: Progress in Protozoology, Proceedings of the First International Conference on Protozoology, Prague 1961, 572–576

Teras J (1966) Differences in the antigenic properties within strains of *Trichomonas vaginalis*. Wiadomości Parazytologiczne T. XII: 357–363

Teras J, Mertens T, Piekarski G (1985) Zur Verbreitung der Urogenitaltrichomoniasis bei HWG-Frauen. Immun Infekt 13: 44–46

Tewfik S, Kassem S A, Aref M K, Awadalla H N, Abadir A (1983) A preliminary report on two cases of visceral leishmaniasis in Egypt. Trans Roy Soc Trop Med Hyg 77: 334–335

Thalhammer O (1981) Toxoplasmose. Dtsch med Wschr 106: 1051–1053

Thiel P H van (1966) The final host of the herringworm *Anisakis marina*. Trop Geogr Med 18: 310–327

Thiel P H van (1984) Maagdarm-granulomen door het eten van rauwe oesters? Ned T Geneesk 128: 449–450

Thiel P H van, Bakker P M (1981) Wormgranulomen in de maag in Nederland en in Japan. Ned T Geneesk 125: 1365–1370

Thomas H, Gönnert R (1977) The efficacy of Praziquantel against cestodes in animals. Z Parasitenkd 52: 117–127

* Tischler W (1969) Grundriß der Humanparasitologie. VEB Gustav Fischer, Jena

Trager W, Jensen J B (1976) Human malaria parasites in continuous culture. Science 193: 673–675

Tzipori S, Smith M, Birch C, Barnes G, Bishop R (1983) Cryptosporidiosis in hospital patients with gastroenteritis. Am J Trop Med Hyg 32: 931–934

Vaněk J, Jirovec O (1952) Parasitäre Pneumonie. ‚Interstitielle' Plasmazellenpneumonie der Frühgeborenen, verursacht durch *Pneumocystis carinii*. Zbl Bakt Orig A 158: 120–127

Vickerman K (1978) Antigenic variation in trypanosomes. Nature 273: 613–617 ...

Voelker J, Vogel H (1965) Zwei neue *Paragonimus*-Arten aus West-Afrika: *Paragonimus africanus* und *Paragonimus uterobilateralis* (Troglotrematidae; Trematoda). Z Tropenmed Parasit 16: 125–148

Voelker J, Sachs R (1984) Experimental infections of rhesus monkeys with *Paragonimus africanus* and *P. uterobilaterlis:* parasitological observations 8–10 years after infection. Zbl Bakt Hyg A 258: 405

Vogel H (1934) Der Entwicklungscyclus von *Opisthorchis felineus*. Far East Assoc Trop Med Nanking 1: 619–624

Vogel H (1957) Über den *Echinococcus multilocularis* Süddeutschlands. I. Das Bandwurmstadium von Stämmen menschlicher und tierischer Herkunft. Z Tropenmed Parasit 8: 404–454

Vogel H (1962) Beobachtungen über die erworbene Immunität von Rhesusaffen gegen *Schistosoma*-Infektionen. Tropenmed Parasit 13: 397–404

Voller A, Cornille-Brögger R, Storey J, Molineaux L (1980) A longitudinal study of *Plasmodium falciparum* malaria in the West Africa savanna using the ELISA technique. WHO Bull 58: 429–438

Vossen M E M H, Beckers P J A, Meuwissen J H E T, Stadhouders A M (1978) Developmental biology of *Pneumocystis carinii*, an alternative view on the life cycle of the parasite. Z Parasitenkd 55: 101–118

Wagner W-H (1986) Malariabekämpfung – Chemotherapie und Immunprophylaxe. ArzneimForsch/Drug Res 36 2–9: 163–175, 409–415

Wahlgren M, Frolov I (1983) Treatment of *Dipetalonema perstans* infections with mebendazole. Trans Roy Soc Trop Med Hyg 77: 422–423

Wahn V, Mehlhorn H (1984) Vier Parasitenarten bei einem achtjährigen Jungen. Kurative Wirkung von Praziquantel gegen *Fasciola hepatica*. Dtsch med Wschr 109: 1486–1488

Walker J (1978) The finding of *Biomphalaria straminea* amongst fish imported into Australia. WHO/Schisto 78.46

Walls K W, Franco E L (1982) Reversed enzyme immunoassay. Lyon Médical 248: 43–49

Walter G, Weber G (1981) Untersuchungen zur Übertragung (transstadial, transovarial) von *Babesia microti*, Stamm ‚Hannover I', in *Ixodes ricinus*. Tropenmed Parasit 32: 228–230

Walzer P D, Perl D P, Krogstad D J, Rawson P G, Schultz M G (1976) *Pneumocystis carinii* pneumonia in the United States: Epidemiologic, diagnostic and clinic features. Natl Cancer Inst Monogr 43: 55–63

Wandmacher G (1979) Untersuchungen über die in-vitro-Empfindlichkeit von *Bacteroides fragilis* gegen Metronidazol, Moxnidazol und Nimorazol. Inaug Diss Med Fak Universität Bonn

Wang S S, Feldman H A (1967) Isolation of *Hartmannella* species from human throats. New Engl J Med 277: 1174–1179

Warhurst D C (1988) Mechanism of chloroquine resistance in malaria. Parasitology Today 4: 211–213

Warren K S (1972) The immunopathogenesis of schistosomiasis: A multidisciplinary approach. Trans Roy Soc Trop Med Hyg 66: 417–432

Wartoń A, Honigberg B M (1983) Analysis of surface saccharides in *Trichomonas vaginalis* strains with various pathogenicity levels by fluorescein-conjugated plant lectins. Z Parasitenkd 69: 149–159

* Watson J M (1960) Medical Helminthology. Baillière Tindall and Cox, London

Wegner D H G (1984) Aktualisierte Daten zur Biltricide-Therapie der Schistosomiasis und anderer Trematoden-Infektionen. In: Boch J (Hrsg) Tropenmedizin Parasitologie. Medizin in Entwicklungsländern Bd 16. Peter Lang, Frankfurt Bern New York, S 193–199

Wegner D H G, Rohwedder R W (1972a) The effect of Nifurtimox in acute Chagas-infection. Arzneim-Forsch (I) Drug Res 22: 1624–1635

Wegner D H G, Rohwedder R W (1972b) Experience with Nifurtimox in chronic Chagas-infection. Arzneim-Forsch/Drug Res 22: 1635–1641

Wegner D H G, Thomas H (1980) Neue Mittel gegen Schistosomiasis. In: Gsell O (Hrsg) Importierte Infektionskrankheiten. Epidemiologie und Therapie. Georg Thieme, Stuttgart New York, S 101–112

Weise H-J (1982) Internationaler Reiseverkehr und meldepflichtige Krankheiten in der Bundesrepublik Deutschland einschl. Berlin (West) 1976–1980. Fortschr Med 100: 771–777

Weise H-J (1984) Entwicklung der Malariaeinschleppungen in die Bundesrepublik Deutschland einschl. Berlin (West) während der letzten fünf Jahre (1978–1982). Bundesgesundheitsblatt 27: 1–10

Weissbrodt H, Albiez E J, Büttner D W (1984) Die Beziehung zwischen der Mikrofilariendichte und dem Lebensalter bei Onchocerciasispatienten in hyperendemischen Dörfern in Liberia. Vorläufige Mitteilung. In: Boch H J (Hrsg) Tropenmedizin Parasitologie. Medizin in Entwicklungsländern Bd 16. Peter Lang, Frankfurt Bern New York, S 213–216

Welch J S, Dobson C, Campbell G R (1980) Immunodiagnosis and seroepidemiology of *Angiostrongylus cantonensis* zoonoses in man. Trans Roy Soc Trop Med Hyg 74: 614–623

Weltgesundheitsorganisation (WHO) (1964) Soil-transmitted helminths. Report of a WHO Expert Committee on helminthiases. Techn Rep Series 277: 15

* WHO (1979) Parasitic Zoonoses. Techn Rep Series 637

* WHO (1980a) Epidemiology and Control of Schistosomiasis. Techn Rep Series 643

* WHO (1980b) Guinea worm disease: Prospects for control. Chronicle 34: 159–160

* WHO (1981) Intestinal Protozoan and Helminthic Infections. Techn Rep Series 666

* WHO (1983) Report of the Steering Committees of the Scientific Working Groups on Malaria 1980–1983. TDR/MAL/SC-SWG (80–83) 83.3, Genf

* WHO (1984a) The Leishmaniases. Techn Rep Series 701

* WHO (1984b) Lymphatic filariasis. Techn Rep Series 702

* WHO (1984c) Advances in malaria chemotherapy. Techn Rep Series 711

* WHO (1985) The control of schistosamiasis. Techn Rep Series 728

* WHO (1986) Epidemiology and control of African trypanosomiasis. Techn Rep Series 739

Werner Hans (1981) Zur Vereinheitlichung der Laboratoriumsdiagnostik der Echinokokkose. Bundesgesundheitsblatt 24: 310–311

Werner Hans, Merks C (1984) Neue Möglichkeiten zum Nachweis einer Chagas-Krankheit mittels künstlicher Xenodiagnose. In: Boch J (Hrsg) Tropenmedizin Parasitologie. Medizin in Entwicklungsländern Bd 16. Peter Lang, Frankfurt Bern New York, S 101–102

Werner Hans, Janitschke K (1985) Aktuelle Probleme der Serodiagnostik der Toxoplasmose unter besonderer Berücksichtigung der Schwangerenvorsorge. Bundesgesundheitsblatt 28: 240–243

Werner Hans, Voss H (1970) Entwicklungsformen von *Toxoplasma gondii* im Säugetierorganismus. Zbl Bakt Hyg I Orig 213: 120–134

* Werner Herbert (1981) Anaerobier-Infektionen: Pathogenese, Klinik, Therapie. Georg Thieme, Stuttgart

Werner Herbert, Krasemann C, Kandler R, Wandmacher G (1980) Metronidazol-Empfindlichkeit von Anaerobiern. Vergleich mit anderen Chemotherapeutica. Münch med Wschr 122: 633–636

Wernsdorfer G (1984) Differentialdiagnose von Malaria und Babesiose beim Menschen. In: Boch J (Hrsg) Tropenmedizin Parasitologie: Medizin in Entwicklungsländern Bd 16. Peter Lang, Frankfurt Bern New York, S 157–160

Westphal A (1969) Amöbendysenterie. In: Grumbach A, Bonin O (Hrsg) Die Infektionskrankheiten des Menschen und ihre Erreger Bd. II, 2. Aufl. Georg Thieme Stuttgart 1830–1838

Whisnant J K, Buckley R H (1976) Successful pyrimethamine-sulfadiazine therapy of *Pneumocystis* pneumonia in infants with X-linked immunodeficiency with hyper-IgM. In: Robbins J B, DeVita V T Jr, Dutz W (Hrsg) Symposium on *Pneumocystis carinii* Infection. Nat Cancer Inst Monogr (ML) 43: 211–216

White A T, Newland H S, Taylor H R, Errtmann K D, Williams P N, Greene B M (1986) Controlled trial and dose finding study of Ivermectin for treatment of onchocerciasis. Proceedings Wellcome Filariasis Seminar '85 in: Trop Med Parasit [Suppl 43] 37: 96

Wielaard F, Gruijthuijsen H van, Duermeyer W, Joss A W L, Skinner L, Williams H, Elven E H van (1983) Diagnosis of acute toxoplasmosis by an enzyme immunoassay for specific immunoglobulin M antibodies. J Clin Microbiol 17: 981–987

Willaert E, Stevens A R (1980) Experimental pneumonitis induced by *Naegleria fowleri* in mice. Trans Roy Soc Trop Med Hyg 74: 779–783

Williams H H (1965) Roundworms in fishes and so-called 'herring-worm disease'. Brit med J I: 164–167

Witassek F (1984) Nachweis von totalem und spezifischem IgE bei Patienten mit Echinokokkose. In: Boch J (Hrsg) Tropenmedizin Parasitologie. Medizin in Entwicklungsländern Bd 16. Peter Lang, Frankfurt Bern New York, S 361

Witschel H, Sundmacher R, Seitz H M (1984) Amöben-Keratitis. Ein klinisch-histopathologischer Fallbericht. Klin Mbl Augenheilk 185: 46–49

Witting P-A (1979) Learning capacity and memory of normal and *Toxoplasma*-infected laboratory rats and mice. Z Parasitenkd 61: 29–51

Woodtli W, Bircher J, Witassek F, Eckert J, Wüthrich B, Ammann R W (1985) Effect of plasma mebendazole concentrations in the treatment of human echinococcosis. Am J Trop Med Hyg 34: 754–760

Wright K A (1979) *Trichinella spiralis:* An intracellular parasite in the intestinal phase. J Parasitol 65: 441–445

Wyneken-Görgen A (1987) *Anisteis*-Infektionen des Menschen und ihre klinische Relevanz in Deutschland. Inaug.-Diss. Med. Fak., Universität Bonn

Yisunsri L, Rieckmann K (1980) In vitro microtechnique for determining the drug susceptibility of cultured parasites of *Plasmodium falciparum*. Trans Roy Soc Trop Med Hyg 74: 809–810

Yokogawa S, Cort W W, Yokogawa M (1960) *Paragonimus* and Paragonimiasis. Exp Parasit 10: 81–205

Yokogawa M, Yoshimura H (1965) *Anisakis*-like larvae causing eosinophilic granulomata in the stomach of man. Am J Trop Med Hyg 14: 770–773

Yoshida Y, Shiota T, Yamada M, Matsumoto Y (1981) Further light microscopic studies on morphology and development of *Pneumocystis carinii.* Zbl Bakt Hyg I Abt Orig A 250: 213–218

Yoshida Y, Matsumoto Y, Yamada M, Okabayashi K, Yoshikawa H, Nakazawa M (1984) *Pneumocystis carinii:* electron microscopic investigation on the interaction of trophozoite and alveolar lining cell. Zbl Bakt Hyg I Abt Orig A 256: 390–399

Young R C, DeVita V T Jr (1976) Treatment of *Pneumocystis carinii* pneumonia: current status of the regimens of pentamidine isethionate and pyrimethamine-sulfadiazine. In: Robbins J B, DeVita V T Jr, Dutz W (Hrsg) Symposium on *Pneumocystis carinii* Infection. Nat Cancer Inst Monogr 43: 193–198

Zeyhle E (1982) Die Verbreitung von *Echinococcus multilocularis* in Südwestdeutschland. In: Bähr R (Hrsg) Aktuelle Probleme der Chirurgie und Orthopädie Bd 23. Hans Huber, Bern Stuttgart Wien, S 26–33

Zeyhle E, Romig T, Macpherson C N L (1984) *Echinococcus granulosus* in the Turkana Area, Northwest Kenya. Zbl Bakt Hyg A 258: 407–408

* Ziefer A, Jacobs T, Seitz H M (1986) *Pneumocystis carinii* – Pneumonie – ein Überblick. Immun Infekt 5: 170–177

Zierdt C H (1988) Blastocystis hominis, a long-misunderstood intestinal parasite. Parasitology Today 4: 15–17

Zierdt C H, Swan J C (1981) Generation, time and growth rate of the human intestinal parasite *Blastocystis hominis.* J Protozool 28: 438–486

Zillmann U, Mehlitz D, Sachs R (1984) Identity of trypanozoon stocks isolated from man and a domestic dog in Liberia. Tropenmed Parasit 35: 105–108

Zwisler O (1989) Praktikum der Parasitosen; tierische Parasitosen des Menschen und ihre Serodiagnose, hrsg. O. Zwisler, 3. Aufl., Verlag H. Hoffmann, Berlin

Subject Index

The **emboldened** page numbers refer to the main chapter which covers the key word in more detail

Printed by Publishers' Graphics LLC